Methods in Molecular Biology™

Series Editor
John M. Walker
School of Life Sciences
University of Hertfordshire
Hatfield, Hertfordshire, AL10 9AB, UK

For further volumes:
http://www.springer.com/series/7651

Rice Protocols

Edited by

Yinong Yang

Department of Plant Pathology and Huck Institutes of Life Sciences, Pennsylvania State University, University Park, PA, USA

Humana Press

Editor
Yinong Yang
Department of Plant Pathology
and Huck Institutes of Life Sciences
Pennsylvania State University
University Park, PA, USA

ISSN 1064-3745 ISSN 1940-6029 (electronic)
ISBN 978-1-4939-5764-4 ISBN 978-1-62703-194-3 (eBook)
DOI 10.1007/978-1-62703-194-3
Springer New York Heidelberg Dordrecht London

Printed on acid-free paper

Humana Press is a brand of Springer
Springer is part of Springer Science+Business Media (www.springer.com)

Preface

Rice, *Oryza sativa*, is one of the most universal food crops for over half of the world's population. It also serves as an excellent model for monocots and economically important cereal crops because of its small genome, extensive genetic resources, relative ease of transformation, and synteny with other cereal genomes. With the completion of a finished genome sequence, increasing efforts have now focused on functional characterization of rice genes, elucidation of the underlying mechanisms involved in major agronomic traits (e.g., high yield, grain quality, abiotic stress tolerance, and disease resistance), and subsequent translation of genomic knowledge into agricultural productivity via molecular breeding and improved cultural practice.

To meet increasing interest in this field, *Rice Protocols* has been compiled to provide a series of core techniques and approaches commonly used in studying rice molecular biology and functional genomics. These include genetic and molecular techniques such as artificial hybridization, fluorescence in situ hybridization, generation and characterization of chemical and T-DNA insertional mutants, quantitative trait loci (QTLs) analysis and map-based cloning, site-specific transgene integration, and artificial microRNA-mediated gene silencing. A number of "omics" techniques have been described to facilitate transcriptome and small RNA profiling, protein identification, and metabolic analysis. In addition, several phenotypic, physiological, and cytological methods have been included for the study of root hair morphology and the evaluation of abiotic and biotic stress responses. The reviews in the book discuss molecular strategies for genetically engineering abiotic stress tolerance, disease resistance, and pharmaceutical production in rice, while the appendix provides many Internet resources related to rice genetic and genomic research. I hope this book is useful for both beginners and experienced researchers, whether they are molecular biologists who want to study rice plants or rice researchers who are interested in learning molecular techniques.

University Park, PA, USA *Yinong Yang*

Contents

Contributors

GANESH KUMAR AGRAWAL • *Research Laboratory for Biotechnology and Biochemistry, Kathmandu, Nepal*
ILLIMAR ALTOSAAR • *Department of Biochemistry, Microbiology and Immunology, Faculty of Medicine, University of Ottawa, Ottawa, ON, Canada; Proteins Easy Corporation, Ottawa, ON, Canada*
MADANA M.R. AMBAVARAM • *Virginia Bioinformatics Institute, Virginia Tech, Blacksburg, VA, USA*
GYNHEUNG AN • *Department of Plant Molecular Systems Biotechnology, Kyung Hee University, Yongin, Republic of Korea; Crop Biotech Institute, Kyung Hee University, Yongin, Republic of Korea*
NIRANJAN BAISAKH • *School of Plant, Environmental, and Soil Sciences, Louisiana State University Agricultural Center, Baton Rouge, LA, USA*
UTLWANG BATLANG • *Virginia Bioinformatics Institute, Virginia Tech, Blacksburg, VA, USA*
ADAM BOGDANOVE • *Department of Plant Pathology, Iowa State University, Ames, IA, USA*
FENG CHEN • *Department of Plant Sciences, University of Tennessee, Knoxville, TN, USA*
HANMIN CHEN • *State Key Laboratory of Plant Physiology and Biochemistry, College of Life Science, Zhejiang University, Hangzhou, China*
KYOUNGWON CHO • *Department of Biotechnology, College of Agriculture and Life Sciences, Chonnam National University, Kwangju, South Korea; Environmental Biology Division, National Institute for Environmental Studies, Tsukuba, Japan*
LUCA COMAI • *UC Davis Genome Center, University of California, Davis, CA, USA*
JENNIFER L. COOPER • *Department of Chemistry, University of Akron, Akron, OH, USA*
JI-PING GAO • *Institute of Plant Physiology and Ecology, Shanghai Institutes for Biological Sciences, Chinese Academy of Sciences, Shanghai, China*
TREVOR GREENHAM • *Department of Biochemistry, Microbiology and Immunology, Faculty of Medicine, University of Ottawa, Ottawa, ON, Canada*
OKSOO HAN • *Department of Biotechnology, College of Agriculture and Life Sciences, Chonnam National University, Kwangju, Republic of Korea*
EMILY H. HELLIWELL • *Department of Plant Pathology, Huck Institutes of Life Sciences, Pennsylvania State University, University Park, PA, USA*
STEVEN HENIKOFF • *HHMI and Basic Sciences Division, Fred Hutchinson Cancer Research Center, Seattle, WA, USA*

VENURA HERATH • *School of Biology and Ecology, University of Maine, Orono, ME, USA*
GURU JAGADEESWARAN • *Department of Biochemistry and Molecular Biology, Oklahoma State University, Stillwater, OK, USA*
YULIN JIA • *USDA-ARS Dale Bumpers National Rice Research Center, Stuttgart, AR, USA*
JIMING JIANG • *Department of Horticulture, University of Wisconsin-Madison, Madison, WI, USA*
KI-HONG JUNG • *Department of Plant Pathology, University of California, Davis, CA, USA; Department of Plant Molecular Systems Biotechnology, Crop Biotech Institute, Kyung Hee University, Yongin, Republic of Korea*
YOUNG-HO JUNG • *Department of Molecular Biology, College of Life Sciences, Sejong University, Seoul, Republic of Korea*
NAM-SOO JWA • *Department of Molecular Biology, College of Life Sciences, Sejong University, Seoul, Republic of Korea*
DEA-WOOK KIM • *National Institute of Crop Science, Rural Development Administration, Suwon, Republic of Korea*
KI YOUNG KIM • *Department of Rice and Winter Cereal Crops, National Institute of Crop Science, Iksan, Republic of Korea*
SUN TAE KIM • *Department of Plant Bioscience, Pusan National University, Busan, Republic of Korea*
AKIHIRO KUBO • *Environmental Biology Division, National Institute for Environmental Studies, Tsukuba, Japan*
JEONG IL LEE • *National Institute of Crop Science, Chuncheon Sub-station, Chuncheon, Republic of Korea*
YONG-FANG LI • *Department of Biochemistry and Molecular Biology, Oklahoma State University, Stillwater, OK, USA*
HONG-XUAN LIN • *Institute of Plant Physiology and Ecology, Shanghai Institutes for Biological Sciences, Chinese Academy of Sciences, Shanghai, China*
GUANGJIE LIU • *USDA-ARS Dale Bumpers National Rice Research Center, Stuttgart, AR, USA*
WENHUA LIU • *Department of Plant Pathology and Huck Institutes of Life Sciences, Pennsylvania State University, University Park, PA, USA*
JUNYA MIZOI • *Laboratory of Plant Molecular Physiology, Graduate School of Agricultural and Life Sciences, University of Tokyo, Tokyo, Japan*
STEPHAN OSSOWSKI • *Department of Molecular Biology, Max Planck Institute for Developmental Biology, Tübingen, Germany*
DONG-SOO PARK • *Rural Development Administration, National Yeongnam Agricultural Research Institute, Milyang, Republic of Korea*
MYOUNG RYOUL PARK • *Department of Crop Science, Chonbuk National University, Jeonju, Republic of Korea; Institute of Agricultural Science and Technology, Chonbuk National University, Jeonju, Republic of Korea*
ANDY PEREIRA • *Department of Crop, Soil, and Environmental Sciences, University of Arkansas, Fayetteville, AR, USA; Virginia Bioinformatics Institute, Virginia Tech, Blacksburg, VA, USA*

RANDEEP RAKWAL • *Research Laboratory for Biotechnology and Biochemistry, Kathmandu, Nepal; Department of Anatomy I, Showa University School of Medicine, Tokyo, Japan; Graduate School of Life and Environmental Sciences, University of Tsukuba, Tsukuba, Japan*
BENILDO G. DE LOS REYES • *School of Biology and Ecology, University of Maine, Orono, ME, USA*
PAMELA C. RONALD • *Department of Plant Pathology, University of California, Davis, CA, USA*
REBECCA SCHWAB • *Department of Molecular Biology, Max Planck Institute for Developmental Biology, Tübingen, Germany*
XUEYAN SHA • *Rice Research Station, Louisiana State University Agricultural Center, Crowley, LA, USA*
JUNKO SHIBATO • *Environmental Biology Division, National Institute for Environmental Studies, Tsukuba, Japan; Department of Anatomy I, Showa University School of Medicine, Tokyo, Japan*
GITIKA SHRIVASTAVA • *Department of Plant Sciences, University of Tennessee, Knoxville, TN, USA*
VIBHA SRIVASTAVA • *Department of Crop, Soil & Environmental Sciences and Department of Horticulture, University of Arkansas, Fayetteville, AR, USA*
RAMANJULU SUNKAR • *Department of Biochemistry and Molecular Biology, Oklahoma State University, Stillwater, OK, USA*
THOMAS H. TAI • *USDA-ARS Crops Pathology and Genetics Research Unit and Department of Plant Sciences, University of California, Davis, CA, USA*
SHIGERU TAMOGAMI • *Laboratory of Biologically Active Compounds, Department of Biological Production, Akita Prefectural University, Akita, Japan*
BRADLEY J. TILL • *Plant Breeding Unit, FAO/IAEA Agricultural & Biotechnology Laboratory, International Atomic Energy Agency, Vienna, Austria*
YANLI TONG • *State Key Laboratory of Plant Physiology and Biochemistry, College of Life Science, Zhejiang University, Hangzhou, China*
JASON G. WALLING • *Department of Horticulture, University of Wisconsin-Madison, Madison, WI, USA*
NORMAN WARTHMANN • *Department of Molecular Biology, Max Planck Institute for Developmental Biology, Tübingen, Germany*
DETLEF WEIGEL • *Department of Molecular Biology, Max Planck Institute for Developmental Biology, Tübingen, Germany*
PING WU • *State Key Laboratory of Plant Physiology and Biochemistry, College of Life Science, Zhejiang University, Hangzhou, China*
KABIN XIE • *Department of Plant Pathology and Huck Institutes of Life Sciences, Pennsylvania State University, University Park, PA, USA*
FUYU XU • *School of Biology and Ecology, University of Maine, Orono, ME, USA*
KAZUKO YAMAGUCHI-SHINOZAKI • *Laboratory of Plant Molecular Physiology, Graduate School of Agricultural and Life Sciences, University of Tokyo, Tokyo, Japan*
BING YANG • *Department of Genetics, Development and Cell Biology, Iowa State University, Ames, IA, USA*

YINONG YANG • *Department of Plant Pathology and Huck Institutes of Life Sciences, Pennsylvania State University, University Park, PA, USA*
ZHIMING YU • *State Key Laboratory of Plant Physiology and Biochemistry, College of Life Science, Zhejiang University, Hangzhou, China*
SONG JOONG YUN • *Department of Crop Science and Institute of Agricultural Science and Technology, Chonbuk National University, Jeonju, Republic of Korea*
WENLI ZHANG • *Department of Horticulture, University of Wisconsin-Madison, Madison, WI, USA*
NAN ZHAO • *Department of Plant Sciences, University of Tennessee, Knoxville, TN, USA*
XIAOFENG ZHUANG • *Department of Plant Sciences, University of Tennessee, Knoxville, TN, USA*

Chapter 1

Rice Artificial Hybridization for Genetic Analysis

Xueyan Sha

Abstract

Artificial hybridization has probably been practiced since ancient time; however, the science of genetics did not initiate until Gregor Mendel conducted a series of crosses between different pure lines of garden pea and made careful observations and systematical analyses of their offspring. Artificial hybridization or crossing between carefully chosen parents has been and still is the primary way to transfer genes from different germplasm for self-pollinated rice. Through gene recombination, novel genetic variation is created by different arrangements of genes existing in parental lines. Procedures of artificial hybridization involve the selection of appropriate panicles from representative plants of the female parents, the emasculation of female parents, and the pollination of emasculated panicles with abundant pollens of selected male parents. Of the numerous proposed methods, hot water and vacuum emasculation have proven to be the most robust and reliable ones. A successful and efficient hybridization program also relies on the knowledge of parental lines or germplasm, the reproductive biology and development of rice, the conditions needed to promote flowering and seed development, and the techniques to synchronize flowering of diverse parents.

Key words: Rice, *Oryza sativa* L., Artificial hybridization, Crossing, Vacuum emasculation, Hot water emasculation, Pollination

1. Introduction

The genus *Oryza* L. is classified under the tribe Oryzeae, subfamily Oryzoideae, of the grass family Poaceae. It contains two cultivated species, the Asian cultivated species *Oryza sativa* and African cultivated species *Oryza glaberrima*, and 21 wild species (1–3). Of the two cultivated species, *O. sativa* is grown globally, while *O. glaberrima* only exists in a restricted area of West Africa (4). Within cultivated Asian species, there are two commonly recognized subspecies, *O. sativa indica* and *O. sativa japonica*. *O. sativa japonica* can be further differentiated into tropical and temperate japonica subpopulations (5–7). Crosses among these different types of rice show varying levels of sterility and do not produce a full spectrum of recombinants (8).

Yinong Yang (ed.), *Rice Protocols*, Methods in Molecular Biology, vol. 956,
DOI 10.1007/978-1-62703-194-3_1, © Springer Science+Business Media, LLC 2013

Rice has a determinate inflorescence, commonly known as a panicle or a head, normally consisting of primary and secondary floral branches. The floral unit of a rice panicle is called a spikelet, which has a pair of glumes that embrace only one floret. Both spikelet and floret have been used in literatures in describing a single rice flower. Rice has the perfect flower, which contains both a pistil and stamens. The reproductive organs of rice flowers are enclosed by glumes, a lemma, and a palea that have to be partially removed during artificial hybridization. The components of the pistil are the stigmas, styles, and ovary. The stigma is plumose, onto which pollen grains are shed. There are six well-developed stamens, each composed of a two-celled anther and filament (9–11).

The process that rice panicle starts exsertion from the flag leaf sheath is called heading (9). In general, anthesis or flowering occurs right after heading, which is a series of events between the opening and closing of a spikelet, and normally lasts about 1–2.5 h. During anthesis, tips of the lemma and palea open, filaments elongate, anthers begin to exsert from the glumes and dehisce, and at last glumes close. Anther dehiscence usually occurs around the time when the lemma and palea open; consequently, pollen grains fall onto the stigma (11–13). Pollen grains are viable for only 5 min after emerging from the anther, whereas the stigma can be fertilized for 3–7 days.

The primary environmental factors considered for the development of suitable flowers and successful seed set are light, temperature, and soil fertility (9). Different measures should be taken to synchronize the flowering of both parents, which include multiple planting dates, short day length treatment, pruning, and exposing parents to different temperature and fertilizer rates (13, 14).

Of various techniques that have been developed to facilitate emasculation, hot water soaking and mechanical removal by vacuum pump are the most common and reliable ones (14–16). The two methods can complement each other by working on different parental lines or during different climate conditions.

Hot water emasculation was first developed for oats (17), and later adopted for rice hybridization (18), which was based on the fact that male-derived pollen grains are more sensitive to high temperatures than female organs (stigma and ovary). When treated with 45°C water for 5 min, rice pollen grains lose viability while the stigma remains active. The hot water method is very robust and efficient in the emasculation of indica and tropical japonica rice (including the US southern long-grain and most medium-grain rice), which undergo anthesis earlier and more concentrated than temperate japonica rice under fair weather conditions (no rain or no overcasting clouds). For temperate japonica rice and under rainy and cloudy conditions, the vacuum emasculation, which was perfected by researchers at the International Rice Research Institute (IRRI) (13, 19), is effective and reliable in removing anthers from female parents by suction force.

The success of the artificial hybridization largely depends on the availability of abundant pollen grains to fertilize the freshly emasculated spikelets of female parents. Proper care needs to be taken to minimize the contamination during and after pollination.

2. Materials

2.1. Parent Acquisition and Maintenance

The International Rice GeneBank at IRRI, the Philippines, has the world's largest rice germplasm collection at about 110,000 accessions (20). National genetic resource centers, such as those in China and India, also maintain large collections of rice germplasm. The current USDA rice collection contains about 20,000 accessions from 116 countries (Yan, W.G. personal communication, 2009). Most recently released varieties are protected by the plant variety protection act (PVP) and/or plant utility patent; therefore, they can only be acquired directly from the breeder(s). In the USA, materials must be introduced through the Plant Germplasm Quarantine Program, USDA-APHIS, Beltsville, MD, before they can be included in field or greenhouse research. Similar regulations also exist in other countries. Special precautions have to be taken, and extra time should be given when dealing with direct plant and seed material exchange with foreign institutes.

Before used for hybridization or crossing, plant introductions may need to be grown for both observation and purification purposes. Physiological and morphological traits, such as maturity, plant height, and pigmentation of different parts of the plant, are frequently used in the purification of parental lines. Genotyping with appropriate molecular markers will serve as a reliable tool to ensure the purity.

2.2. Tools for Emasculation

1. Vacuum pump: Model 1531-320-0557X, 0.75 hp, 600 mmHg VAC (GAST MFG Corp., Benton Harbor, MI).
2. Vinyl tube with a 6.35 mm inside diameter (VWR).
3. Disposable Borosilicate Glass Pasteur Pipettes, 146 mm in length (Fisher).
4. Thermometer (0–100°C) (Fisher).
5. Surgical scissors: 12 cm length, sharp, and pointed (Fisher).
6. Glassine cross bags: About 5×22 cm (Lawson Bags Inc., Northfield, IL).
7. Nursery markers: Black.
8. Paper clips: No. 1, nonskid.
9. Dishpans: 10.7 L, 33.5 cm L×26 cm W×13.3 cm D (Rubbermaid).
10. Timer: 0–60 min (VWR).

11. Wooden stakes: Treated, 30 cm in length (Hummert International, Earth City, MO).
12. Coin envelope: #3 (Spear Envelope, Minnetonka, MN).
13. Plastic pots: 3.76 L, 15–17 cm diameter, and 17–19 cm deep, proliferated (Hummert International, Earth City, MO).
14. Lighted magnifier (Staples).

3. Methods

It is critical to obtain vigorous and healthy plants with normal flower development, pollen shed, and seed set to enhance the success of artificial hybridization. Both male and female parental lines must flower at the same time for a successful hybridization to occur. To improve the efficiency, appropriate plants should be selected and trimmed to facilitate the emasculation process.

3.1. Plant Culture in the Greenhouse

Since conditions in the greenhouse are different from the field, specific attention must be given to temperature, light, humidity, fertilizer, and pest control in order to obtain healthy and productive plants for crossing. The greenhouse temperature should be kept at 29–35°C during the day and 22–25°C during the night by proper ventilation and/or heaters. Rice plants can be exposed to either full natural or artificial light with the intensity in the range of 600–1,000 μmol/m^2 s and duration of 12–14 h (9, 21). It is recommended that the relative humidity be kept at 60–70 %, which will improve seed set by maximizing anther dehiscence (see Note 1).

1. Obtain medium-textured soil by mixing pulverized clay and fine river sand with well-decomposed organic matter such as peat moss using a volumetric ratio of 1:2:1.
2. Fill 3.76 L (with an inside diameter of 15–17 cm, and depth of 17–19 cm) perforated plastic pots with the soil mixture. Add the base fertilizer at the rate of 2.2 g N, 6.7 g P_2O_5, and 6.7 g K_2O/m^2 before sowing the pot with five dry or pre-germinated seeds. Keep all pots in wooden or concrete benches covered with plastic film. Tap water should be added and kept at 10 cm below the soil surface.
3. Thin to two to three plants in each pot 15–20 days after sowing and topdress with nitrogen fertilizer at the rate of 15–20 g m^{-2}. Raise the water level to maintain to about 2–4 cm above the top of the pot (see Note 2).
4. During the growing season, the parental lines should be checked periodically and carefully for any off-type plants.

3.2. Synchronization of Flowering

In order to have a successful hybridization, it is essential to synchronize the flowering of both male and female parents so that large quantities of fertile pollen from the male parents will be readily available to fertilize the receptive flowers of the female parent. Synchronization can be achieved by planting seedings at multiple dates, providing a shorter critical day length for photoperiod-sensitive parents, and delaying or hastening flowering of the parents.

1. If the maturity of all parental lines is known, adjust the planting date for late maturing parents. Otherwise, four plantings of each parent at 14-day intervals are necessary to ensure a perfect match.
2. To enable crossing of parental lines with strong photoperiod sensitivity, artificially shorten the day length to 10–12 h for one or both parents by covering plants (normally from 30 days after seeding to heading) with lightproof material.
3. In subtropical or temperate regions, the planting of both parents in a greenhouse during the off-season (winter) may result in simultaneous flowering when the day length is shorter than the critical photoperiod for all genotypes.
4. The number of growing-degree-days, which determines the flowering of rice varieties in the specific environment, can be artificially altered by maintaining the temperature of some parents higher or lower than that of others in separate greenhouses or growth chambers.
5. Removal of the main or primary tillers from plants or space planting at one plant for a 3.76 L pot can delay the flowering of the early maturing parent. The onset of flowering on the new or tertiary tillers that are stimulated by the abovementioned practices will be delayed to achieve the synchronization of the flowering of both parents.
6. Cut off the early parent at 10–15 cm from the base of the plant and then apply N fertilizer to produce a ratoon crop, which will result in significant delay in flowering.
7. As common practices in hybrid rice seed production, flowering time of early parents can be delayed by applying an extra dose of nitrogen fertilizer at the vegetative stage while flowering of late parents can be expedited by spraying 1 % solution of phosphate fertilizer immediately after the panicle initiation stage (22).

3.3. Parent Selection

1. Early-flowering lines that have either large anthers or small panicle size are often used as males, while late-flowering lines frequently are used as females.
2. Parental lines that possess recessive alleles for qualitative marker traits such as leaf or glume pubescence, auricle color, and stigma color are generally used as females. This allows the differentiation of true hybrids from those that are the result of

accidental selfing. Since true hybrids show the dominant trait of the male, those exhibiting the recessive trait of the female are the result of accidental self-pollination.

3. For crosses between species that have different polyploidy levels, the parent with the highest chromosome number is usually used as the female parent.
4. For the development of mapping populations, crossing parents that may cause excessive segregation in height and maturity should be avoided, while lines that have good tissue culture ability should be included.

3.4. Emasculation

For the hot water method, it is best to carry out the emasculation between 8:00 AM and 9:30 AM, while the vacuum suction approach can be done either in the afternoon after 3:30 PM or in the early morning before 9:00 AM. The emasculated female plant remains receptive for 3–4 days. Female flowers used for artificial hybridization generally are labeled on glassine cross bags. Information on the label may include female name, greenhouse/field designation, pedigree, date of emasculation, male information, and pollination date. The cross bags should be placed on the female panicle right after the emasculation and immediately following pollination.

3.4.1. Hot Water Emasculation

Hot water emasculation consists of soaking flowers at a water temperature high enough to kill the pollen without injuring the stigma and ovary. When or after the panicle is submerged into the hot water, the glumes of the spikelets, which are anticipated to flower later, will open in advance. Pollens from those spikelets will have the maximum chance to be killed by the hot water.

1. Select representative female plants with multiple panicles that have undergone anthesis in only the upper 15–35 % of spikelets (see Note 3).
2. Remove the plants that will be used as females in the next morning out of the greenhouse bench and let them drain (see Note 4).
3. Prepare 45°C water in a dishpan by mixing the appropriate amount of the hot tap water with cold tap water (see Note 5).
4. Submerge the entire selected panicle(s) into the water and set the timer for 5 min. The pot that the rice is planted into should be raised above the submerged panicles to avoid breaking the culm. The culms of the plant should be held steady by a metal rod lying on top of the dishpan. Hot tap water is periodically added to keep the water in the dishpan at 45°C for the entire period. The procedures are shown in Fig. 1.
5. After 5 min, remove the plant from the water. Tap the panicle to get rid of the water. The glumes of the spikelets that are due

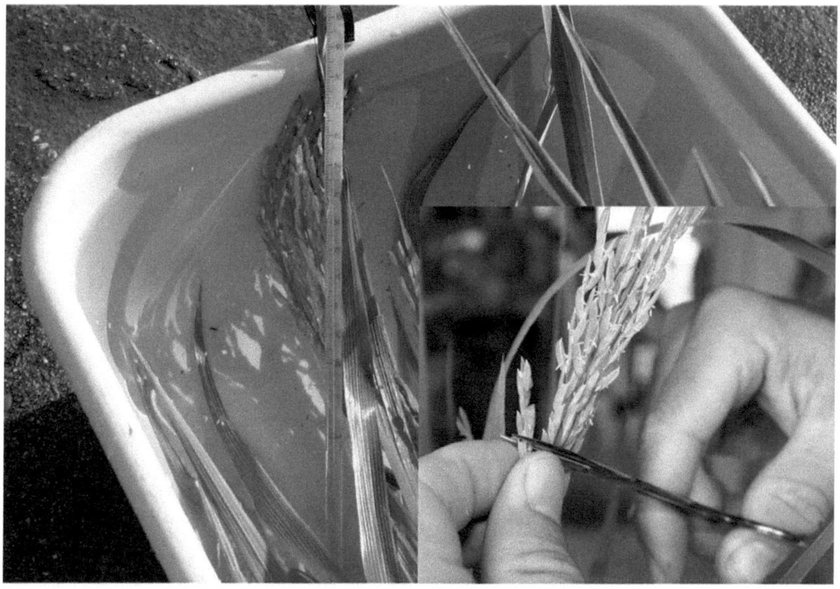

Fig. 1. Hot water emasculation. (**a**) Prepare the 45°C water in a dishpan after 8:00 AM; (**b**) cut off flag leaves of all headed panicles of selected female plants and soak them in the water for 5 min; (**c**) maintain the water temperature at or slightly above 45°C during the soaking period by adjusting with hot tap water every 2 min; (**d**) after 5 min, take the plants out of the water and tap the panicles to get rid of the water; (**e**) by working from bottom up, cut off all spikelets whose glumes are not open, and meanwhile clip off upper 1/3 glumes of each opened spikelet; and (**f**) place one glassine cross bag, which contains the information of female identification number or name and date, emasculated panicle, and the bag. The male information will be added to the bag after pollination.

to anthesis later that day will open in advance either during or shortly after the soaking.

6. Clip off the upper third of the glumes of each opened spikelet with a scissors. All remaining spikelets whose glumes have not opened within 20 min after the soaking should be eliminated. At least 20–30 emasculated spikelets are needed for a female panicle. The procedures are shown in the inset at the bottom right corner of Fig. 1.
7. Once emasculation is finished, place a glassine crossing bag over the panicle with the female label on the bag, along with the date. Leave the panicle about 1 in. from the top and fold the bag at a 45° angle from the bottom right corner, and place a paper clip to hold the corner to the stem. Place the plant back onto the greenhouse bench.

3.4.2. Vacuum Emasculation

A vacuum emasculator (Fig. 2) contains a small electric pump that was connected to filter(s) for pollen grain and incoming air. The filter is linked to a glass flask that is sealed with a rubber stopper by glass and vinyl tubes. Another 1-m-long vinyl tube is then connected to the glass tube out of the flask in one end and a glass pipette in the other end. The narrow nozzle of the glass pipette is

Fig. 2. Vacuum emasculation. (**a**) Select panicles that just emerged from flag leaf sheath the previous day; (**b**) cut off the flag leaf and portion of the sheath to expose the entire panicle; (**c**) trim the panicle by cutting off those that already went through the anthesis or the ones that are too young (*white colored*) and leave 25–35 well-spaced spikelets in three to four branches; (**d**) clip off upper 1/3 glumes of each remaining spikelet to expose its anther for vacuum suction. The cut opening should be made as wide as possible by applying a slant cutting or reverse L-shaped cutting, in which the scissors go upward first in the right side of the glumes and after they reach the top, cut from right to left horizontally; (**e**) by working upward from the base of each branch, emasculation is carried out by inserting the pipette tip to each partially clipped spikelet by wiggling the top of the anthers until an air disruption sound is heard; and (**f**) a glassine cross bag, which is labeled with female identification or name and date, is placed on the panicle.

used for sucking anthers from the clipped spikelets. The glass pipette should be checked frequently for pollen buildup that will hinder the proper suction of anthers and changed if necessary. The suction pressure of the vacuum emasculator is pre-calibrated and set at 500–600 mmHg so that only the anthers are torn away but the stigmas are not damaged (12, 13, 23).

1. Panicles that have just emerged out of the flag leaf sheath and have not undergone anthesis or those that have just flowered the day before with only a few spikelets at the tip of the panicle should be selected and marked by cutting off the flag leaf.
2. Cut off all spikelets that have undergone anthesis with scissors. Remove young branches on the bottom of the panicle and leave the branches just below the flowered portion, which will flower the following days (see Note 6).
3. By cutting off spikelets that are close to the rachis or face the panicle axis, about 25–35 spikelets that remain uniformly spaced along the panicle branch and face the same direction should be selected for clipping.

4. Cut off about one-third of each selected spikelet with a pair of shape scissors to expose anthers for vacuuming. By grasping the spikelet between the thumb and forefinger and holding the pedicled spikelet away from the operator with one hand, the cutting is carried out by the other hand. For easy removal of the anthers, the cutting should be as oblique as possible at a wide angle or in a reverse "L" shape without injuring the stigma.
5. Immediately after clipping, emasculation is carried out by touching the pipette tip to the exposed anthers inside the spikelet. Gently waggle the tip until you hear the airflow interruption sound caused by the extracted anthers. This procedure is shown in the inset at the upper right corner of Fig. 2.
6. The spikelets can be reviewed and rechecked with the aid of a lighted magnifier to ensure that there are no anthers left in the emasculated spikelets. If an anther breaks, the spikelet should be removed to avoid selfing.
7. It is preferred to emasculate as quickly as possible in order to avoid desiccation injury to any part of floret including stigma.
8. Immediately cover the emasculated panicles with glassine paper bags to prevent stigmas from being desiccated or contaminated. After emasculation, put female plants back on the greenhouse bench for pollination.

3.5. Pollination

Pollination is the process to transfer pollen from the male parent to the emasculated female spikelets. It may be done immediately after emasculation or may be delayed for up to 4 days.

1. Before pollination, separate the male plants from each other (not only for plants but also for panicles). Since the pollen dust can be easily blown away by any disturbance, efforts should be attempted to minimize the air turbulence inside the greenhouse during the pollination period.
2. In a normal sunny day, male plants bloom from 10:30 AM to 1:00 PM central time in Crowley, LA, USA (30° 12′ 49″ N, 92° 22′ 25″ W) and are ready to be used to fertilize the emasculated female panicles. During the anticipated blooming period, check male plants every 15 min for possible pollen shedding. Dusty pollen grains shed from anthers have a light yellow color that can be observed visually or by flicking the anthers on finger nails (see Note 7).
3. At the peak of blooming, cut off the entire or part of the flowering panicle and take it to the female parent. Next, cut off the top of the glassine bag protecting the emasculated spikelets f and insert and rotate the male panicle vigorously to shed pollen on the female panicle. Reseal the bag and repeat this process two to three times until the female panicle receives enough pollen (see Note 8).

4. After pollinating the female, add the male label and/or name to the crossing bag.
5. When the maximum number of spikelets undergo anthesis, dusting or pollination should be done quickly because the survival of pollen grains under natural conditions is very short. While dusting, appropriate care should be taken to avoid any contamination from foreign pollen.
6. About 5–7 days after pollination, the result of the crossing can be determined by checking the seed set of the female panicle. If it fails to produce enough seeds, additional hybridization should be attempted (see Note 9).
7. The mature seeds from the cross (F_1), along with its corresponding male and female parents, can be harvested 28 days after pollination.

4. Notes

1. In general, rice plants are more vigorous and more productive under natural field conditions than those under controlled greenhouse conditions. If parental lines are grown in the field, they can be directly seeded in the field or germinated in the greenhouse and later transplanted into the field. Hybridization nurseries should be space-planted (at 60×25 cm) in easily accessible locations, preferably on the side of the field near a levee or a driveway.
2. During the cold season when the tap water is cold, the water level of the bench should be lowered to 5–10 cm below the soil surface to let rice roots and the crown be exposed to the warm air.
3. It takes 7–10 days for all the spikelets within the same panicle to complete anthesis. Normally the peak of anthesis occurs at the 3rd or the 4th day.
4. If parental lines are grown in the field, they must be brought in at the late booting stage, transferred into gallon-size perforated plastic pots, and placed in benches filled with water. Such transplanting is normally carried out in the early morning on raining or cloudy days to minimize plant shock and wilting. Each pot is fertilized with about 2 g of 13-13-13 $N/P_2O_5/K_2O$ complete fertilizer and properly labeled with a wooden stake, which has the identification number and variety name/pedigree of the plants. Since pollen abundance is the determining factor for the seed set of artificial hybridization, more male parental plants should be used as compared to female plants.
5. The thermostat of the water heater generally is preset at about 50°C; it must be adjusted to 55°C to provide water hot enough to be used for crossing.

6. Within the same panicle, spikelets on the upper branches undergo anthesis earlier than those on the lower branches; within a branch, the top spikelet flowers first, followed by the one at the bottom, and then moves upward. The one below the top spikelet normally flowers last in a branch.
7. When the male parents are located in a different place from the female, panicles can be clipped from the male plants and placed in a pot filled with water, which is kept in a protected area until pollen is shed naturally. If anthers from the male parents fail to shed pollen at an anticipated time, the upper one-third of each spikelet should be clipped to facilitate pollen dispersal.
8. Bag a flowering male inflorescence above the emasculated female panicle in a glassine cross bag. Over a period of time, pollen shed from the male falls onto the stigma of the female parent. This can be aided by tapping the bag during blooming time. After 4 days, the male panicle needs to be taken out of the bag.
9. For some parental lines that may be difficult to mate, reciprocal crosses should be tried.

Acknowledgments

The author wishes to thank Steve Linscombe, Kris De Pauw, Shane J. Theunissen, and Blake J. Henry for their help with various aspects of this study. Financial support from the Louisiana Rice Research Board is gratefully acknowledged.

References

1. Lu BR (1999) Taxonomy of the genus Oryza (Poaceae): historical perspective and current status. Int Rice Res Notes 24:4–8
2. Khush GS (1997) Origin, dispersal, cultivation and variation of rice. Plant Mol Biol 35:25–34
3. Vaughan DA, Morishima H, Kadowaki K (2003) Diversity in the Oryza genus. Curr Opin Plant Mol Biol 6:139–146
4. Carney JA (2001) Black rice. Harvard University Press, Cambridge, MA
5. Sweeney M, McCouch S (2007) The complex history of the domestication of rice. Ann Bot 100:951–957
6. Nia J, Colowit PM, Mackill DJ (2002) Evaluation of genetic diversity in rice subspecies using microsatellite markers. Crop Sci 42:601–607
7. Garris AJ, Tai TT, Coburn J, Kresovich S, McCouch SR (2005) Genetic structure and diversity in *Oryza sativa* L. Genetics 169:1631–1638
8. Singh RK, Khush GS, Singh US, Singh AK, Singh S (2000) Breeding aromatic rice for high yield, improved aroma and grain quality. In: Singh RK, Singh US, Khush GS (eds) Aromatic rices. Oxford & IBH Publishing Co. Pvt. Ltd, New Delhi, India, pp 71–105
9. Yoshida S (1981) Fundamentals of rice crop science. International Rice Research Institute (IRRI), Los Banos, Laguna, Philippines
10. Datta D (1981) Principles and practices of rice production. IRRI, Los Banos, Philippines
11. Ikeda K, Sunohara H, Nagato Y (2004) Developmental course of inflorescence and spikelet in rice. Breed Sci 54:147–156
12. Coffman WR, Herrera RM (1980) Rice. In: Harley HH, Fehr WR (eds) Hybridization of

crop plants. Crop Science Society of America, Madison, WI, pp 511–522
13. Jennings PR, Coffman WR, Kauffman HE (1979) Rice improvement. IRRI, Los Banos, Laguna, Philippines
14. Fehr WR (1987) Principles of cultivar development, vol 1. Theory and technique. Macmillan Publishing Company, New York
15. Mackill DJ, Coffman WR, Garrity DP (1996) Rainfed lowland rice improvement. IRRI, Los Banos, Philippines
16. House LR (1985) A guide to sorghum breeding, 2nd edn. International Crops Research Institute for the Semi-Arid Tropics (ICRISAT), Patancheru 502324, Andhra Pradesh, India
17. Stephens JC, Quinby JR (1933) Bulk emasculation of sorghum flowers. Agron J 25:233–234
18. Jodon NE (1938) Experiments on artificial hybridization of rice. Agron J 30:294–305
19. Hargrove TR (1974) Vacuum emasculation streamlines hybridization. IRRI Report 3: 3–4
20. Jackson MT (1997) Conservation of rice genetic resources: the role of the International Rice Genebank at IRRI. Plant Mol Biol 35:61–67
21. Owen PC (1971) The effects of temperature on the growth and development of rice. Field Crop Abstr 24:1–8
22. Virmani SS, Sharma HL (1993) Manual for hybrid rice seed production. IRRI, Los Banos, Philippines
23. Collard BCY, Iftekharuddaula K, Thomson MJ, Pamplona A, Mackill DJ (2008) An electronic manual on marker assisted backcrossing in rice: theory and applications, 1st edn. Available at http://mcclintock.generationcp.org/index.php?option=com_content&task=view&id=92&Itemid=114

Chapter 2

Fluorescence in Situ Hybridization Techniques for Cytogenetic and Genomic Analyses

Jason G. Walling, Wenli Zhang, and Jiming Jiang

Abstract

Fluorescent in situ hybridization (FISH) is a powerful method to visualize DNA sequences in the context of the whole chromosome. Yet despite the value of FISH analysis for cytogenetic studies, there are surprisingly few labs that are able to adapt the technique for their experiments in chromosomal and genome biology. Here we present a comprehensive FISH protocol acquired from over 20 years of collective experience using different plant species. Our description uses rice as a model for performing a complete FISH procedure, but the protocol can be readily adapted for other plant species. We have provided more specialized instruction beyond routine FISH, which includes the preparation of meiotic and mitotic samples suitable for FISH analysis, procedures for direct and indirect labeling of DNA probes, and techniques for increasing signal strength using layers of antibodies.

Key words: Fluorescent in situ hybridization, Mitotic metaphase chromosomes, Pachytene chromosomes

1. Introduction

Currently there are two popular methodologies for developing physical maps in plant genome research. The first is to develop a bacterial artificial chromosome (BAC) library in which all clones are fingerprinted and contiged (1). This approach generates a high-resolution genome-wide physical map. However, it also requires genetic and genomic resources to anchor BAC contigs to individual chromosomes. The second methodology is to map DNA clones, such as BACs, to individual chromosomes by fluorescence in situ hybridization (FISH) (2, 3). FISH essentially is a manual technique with inherently low throughput. FISH-based physical maps have a limited and sometimes variable resolution depending on the cytological targets (somatic metaphase chromosomes or meiotic pachytene chromosomes). However, FISH as a physical mapping

Yinong Yang (ed.), *Rice Protocols*, Methods in Molecular Biology, vol. 956,
DOI 10.1007/978-1-62703-194-3_2, © Springer Science+Business Media, LLC 2013

tool has several beneficial advantages that cannot be replaced by other techniques. First, DNA sequences can be directly mapped to specific chromosomes. Second, two distinct types of chromatin, euchromatin and heterochromatin, can be differentially visualized at mitotic interphase and/or at the meiotic pachytene stage. Thus, FISH can anchor DNA sequences to a specific euchromatin/heterochromatin domain.

Rice has been established as one of the best model species for cytogenetic research. The 12 rice pachytene chromosomes are well differentiated from each other (4–7) and provide excellent cytological targets for FISH mapping (3, 8). The rice genome is relatively small and contains a limited amount of repetitive DNA sequences compared to other cereal crops. Therefore, most large-insert genomic DNA clones, such as BACs, can be readily used as FISH and/or fiber-FISH probes (2, 9). In contrast, a significant proportion of genomic clones from species with larger and more complex genomes, such as maize and wheat, often cannot be used as FISH probes because the repeats within the clones generate cross-hybridization signals that mask the true locations of the FISH probes.

FISH as a tool for cytogenetic and genomic research has been extensively used in both rice cytogenetic and genomic research. Cytological targets used in rice FISH mapping include interphase nuclei (2, 10), somatic metaphase chromosomes (2, 11, 12), pachytene chromosomes (3, 7, 8), and DNA fibers (13, 14). In this chapter we provide a comprehensive FISH protocol that can be applied to all of these cytological targets. Procedures including probe labeling, FISH, FISH detection, and signal amplification using layered antibodies are outlined. Furthermore we provide in-house technical notes that are key to the success of FISH experiments. Such technical notes are usually not provided in standard research papers.

2. Materials

2.1. Chromosome Preparation

1. Rice Seeds (*Oryza sativa* ssp. *japonica* cv. Nipponbare).
2. 2 mM 8-hydroxyquinoline: 0.15 g 8-hydroxyquinoline, 500 ml ddH_2O. Store in the dark at 4°C.
3. Fixation Solution: 3 parts 100% EtOH, 1 part glacial acetic acid. Make fresh.
4. 45% Acetic Acid.
5. Digestion Cocktail: 1% (w/v) Pectinase from *Aspergillus niger* (Fluka), 2% (w/v) Cellulase "Onozuka" R-10 (Yakult), in 1× PBS. Store at 4°C.
6. Microscope Slides (Fisherbrand Premium).
7. 18 × 18 or 22 × 22 mm glass coverslips (Fisherbrand).

8. Fine Forceps.
9. Alcohol burner.
10. Acetocarmine.
11. Phase contrast microscope.

2.2. Probe Labeling

1. Labeled dNTP solution: 0.166 mM dUTP (conjugated with biotin, dig, or fluorophore), 0.333 mM dTTP. Store at −20°C.
2. dNTP solution: 0.5 mM dATP, 0.5 mM dGTP, 0.5 mM dCTP. Store at −20°C.
3. 10× Nick Translation buffer: 0.5 M Tris–HCl pH 7.5, 50 mM $MgCl_2$. Store at −20°C.
4. DNase I stock buffer: 10 mM HEPES, pH 7.5, 50% (V/V) glycerol, 10 mM $CaCl_2$, 10 mM $MgCl_2$. Store at −20°C.
5. DNase I dilution buffer: 10 mM HEPES, pH 7.5, 15% (V/V) glycerol, 10 mM $CaCl_2$, 10 mM $MgCl_2$. Store at −20°C.
6. DNase I.
7. DNA Polymerase I.
8. Plasmid or Genomic DNA to be labeled.
9. 0.5 M Ethylenediaminetetraacetic Acid (EDTA).
10. 1% agarose gel (small).
11. 100 bp ladder.
12. Thermocycler or water bath set to 15°C.

2.3. Purification of Labeled Probes

1. Sephadex G-50 (Sigma-Molecular biology grade).
2. Glass wool (Fisher).
3. 1 ml Tuberculin syringe.
4. 15 ml Corex tube.
5. 1× TE: 10 mM Tris-HCl (pH 7.6), 1 mM EDTA (pH 8.0). Make from 10× stock.

2.4. FISH

1. 70%, 90%, 100% Ethanol. Keep one set at room temp and another set at −20°C.
2. 20× SSC: 3 M NaCl, 0.3 M $Na_3C_6H_5O_7$ (trisodium citrate 2 hydrate), pH 7.0. Also make 2× dilution. Store both at room temperature.
3. 70% Formamide in 2× SSC (10 ml): 7 ml deionized formamide, 1 ml 20× SSC, 2 ml ddH_2O. Store at 4°C.
4. Deionized Formamide (Amersham).
5. 50% Dextran sulfate: 5 g. Dextran sulfate in 10 ml ddH_2O. Viscous. Aliquot in 1 ml tubes and store at −20°C.
6. Directly or indirectly labeled DNA probes.
7. Rubber cement.

8. 22×22 and 22×40 mm glass coverslips.
9. Moist Chamber: Tupperware style container, lined with moist towel, and covered with aluminum foil.
10. Coplin Jar.

2.5. FISH Detection

2.5.1. Direct Detection

1. 2× SSC: Dilute from 20× (see Subheading 2.4).
2. 1× PBS: 0.13 M NaCl, 0.007 M Na_2HPO_4, 0.003 M NaH_2PO_4, pH 7.4 (make 10× stock). Store at room temperature.
3. Vectashield Mounting Medium: Supplemented with 4′, 6-diamidino-2-phenylindole (DAPI) or Propidium Iodide (PI) (Vector Laboratories Inc).

2.5.2. Indirect Detection/ Antibody Layering

1. 2× SSC: Dilute from 20× (see Subheading 2.4).
2. 1× PBS: 0.13 M NaCl, 0.007 M Na_2HPO_4, 0.003 M NaH_2PO_4, pH 7.4 (make 10× stock). Store at room temperature.
3. Secondary antibodies: An exemplary set of antibodies that can be used in series for indirect signal amplification is outlined in Subheading 3 (potential sources: Invitrogen, Roche, Jackson ImmunoResearch, etc.) (also see Note 15).
4. 1× TNT: 0.1 M Tris–HCl, 0.15 M NaCl, 0.05% Tween 20, pH 7.5. Make from 10× stock.
5. Vectashield Mounting Medium: Supplemented with DAPI or PI (Vector Laboratories Inc), 5× TNB: 0.5 M Tris–HCl (pH 7.5), 0.75 M NaCl, 2.5% (w/v) Blocking Reagent (Roche). Store at −20°C.
6. Epifluorescence microscope with appropriate filters and image acquisition hardware.
7. Coplin Jars.

3. Methods

FISH in rice includes two major steps: first, preparation of quality slides containing either meiotically or mitotically derived chromosome spreads and second, the actual probe hybridization and detection steps. The importance of obtaining quality chromosome spreads for producing robust FISH results is often overlooked. For this reason we provide an in-depth description of how to obtain either mitotic or meiotic chromosomes from their respective tissue sources. Tissue collection and chromosome preparation may take several weeks before satisfactory slides are produced. The entire FISH procedure however can be completed in two separate days. The first day consists of probe labeling, denaturing of the probe and target (chromosome) DNA, and an overnight incubation to allow for probe–target heteroduplex formation.

We prefer to use nick translation to label DNA probes. A major advantage of using nick-translation is that the average size of the labeled DNA fragments can be readily controlled. Generally any type of DNA sequence can be used for labeling, including genomic DNA, PCR-amplified DNA fragments, plasmid clones, and BAC clones. However, the PCR fragments or inserts or plasmid clones should be at least 200 bp long. DNA with initial fragments of <200 bp often do not generate good probes.

3.1. Preparation of Somatic Chromosomes from Rice Root Tips

3.1.1. Collection of Root Meristem Tissue for Chromosome Preparations

Root Tips Harvested from Germinating Seeds of *Oryza sativa*

1. Place 25–50 seeds on moist filter paper in a Petri dish and incubate at room temperature or slightly above (20–30°C). Setting them on the top of a warm incubator works well (see Note 1).
2. Make sure that filter paper does not dry out over the course of the germination. Change filter paper after a few days.
3. The seeds should germinate in 2–5 days.
4. Harvest the root tips when they are approximately 3–4 cm long.

Root Tips Harvested from Potted Plants (See Note 2)

1. Plant an appropriate number of seeds (5–10) in 4 or 6 in. pots. Field soil is the best planting medium for long-term rice growth. Artificial soil can be used if only growing the rice as an immediate source for roots tips and/or short-term use.
2. When roots start to emerge from the bottom of the container, the plants are ready for root sampling.
3. Remove plants from pots by inverting the containers.
4. Using a scissors or forceps, sample actively growing root tips. Healthy, actively growing root tips will be mostly pasty white in color with a slight yellowing toward the tip. Both primary and secondary roots are good sources for mitotic chromosomes.
5. Collect root tips into a container filled with ddH_2O on ice. Keep the container on ice until sampling is complete.
6. When collection has finished, promptly start the root tip treatment procedures described at the beginning of this section under Subheading 3.1.

3.1.2. Pretreatment of Root Tips and Slide Preparation

1. Harvest fresh roots 3–4 cm long from potted plants or germinating seedlings as described above.
2. Pretreat the roots in 2 mM 8-hydroxyquinoline at room temperature for 2–4 h to accumulate metaphase chromosome. This time varies depending on species.
3. Fix the roots with fixation solution for 24 h at room temperature and then keep at -20°C. The specimens can be stored in the fixative at -20°C for several months.

4. Wash the roots two times with distilled water for 5 min each (see Note 3).
5. Carefully dissect the meristem tissue away from the rest of the root and place it in a 1.5 ml microfuge tube containing 50 μl of digestion cocktail.
6. Incubate at 37°C in water bath for 30–60 min (see Note 4).
7. Carefully remove the digestion cocktail with a pipette. Replace the cocktail with 100 μl ddH_2O and leave tube on ice for 2 min. Repeat this wash procedure two more times.
8. Fix the root tips again by replacing the ddH_2O with 200 μl of fixation solution. Place the sample at −20°C for at least 30 min. Samples can be stored like this for several days if needed.
9. Using a pipette with a cut pipette tip, transfer one digested root tip to an ice-cold and clean microscope slide. Leave approximately 20–30 μl of fixation solution surrounding the specimen on the slide.
10. Using a forceps or a pair of probes, quickly macerate the specimen in the fixation solution. Make sure that the fixative does not evaporate. Add more drops if necessary.
11. Upon completely macerating the sample, apply an additional 20 μl of the fixation solution to flank each side of the specimen area.
12. Gently heat the bottom of the slide over an alcohol burner. While doing so, briefly expose the specimen and fixation solution to the flame by slightly tilting the slide into the flame. Only expose the sample to the flame long enough for the solution to ignite.
13. Pull the slide away from the flame and allow the flame to extinguish itself by burning up the rest of the fixative. Steps 12 and 13 should not take more than about 5 s.
14. Let the slide dry and cool at room temperature.
15. Slides can be used immediately for FISH. Alternatively, prepared slides can be stored, dry and without a coverslip, in slide box at room temperature or at 4°C.

3.2. Preparation of Meiotic Chromosomes at Pachytene Stage from Rice Anthers

3.2.1. Panicles Containing Anthers at Pachytene Stage of Meiosis

1. Rice panicles should be harvested from actively growing plants during the booting stage or prior to their emergence from the leaf sheath (see Note 5).
2. Remove the entire panicle from the boot by peeling away the sheath leaves and breaking the panicle away from the stem.
3. Immerse the panicle in fixation solution for 1 day at room temperature, and then store it at −20°C until use.
4. (Optional) It has been suggested that using a vacuum to infiltrate the fixative into the tissue provides superior fixation results. Some members of our lab have found merit in this.

3.2.2. Preparation of Pachytene Chromosome Slides from Collected Anthers

1. Remove a panicle from the fixation solution. Remove one floret and place it on a clean microscope slide. Place one drop of fixation solution on specimen to prevent from drying. The rest of the panicle can be replaced back into the fixation solution (see Note 6).
2. Carefully dissect away the anthers from the glume by cutting the filaments attached to the anthers. Do not let the anthers dry.
3. Use a fine needle to gently but thoroughly smash the anther to release the meiocytes. Place one drop of 1% acetocarmine on the slide and mix with cell suspension.
4. Remove extra tissue debris and cover the solution with an 18 × 18 mm or 22 × 22 mm coverslip.
5. Briefly heat the slide using an alcohol burner. Remove excess acetocarmine, by adding one or two drops of 45% acetic acid along one side of coverslip and use a piece of filter paper to wick the fluid from the other side. This will destain the aceto-carmine-stained cells.
6. Gently heat the slide using an alcohol burner to just short of the boiling point. It should feel uncomfortably warm to the touch, but not so hot as to boil the solution.
7. Quickly flip the slide upside down and lay it on a piece of filter paper. Gently apply hand pressure to help flatten and spread the cells on the slide.
8. Store the slides at –80°C.

3.3. Probe Labeling (See Note 7)

1. Prepare the following reaction solution:

Nick Translation MasterMix components	μl of reagent
10× Nick Translation Buffer	5
Unlabeled dNTP solution (+A,C,G) (–T)	2
Labeled dNTP solution (fluorescent-direct) (biotin- or digoxigenin-indirect)	2
Plasmid or Genomic DNA (~1 microgram)	X
DNA Polymerase I	X (12 units)
DNase I (diluted)* (see Note 8)	1
ddH_2O	X
Total volume	50

2. Gently mix after each step of adding solutions and spin down at the end.
3. Incubate reaction solution in water bath or thermocycler set to 15°C for 1.5 h.
4. Temporarily stop the reaction by placing the tube on ice.

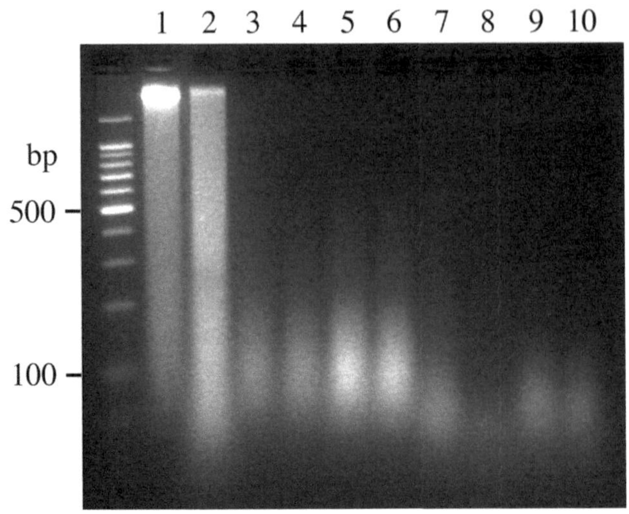

Fig. 1. Gel electrophoresis of nick translated BAC DNA. In each lane, approximately 1 μg of DNA from a BAC clone was nick translated. Concentrations of DNase I in nick reactions decrease from left to right across the lanes. Lanes 1 and 2 are considered underdigested; lanes 3–6 are perfectly digested and lanes 7–10 are overdigested.

5. Check the size of the labeled products by running a 4 μl aliquot on a 2% agarose gel next to a 100 bp ladder (see Note 9).
6. After staining the gel with ethidium bromide for 10 min, view the DNA size under UV light. If the smear from the product is less than 750 bp then the probe is suitable for FISH and you should go directly to step 8 (see Fig. 1).
7. If the upper limit of the size range is significantly larger than 750 bp (>1 kb), add more DNase I (1–3 μl), place the tube at 15°C for another 20–40 min, and then run another 2% agarose gel to check the DNA size.
8. Add 5 μl of 0.2 mM EDTA to stop the reaction. At this point, the probe can either be further purified (see Subheading 3.4) or used directly for FISH.

3.4. Purification of Labeled Probes (Optional)

1. Plug the bottom of a 1 ml tuberculin syringe with siliconized glass wool.
2. Fill the syringe with Sephadex G-50 up to the top using a sterilized Pasteur pipette.
3. Place the sephadex-filled syringe into a 15 ml Corex tube.
4. Centrifuge at 1,500rpm (300×*g*) for 4 min to pack down the sephadex.
5. Repeat steps 2–4 so that the packed sephadex column has a total volume of about 0.9 ml.
6. Wash the column twice with 55 μl of 1× TE and centrifuge at 1,500 rpm for 4 min each.

7. Place a sterilized 1.5 ml tube on the bottom of the 15 ml Corex tube; insert the packed and washed syringe into the Corex tube such that the tip of the syringe is fitted into the 1.5 ml tube. Load the nick translation product (55 μl, with stop buffer) and centrifuge at 1,500 rpm for 4 min.
8. The labeled probe DNA is collected in the 1.5 ml tube.

3.5. Hybridization of the Probe to Chromosomes

3.5.1. Target/Chromosome Denaturation

1. Remove the coverslips from the frozen slides (stored in a −80°C freezer) with a razor blade by catching an edge of the razor blade under a corner of the coverslip and smartly lifting up (see Note 10).
2. Immediately immerse the slides in an ethanol series (70%, 95%, 100% ethanol) for 5 min each at room temperature. Allow slide to completely dry for 5 min on bench.
3. Add 100 μl of the 70% formamide in 2× SSC solution on the dried slide and cover the slide with a 22×40 mm coverslip. Denature slide on a hot plate set to 80°C for approximately 60–90 s. Do not exceed 2 min.
4. Let coverslips fall off and immediately immerse into a cold ethanol series (70%, 90%, and 100%) 5 min each at −20°C and air-dry the slide.

3.5.2. Probe Mix and Denaturation

1. Prepare the hybridization mixture as follows (see Note 11):

FISH MasterMix components	μl of reagent
Deionized Formamide	5
50% Dextran Sulfate	2
20× SSC	1
Indirect or direct labeled probe	1–2
Total volume	~10/slide

2. Denature this mixture by placing sample in an 80°C heat block for 5 min. After denaturation, briefly centrifuge and immerse tube into an ice bath.

3.5.3. Probe/Target Hybridization

1. Apply 10 μl of hybridization mixture to the slide and cover with an 18×18 mm or 22×22 mm coverslip. Seal the coverslip by applying a line of rubber cement around the edges and place the slide in a moist chamber (see Note 12).
2. Incubate the wet chamber at 37°C for a minimum of 8 h or overnight. Slides can be kept in a sealed moist chamber for 3–4 days.

3.6. Detection of Directly Labeled DNA Probes

1. Using a forceps, carefully peel off the rubber cement without allowing the coverslip to move. Immerse the slides, still with coverslip, in a coplin jar containing 2× SSC. Slow shaking of the coplin jar will help the coverslips fall from the slides.

2. Place slides in a new coplin jar and wash the slides with the following steps (see Note 13):

2× SSC, room temperature	5 min
2× SSC, 42°C	10 min
2× SSC, room temperature	5 min
1× PBS, room temperature	5 min

3. Shake off excess PBS and mount the coverslip by placing one drop (approximately 20 μl) of an antifade-mounting medium (Vectashield) supplemented with a counterstain such as DAPI or PI (see Note 14).
4. The slide can now be viewed using fluorescent microscopy. Store the slide in a dark box at –20°C. Slides stored under these conditions will be stable for several months.

3.7. Detection of Two Indirectly Labeled DNA Probes with a Single Layer of Antibodies

1. Employ the stringency washes as described in Subheading 3.6.
2. Prepare master mix for detecting two labeled probes as follows:

Detection MasterMix components	μl of reagent
5× TNB Buffer	20
ddH_2O	80
Rhodamine anti-digoxigenin (see Note 15)	1
FITC (Strept)Avidin (see Note 15)	1
Total volume	~100/slide

3. Place 100 μl of antibody detection mix on the slide and cover with a 22×40 mm coverslip. Parafilm cut to size can also be used to cover the antibody solution on the slide.
4. Place the slide in a sealed moist chamber and incubate at 37°C for 30–60 min.
5. Remove the coverslip by tilting the slide and carefully shaking it off.
6. Wash the slides using the following steps:

1× TNT	5 min, room temperature (repeat 2×)
1× PBS	5 min, room temperature

7. Shake off excess 1× PBS and mount the coverslip by placing one drop (approximately 20 μl) of an antifade mounting medium (Vectashield) supplemented with a counterstain such as DAPI or PI onto the target area. Lay a single 18×18 mm or

22 × 22 mm coverslip over the solution and seat it using light hand pressure (see Note 14).

8. The slide can now be viewed using fluorescent microscopy. Store the slide in a dark box at −20°C. Slides stored under these conditions will be stable for several months.

3.8. Detection and Signal Amplification of Two Indirectly Labeled DNA Probes Using Multiple Layers of Antibodies

1. Employ the stringency washes outlined in Subheading 3.6.
2. Prepare *layer 1* master mix for detecting two probes:

Multilayer detection MasterMix components: Layer 1	μl of reagent
5× TNB Buffer	20
ddH_2O	80
Mouse anti-digoxigenin (see Note 15)	1
FITC streptavidin (see Note 15)	1
Total volume	~100/slide

3. Place 100 μl of antibody detection mix on the slide and cover with a 22 × 40 mm coverslip. Parafilm cut to size can also be used to cover the antibody solution on the slide.
4. Place the slide in a moist chamber and incubate at 37°C for 30–60 min.
5. Remove the coverslip by tilting the slide and carefully shaking it off.
6. Place the slide in a coplin jar and wash in 1× TNT for 5 min. Repeat this wash two more times for a total of three washes in 1× TNT.
7. Decant 1× TNT from coplin jar and replace it with 1× PBS. Rinse for 3 min.
8. Prepare *Layer 2* master mix for detecting two probes (see Note 16):

Multilayer detection MasterMix components: Layer 2	μl of reagent
5× TNB buffer	20
ddH_2O	80
Goat anti-streptavidin conjugated with biotin (see Note 15)	1
Total volume	~100/slide

9. Repeat steps 3–7.

10. Prepare *Layer 3* master mix for detecting two probes:

Multilayer detection MasterMix components: Layer 3	μl of reagent
5× TNB Buffer	20
ddH_2O	80
FITC streptavidin (see Note 15)	1
Rhodamine (or Texas red) anti-mouse (see Note 15)	1
Total volume	~100/slide

11. Repeat steps 3–7.
12. Shake off excess 1× PBS and mount the coverslip by placing one drop (approximately 20 μl) of an antifade mounting medium (Vectashield) supplemented with a counterstain such as DAPI or PI onto the target area. Lay a single 18 × 18 mm or 22 × 22 mm coverslip over the solution and seat it using light hand pressure (see Note 14).
13. The slide can now be viewed using fluorescent microscopy. Store the slide in a dark box at −20°C. Slides stored under these conditions will be stable for several months.

4. Notes

1. Seeds of wild rice (wild *Oryza* species) species generally take longer to germinate, require a presterilization step, and can also have a lower germination rate. However, seeds of cultivated rice (*Oryza sativa*) should readily germinate on moist/sterile filter paper without further manipulation.
2. Root samples from actively growing plants provide good source material for somatic chromosome preparation. Collect sample tissue from plants that are actively growing and whose roots have not become brown and pot bound. Light pruning of roots and repotting into larger pots can entice older plants to produce new root growth.
3. Most preparations can be readily viewed under phase contrast without staining. If a level of chromosome staining is preferred, we suggest using acetocarmine to stain the root tips prior to slide preparation.
4. The digestion time varies depending on the species and size of the root tip samples. Typically, the root tips are sufficiently digested when upon flicking the tubes, you can observe pieces of root material falling off the otherwise intact root tip.

5. Generally, once panicles have emerged from the leaf sheath meiosis is complete and the materials will no longer yield meiotic chromosomes at pachytene stage. The initiation of meiotic division varies depending on growth conditions and the time of harvest. Initially, panicles at different stages of booting should be sampled to determine optimal timing under your conditions.
6. Because the stage of floret development on a given panicle is variable, different sizes of florets should be checked for the proper stage of meiotic division. Typically slides should be prepared from the smallest florets first and checked under a microscope for the presence of pachytene chromosomes. Based on the cell morphology and stage of meiosis observed, you may need to target slightly more or less developed florets on a given panicle.
7. DNA can be labeled using a "direct labeling" approach in which the DNA itself is labeled with fluorescence-tagged dUTPs such as Texas red-dUTPs or FITC-dUTPs, or using an "indirect labeling" approach in which the DNA is labeled with a nonfluorescent tag using biotin-dUTP and digoxigenin-dUTP. In the latter case the primary DNA tag is subsequently detected with a fluorescently labeled secondary antibody. For FISH, the labeled DNA fragments used in hybridization need to be between ~100 and 500 bp long; otherwise, relatively high levels of background becomes visible on the slide, and the hybridization signal itself becomes less punctuated.
8. DNase I can be purchased as a powder and reconstituted using the DNase I stock buffer. The stock solution needs to be further diluted using the DNase I dilution buffer to reduce the enzyme's activity. Adjust the concentration of DNase I in the labeling reaction to control the final size of the DNA product such that the average size is 100–500 bp. Since fluorescence tags are light sensitive, all the steps requiring handling of them should be completed in the dark.
9. Next to FISH, the best way to confirm the quality of your labeled product is to check the size of the final nick translation product by running a small agarose gel. Although we introduce 100–500 bp as an optimum size range, a slightly larger size range (100–1,000 bp) will probably work fine. The random priming technique can also be used for labeling. However, since the size of the probe is relatively easier to control by nick translation, we prefer to use the nick translation technique. If you plan on doing only a few FISH experiments you may save time by purchasing one of the many commercial nick translation kits available. However, we often find the consistency of the labeling results of these kits to be variable.

10. If the slides were made using the flame dry method as described above, then steps 1 and 2 of "Target/Chromosome denaturation" can be skipped.
11. If using more than two probes, combine them (0.5–1 μl/slide) into one tube and dry them down in a vacuum centrifuge prior to adding the rest of the ingredients. 50% dextran sulfate is thick, viscous, and best pipetted slowly and with a cut pipette tip. Mix the probe mixture well using a pipette. For probing a larger target area the master mix can be scaled to 20 μl.
12. Plastic boxes with tight closing lids work well as moist chambers (e.g., Tupperware style boxes). Line the bottom of the box with several layers of clean filter paper. Suspend the slides over the filter paper by placing two rods or scoopulas in parallel in the box to set your slides on. Keep the filter paper moist.
13. Protect the slides from light during the washes. You may increase the temperature and time of the 2× SSC wash if you want to reduce background. Higher stringency washes can be achieved by completing a 10-min wash in a solution of 50% formamide in 2× SSC at 42°C; however, the washing series outlined above works fine for the majority of FISH experiments.
14. The slide can be stained prior to antifade mounting in a solution of 1 μg/ml of DAPI in 1× PBS for 15 min followed by two washes in 1× PBS. Intuitively, if the probe is labeled with rhodamine or a rhodamine analog, do not use propidium iodide as a counterstain since both the probe and counterstain will fluoresce red rendering the probe signal indistinguishable.
15. Many different antibodies conjugated with fluoriphors are commercially available and are too numerous to list here. We suggest choosing fluorophores with excitation/emission values that most closely coincide with the capabilities of your microscope filters. In our hands both avidin and streptavidin work well for detecting biotin-labeled probes. Fluorescent antibody conjugates should be used at the manufacturer's recommended concentration for in situ detection of probes. Typically a 1:10 to 1:20 dilution (in 1× PBS) is acceptable for our working stocks.
16. Our procedure for signal amplification only includes two layers of antibodies for detection of digoxigenin-labeled probes, while our detection of the biotin-labeled probes uses three layers. In our hands, this scheme yields robust and repeatable results; however, this is just one example and experimentation with other layering schemes may be required to detect recalcitrant probes. Finally, we find that using more than three layers to amplify a signal results in elevated levels of background noise and thus applying more than three layers is generally not recommended.

Acknowledgments

Cytogenetic studies of *Oryza* species in our lab have been supported by grant 2006-35604-16649 from the US Department of Agriculture Cooperative State Research, Education, and Extension Service (CSREES) and grant DBI-0603927 from the National Science Foundation.

References

1. Chen MS, Presting G, Barbazuk WB, Goicoechea JL, Blackmon B, Fang FC, Kim H, Frisch D, Yu YS, Sun SH, Higingbottom S, Phimphilai J, Phimphilai D, Thurmond S, Gaudette B, Li P, Liu JD, Hatfield J, Main D, Farrar K, Henderson C, Barnett L, Costa R, Williams B, Walser S, Atkins M, Hall C, Budiman MA, Tomkins JP, Luo MZ, Bancroft I, Salse J, Regad F, Mohapatra T, Singh NK, Tyagi AK, Soderlund C, Dean RA, Wing RA (2002) An integrated physical and genetic map of the rice genome. Plant Cell 14:537–545
2. Jiang JM, Gill BS, Wang GL, Ronald PC, Ward DC (1995) Metaphase and interphase fluorescence in situ hybridization mapping of the rice genome with bacterial artificial chromosomes. Proc Natl Acad Sci USA 92:4487–4491
3. Cheng Z, Presting GG, Buell CR, Wing RA, Jiang JM (2001) High-resolution pachytene chromosome mapping of bacterial artificial chromosomes anchored by genetic markers reveals the centromere location and the distribution of genetic recombination along chromosome 10 of rice. Genetics 157:1749–1757
4. Kurata N, Omura T, Iwata N (1981) Studies on centromere, chromomere and nucleolus in pachytene nuclei of rice, Oryza sativa, microsporocytes. Cytologia 46:791–800
5. Khush GS, Singh RJ, Sur SC, Librojo AL (1984) Primary trisomics of rice: origin, morphology, cytology and use in linkage mapping. Genetics 107:141–163
6. Chung MC, Wu HK (1987) Karyotype analysis of 'IR36' and two trisomic lines of rice. Bot Bull Acad Sin 28:289–304
7. Cheng Z, Buell CR, Wing RA, Gu M, Jiang JM (2001) Toward a cytological characterization of the rice genome. Genome Res 11: 2133–2141
8. Kao FI, Cheng YY, Chow TY, Chen HH, Liu SM, Cheng CH, Chung MC (2006) An integrated map of Oryza sativa L. chromosome 5. Theor Appl Genet 112:891–902
9. Cheng ZK, Buell CR, Wing RA, Jiang JM (2002) Resolution of fluorescence in-situ hybridization mapping on rice mitotic prometaphase chromosomes, meiotic pachytene chromosomes and extended DNA fibers. Chromosome Res 10:379–387
10. Dong F, Jiang JM (1998) Non-Rabl patterns of centromere and telomere distribution in the interphase nuclei of plant cells. Chromosome Res 6:551–558
11. Fukui K, Ohmido N, Khush GS (1994) Variability in rDNA loci in the genus Oryza detected trough fluorescence in-situ hybridization. Theor Appl Genet 87:893–899
12. Ohmido N, Fukui K (1997) Visual verification of close disposition between a rice A genome-specific DNA sequence (TrsA) and the telomere sequence. Plant Mol Biol 35: 963–968
13. Ohmido N, Kijima K, Akiyama Y, de Jong JH, Fukui K (2000) Quantification of total genomic DNA and selected repetitive sequences reveals concurrent changes in different DNA families in *indica* and *japonica* rice. Mol Gen Genet 263:388–394
14. Cheng Z, Stupar RM, Gu M, Jiang JM (2001) A tandemly repeated DNA sequence is associated with both knob-like heterochromatin and a highly decondensed structure in the meiotic pachytene chromosomes of rice. Chromosoma 110:24–31

Chapter 3

Generation of Rice Mutants by Chemical Mutagenesis

Thomas H. Tai

Abstract

Chemical mutagenesis of rice has been used extensively to generate useful genetic variation for the purpose of breeding improved varieties. More recently, advances in high-throughput genotyping platforms have enabled the efficient detection of point mutations generated by chemical agents. This in turn has renewed interest in using traditional chemical mutagenesis to generate mutant populations for gene discovery and functional characterization. Targeting of Induced Local Lesions in Genomes (TILLING) is a powerful reverse genetics method which combines chemical mutagenesis with the high-throughput discovery of point mutations. Numerous chemical mutagens have been shown to be effective in generating point mutations and small deletions in rice. This chapter describes the use of a combination of sodium azide (NaN_3) and N-nitroso-N-methylurea to generate populations that are suitable for TILLING as well as forward genetics and mutation breeding.

Key words: Chemical mutagenesis, Reverse genetics, Forward genetics, TILLING, Mutation breeding

1. Introduction

The generation and isolation of induced mutants is one of the most powerful approaches to identifying genes and characterizing their function. In plants, mutagenesis has long been an important tool for crop improvement and over the past 50 years, more than 500 rice varieties have been developed that were either direct releases of induced mutants or the product of crossbreeding with induced mutant germplasm (1, 2). The advent of molecular biology has established the importance of mutagenesis in helping to unlock the molecular basis of gene function through both forward and reverse genetic screens. While the relative ease and low cost of classical mutagenesis using chemical and physical agents make these methods particularly suited for generating variation for crop improvement, insertional mutagenesis using biological means (i.e., T-DNAs, transposons, retrotransposons) has been the primary tool

Yinong Yang (ed.), *Rice Protocols*, Methods in Molecular Biology, vol. 956,
DOI 10.1007/978-1-62703-194-3_3, © Springer Science+Business Media, LLC 2013

for gene cloning and functional analysis in species amenable to genetic transformation such as rice.

The primary advantage of insertional mutagenesis is the presence of a molecular tag that can facilitate efficient cloning of genes identified via forward genetic screens as well as the efficient identification of lines containing insertions in target genes. In contrast, chemical mutagens typically generate small lesions (i.e., base pair substitutions or small deletions) that are more difficult to detect. However, the nature of the lesions generated by chemical mutagens (i.e., point mutations) is also the basis for one of the advantages of chemical mutagenesis over insertional methods for functional analysis of genes. Unlike insertion events which generally result in loss of function alleles, point mutations generated by chemical mutagens often result in varying degrees of functionality (i.e., an allelic series). The availability of several mutant alleles provides more information on function than single knockout mutations. In addition, chemical mutagens can be used to produce a high density of randomly distributed mutations; thus saturation mutagenesis can be achieved with a relatively small mutant population (3). This contrasts sharply with insertional mutant populations in rice which average about two insertions per line. Based on the current rice genome annotation of about 32,000 genes and a genome size of 389 Mb, over 580,000 independent insertions (~ 300,000 insertion lines) are needed for one insertion in each gene at a 99% probability (4).

The development of the reverse genetics method known as Targeting of Induced Local Lesions in Genomes (TILLING) has sparked a renewed interest in the use of chemically induced point mutations for functional genomics. TILLING combines traditional chemical mutagenesis with high-throughput detection of point mutations. Originally, point mutation discovery was performed using denaturing high-performance liquid chromatography followed by a high-throughput platform consisting of enzymatic mismatch cleavage and semi-automated slab gel detection of labeled DNA fragments (5). More recently, next-generation sequencing platforms have been employed for even more efficient detection of point mutations (6) and further solidifying the utility of chemically induced mutant populations as a resource for functional genomics.

Like *Arabidopsis* and tomato, rice is a self-fertilizing species. As such, only dominant mutations are visible in plants derived from mutagenized seeds (i.e., M1 plants) and recessive mutations can only be observed in the progeny of M1 plants (i.e., M2 seeds and plants). Exceptions to this are recessive mutant phenotypes involving maternal tissue which can only be detected in the M3 seed (7). Although mutagens can be used to treat various materials including pollen and fertilized eggs (i.e., single-cell zygotes), seeds are the most commonly mutagenized materials as they are readily available and treatment is fairly straightforward (particularly important given the toxic and often carcinogenic properties of chemical mutagens). Seeds contain multicellular embryos and each cell in these embryos will be affected independently by the mutagen, resulting in a chimeric M1 plant. In fact, the appearance of foliar chimeras at a frequency ranging from 0.1 to 1% of the M1 plants is typically taken as a positive indication of an effective mutagenesis. Only those mutations that occur in cells that contribute to the germline will be passed on to the M1 progeny and the number of these cells in the embryo is known as the genetically effective cell number, GECN (8). In rice, analysis of M1 panicles from irradiation mutagenesis indicated that the GECN ranged from one to several initial cells (9) and that the number of M1 panicles appearing to be derived from a single cell increased with increasing doses of irradiation, suggesting that survival of an initial group of cells until gametogenesis determines the segregation ratios observed in the progeny.

Traditionally, chemical mutagenesis of higher plants involves the treatment of seeds with alkylating agents such as ethyl methanesulfonate (EMS) which modify nucleotides in DNA and result in mispairing and subsequent base changes due to DNA repair. Because of its effectiveness in generating point mutations (primarily G/C to A/T transitions) and relatively low mortality, EMS is one of the most widely used chemical mutagens in higher plants. Several protocols describing EMS mutagenesis of *Arabidopsis* seeds are generally applicable to other higher plants and have been published along with excellent descriptions concerning the general theories and considerations when performing seed treatments and conducting mutant screens (7, 10–12). While EMS has been used in rice mutagenesis, in our experience, concentrations (0.5–0.75% v/v) that produce significant number of mutant phenotypes in the M2 generation while producing sufficient M3 seeds to serve as a TILLING population resource do not have a sufficiently high enough mutation density for cost-effective TILLING (Bradley Till, personal communication). We have generated populations in japonica rice with sufficient mutation densities using 1.5% EMS (13), but treatments at these concentrations are quite toxic to the seeds and generating populations of the necessary size requires significantly greater number of seeds than the alternative mutagenesis protocol presented here.

A number of chemical mutagens have proven to be effective in inducing mutations in rice including EMS (14), sodium azide (NaN_3; 15), and N-methyl-N-nitrosourea (MNU; 16). NaN_3 is a potent respiration inhibitor which is metabolized in vivo to the mutagenic compound azidoalanine (17) that predominantly leads to base transitions (18, 19). Like EMS, MNU is an alkylating agent. In rice, the majority of reported mutations from MNU mutagenesis are G/C to A/T transitions and there may be a neighboring nucleotide bias associated with MNU-induced mutations (20).

The protocol described in the chapter involves a double mutagenic treatment of rice seeds with NaN_3 followed by MNU and is a modification of a highly effective protocol established in barley

(21). In brief, dry rice seeds are subjected to presoaking to enhance the mutagenic effect of NaN_3. After treatment with NaN_3, the seeds are washed and subjected to an inter-incubation germination period. MNU treatment is then performed, which is followed by thorough washing and planting of mutagenized seeds. Using this protocol, mutation rates on the order of 1/300 kb have been obtained (13) and populations for TILLING have been produced (http://www.tilling.ucdavis.edu/index.php/Rice_Tilling).

2. Materials

1. Rice seeds of the desired genotype (dried-moisture content of about 12–14% at room temperature, hulls free of discoloration and other obvious signs of damage or microbial contamination). Seeds should be free of debris (e.g., leaves, stems, empty hulls). For genetic studies, seeds originating from a single plant are typically used to generate mutant populations. This usually involves planting 20–50 seeds from a single, true breeding plant and bulking the subsequent generation of seeds in order to obtain enough material for mutagenesis and a reserve for additional treatments or seed amplification.
2. 0.1 M sodium phosphate buffer, pH 3: Buffer is made by dissolving monosodium phosphate in ultrapure water and adjusting to pH 3 with phosphoric acid. This solution can be made in advance and stored at room temperature.
3. 1 mM sodium azide: Sodium azide (F.W. 65.01) comes as a white powder and is highly toxic. Before use read Material Safety Data Sheet. Solution should be prepared just prior to use by dissolving sodium azide in 0.1 M sodium phosphate buffer, pH 3. Any remaining solution should be disposed of as hazardous chemical waste. All manipulations should be carried out in a chemical safety fume hood and while wearing appropriate personal protective equipment to prevent accidental inhalation, ingestion, and skin/eye contact.
4. 15 mM MNU: MNU (F.W. 103.08) comes as pale yellow crystals or light yellow powder and is toxic and carcinogenic. Before use read Material Safety Data Sheet. Solution should be prepared just prior to use by dissolving MNU in ultrapure water. Any remaining solutions should be disposed of as hazardous chemical waste. All manipulations should be carried out in a chemical safety fume hood and while wearing appropriate personal protective equipment to prevent accidental inhalation, ingestion, and skin/eye contact.
5. Trays, inserts/plug flats, and humidity domes are available from all greenhouse supply stores.

3. Methods

1. Rice seeds should be tested for germination prior to mutagenesis. Imbibe 100 seeds in an excess of water in a 50 ml tube for 18–24 h at 30–37°C. Decant water and leave seeds in the tube in a horizontal position in the incubator for an additional 24–48 h. Remove seeds from the tube and assess germination. Generally, a germination rate of about 85–95% is observed. In preparation for large-scale mutagenesis, the 1,000 seed-weight of the rice should also be determined (see Note 1).
2. Based on the 1,000 seed-weight of your rice accession, weigh out the equivalent of 10,000 seeds (~240–340 × *g*) and place in a 1 L flask. In addition, 100–200 seeds should be placed in a smaller flask to serve as a control. Add 400 ml of deionized water to the 1 L flask and a proportional amount of water to the seeds in the smaller flask. Flasks may be sealed with rubber stoppers or parafilm. Soak seeds overnight at room temperature (~14–18 h at 22–25°C) (see Note 2).
3. Decant water from imbibed rice seeds. Remove excess water by inserting a wad of paper towels into the neck of the flask and leaving it inverted for a few minutes. From this point on, all manipulations should be performed in a chemical safety hood while wearing personal protective equipment as specified by Material Safety Data Sheets for sodium azide and MNU. Prepare 1 mM sodium azide by adding 21.1 mg of sodium azide to 325 ml of 0.1 M sodium phosphate buffer, pH 3 (see Note 3). Add NaN_3 solution to the seeds and seal flask (see Subheading 2, item 3, and Note 3). Swirl to mix seeds and NaN_3 solution and leave in the hood at room temperature for 3 h with occasional swirling (3–5 times). Control seeds should be decanted and treated with a proportional amount of phosphate buffer without sodium azide. These seeds should be handled outside of the fume hood to prevent contamination with hydrazoic acid evolving from the NaN_3 solution.
4. After 3 h, decant NaN_3 solution into a hazardous waste container for disposal according to your institutional guidelines. Rinse seeds three times by adding 400 ml of deionized water, briefly swirling, and incubating for 5 min each time before decanting rinses into the waste container. After the final rinse, add 400 ml of deionized water to the seeds and leave overnight (~16–18 h) at room temperature. Repeat process with control seeds, disposing of the phosphate buffer and rinses as nonhazardous waste.
5. Decant overnight soak solution into hazardous waste container. Prepare 15 mM MNU solution by adding 618 mg MNU to 250 ml ultrapure water, mixing with a magnetic stirrer

until dissolved, and then bringing to a final volume of 400 ml with deionized water (see Note 4). Add MNU solution to seeds, swirl to mix, and seal flask. Incubate for 3 h at room temperature with occasional swirling. Control seeds may be taken through the same steps without addition of MNU and using proportional volumes for soaking.

6. Following incubation, decant MNU solution to hazardous waste container and rinse seeds three times as in step 4. After the third rinse, remove seeds from fume hood and rinse by running tap water through the flask for 1 hour. Drain seeds in a large kitchen strainer/colander to remove excess water and allow them to dry briefly on a tray lined with paper towels. This will aid in sowing, but the seeds should not be allowed to dry out completely (see Note 5). Control seeds may be taken through the same steps.
7. In a greenhouse, prepare plastic trays (e.g., 1020 trays) without holes by filling with soil to a depth of 2–3 cm. Water thoroughly (some standing water, <1 cm, is ok). Surface of soil should be as level as possible. Sow 250–500 mutagenized seeds per tray either by broadcasting by hand (wear gloves) or with a spoon, spatula, or a similar tool. Distribute seeds as evenly as possible, but do not over-manipulate them and damage any emerging coleoptiles/radicles. Cover trays with clear plastic humidity domes and place at around 25–30°C and avoid direct sunlight (see Note 6). One tray should be sown with a known number of mutagenized seeds for comparison of germination rates with control seeds sown in a separate tray. A reduction in germination of treated versus control seeds of 15–20% is typical. Once seedlings are 3–5 cm tall begin removing humidity domes by propping open for one day before completely removing. Transplant two to three leaf-stage seedlings into 50- or 98-cell inserts/plug flats filled with the appropriate soil (see Note 7).
8. About 30–45 days after flowering (anthesis), harvest mature panicles from each M1 plant individually and store in an appropriate sized envelope. Each plant should produce one to three panicles (see Note 8). At this time, the fertility can be assessed on a random subset of plants derived from the mutagenized and control seeds by counting empty and filled grains. Panicles should be left to dry for 2 or 3 days at ambient temperature and then placed in a drying oven at 45–50°C for 5–7 days. Select an individual panicle and thresh the M2 seeds (i.e., remove from panicles) by hand. Select three to five seeds randomly from each M1 plant and sow seeds in a large pot of soil for greenhouse seed production. After seedlings reach the three to four leaf stage, thin to one M2 plant per pot. Tissue for DNA isolation should be harvested from

each plant at or before mid-tillering stage. Harvest panicles (i.e., M3 seeds) of each M2 plant and process as before (see Note 9). For moderate-term storage (10–15 years), seeds should be placed at 4–10°C.

4. Notes

1. Temperatures reflect those at which lab incubators are commonly set. Seeds can be germinated at room temperature, but will germinate more rapidly at ≥30°C. The 1,000-seed weight varies greatly depending on genotype and is simply determined by counting out and weighing 1,000 seeds.
2. Volumes are based on treatment of 10,000 seeds and should be adjusted proportionally if fewer seeds are treated. We typically perform mutagenesis of more than 10,000 seeds in batches of 10,000 or less. Presoaking rice seeds prior to mutagen treatment increases mutation frequency due to the initiation of DNA synthesis and the more rapid infusion of mutagen into hydrated tissues.
3. For weighing out toxic chemicals such as NaN_3 and MNU, an analytical balance should be placed inside the chemical safety fume hood. If moving a balance is not practical or desirable, a disposable 15 or 50 ml screw cap conical tube (held upright in a small glass beaker) may be tared prior to the addition of the NaN_3 to the tube in the fume hood. The capped tube is then weighed outside of the hood and the amount of NaN_3 may be adjusted as needed. Ten to 20 ml of buffer or water is added to suspend the NaN_3 and the required amount transferred by pipeting to a larger vessel containing more buffer/water. In this case, all manipulations except weighing of the NaN_3 in sealed tube should be done in fume hood.
4. At the concentrations used in this protocol, MNU is soluble in water, but it may take 15–20 min to dissolve. As with NaN_3, MNU may be weighed in a pre-tared tube (we recommend using 50 ml conical tube and 20–40 ml of water). At this concentration, the MNU may not spontaneously dissolve and should be mixed well by gentle vortexing or pipeting before immediately transferring the appropriate volume to the larger vessel for preparation of the final solution. All manipulations except weighing of chemical in sealed tube should be done in fume hood.
5. Control seeds will have germinated and emerging coleoptiles and radicles may be observed at this stage of the protocol. The treated seeds will be somewhat delayed; however, some coleoptiles and radicles may have emerged (although generally these

will not be as long as the control seeds). Germinated seeds should be treated gently to prevent tissue damage and should not be dried out completely.

6. We prefer to sow seeds in batches in flat-bottom trays and then transplant those that survive to 50- or 98-cell inserts/plug flats. Germination trays should be placed at warmer temperatures to promote seedling growth. The humidity domes help maintain temperatures although direct sunlight should be avoided as the temperatures may become too high (we often cover the domes with some newspaper or paper grocery bags until coleoptiles of≥2–3 cm are observed). Lower temperatures (20–25°C) favor the growth of some fungi and water molds that normally would not affect seedling mortality, but can pose a significant problem with treated seeds which generally develop more slowly. Transplanting maximizes available space and is performed by using fine-tipped forceps to grasp the seed (avoid soft tissue of the seedling) and insert it 1–2 cm into soil (sandy or clay loam) that has been softened by thoroughly watering. If space is not an issue, sowing directly into cells is more efficient.
7. Cell/plug size generally determines the number of panicles produced although the genotype may also be a factor. In our experience, plants grown in the 98 plug flats typically produce one or two panicles. For developing TILLING populations, a single panicle is selected although we usually harvest all the panicles for possible forward genetic screens. In about one-third of the cases where morphological mutants are observed from one panicle at least one other panicle from that plant produces plants with the same mutant phenotype.
8. For TILLING populations, tissue for DNA isolation is harvested and lyophilized prior to flowering. Once the fertility of the M2 plant can be assessed, a decision is made whether to include this plant in the TILLING mutant library.
9. Using *Oryza sativa* ssp. japonica (cv. Nipponbare), mutagenesis of ~10,000 seeds yields 2,500–3,000 M2 plants (one M2 seed per M1 plant) that produce sufficient quantities of M3 seed to serve as a TILLING resource.

Acknowledgements

This protocol was developed with the technical assistance of P.M. Colowit and supported by the USDA Agricultural Research Service CRIS Project 5306-21000-016/17-00D and grant 2004-353604-14265 from the USDA Cooperative State Research, Education, and Extension Service, NRI Plant Genome Program.

References

1. Rutger JN (1983) Applications of induced and spontaneous mutation in rice breeding and genetics. In: Brady NC (ed) Advances in agronomy. Academic, New York, NY, pp 383–413
2. Tai TH (2007) Induced mutation in rice (*Oryza sativa* L.). Israel J Plant Sci 55: 137–145
3. Henikoff S, Comai L (2003) Single-nucleotide mutations for plant functional genomics. Annu Rev Plant Biol 54:37–401
4. Zhang Q, Li J, Xue Y, Han B, Deng XW (2008) Rice 202: a call for an international coordinated effort in rice functional genomics. Mol Plant 1:715–719
5. Henikoff S, Till BJ, Comai L (2004) TILLING. Traditional mutagenesis meets functional genomics. Plant Physiol 135:630–636
6. Rigola D, van Oeveren J, Janssen A, Bonne A, Schneiders H, van der Poel HJA, van Orsouw NJ, Hogers RCJ, de Both MTJ, van Eijk MJT (2009) High-throughput detection of induced mutations and natural variation using KeyPoint™ technology. PLoS One 4:e4761
7. Koornneef M (2002) Classical mutagenesis in higher plants. In: Gilmartin PM, Bowler C (eds) Molecular plant biology: a practical approach, vol 1. Oxford University, Oxford, UK, pp 1–11
8. Li SL, Redei GP (1969) Estimation of mutation rate in autogamous diploids. Radiat Bot 9:125–131
9. Yamaguchi H (1962) The chimaeric formation in an X1 panicle after irradiation of dormant rice seed. Radiat Bot 2:71–77
10. Leyser O (2000) Mutagenesis. In: Tucker GA, Roberts JA (eds) Methods in molecular biology: plant hormone protocols. Humana, Totowa, NJ, pp 133–144
11. Kim Y, Schumaker KS, Zhu J (2006) EMS mutagenesis of Arabidopsis. In: Salinas J, Sanchez-Serrano JJ (eds) Methods in Molecular Biology: Arabidopsis Protocols, 2nd edn. Humana, Totowa, NJ, pp 101–103
12. Maple J, Møller SG (2007) Mutagenesis in Arabidopsis. In: Rosato E (ed) Methods in Molecular Biology: Circadian Rhythmns: Methods and Protocols. Humana, Totowa, NJ, pp 197–206
13. Till BJ, Cooper J, Tai TH, Colowit P, Greene EA, Henikoff S, Comai L (2007) Discovery of chemically induced mutations in rice by TILLING. BMC Plant Biol 7:19
14. Wu JL, Wu C, Lei C, Baraoidan M, Bordeos A, Madamba MR, Ramos-Pamplona M, Mauleon R, Portugal A, Ulat VJ, Bruskiewich R, Wang G, Jan Leach J, Khush G, Leung H (2005) Chemical-and irradiation-induced mutants of indica rice IR64 for forward and reverse genetics. Plant Mol Biol 59:85–97
15. Awan MA, Konzak CF, Rutger JN, Nilan RA (1980) Mutagenic effects of sodium azide in rice. Crop Sci 20:663–668
16. Suzuki T, Eiguchi M, Kumamaru T, Satoh H, Matsusaka H, Moriguchi K, Nagato Y, Kurata N (2008) MNU-induced mutant pools and high performance TILLING enable finding any gene mutation in rice. Mol Genet Genomics 279:213–223
17. Owais WM, Kleinhofs A (1988) Metabolic activation of the mutagen azide in biological systems. Mutat Res 197:31–323
18. Olsen O, Wang X, von Wettstein D (1993) Sodium azide mutagenesis: preferential generation of A·T→G·C transitions in the barley Ant18 gene. Proc Natl Acad Sci USA 90: 8043–8047
19. Talame V, Bovina R, Sanguineti MC, Tuberosa R, Lundqvist U, Salvi S (2008) TILLMore, a resource for the discovery of chemically induced mutants in barley. Plant Biotech J 6:477–485
20. Suzuki T, Moriguchi K, Tsuda K, Eiguchi M, Kumamaru T, Satoh H, Kurata N (2010) Neighboring nucleotide bias around MNU-induced mutations in rice. Rice Genet Newsl 25:90–91
21. Szarejko I, Maluszynski M (1999) High frequency of mutations after mutagenic treatment of barley seeds with NaN_3 and MNH with application of inter-incubation germination period. Mutat Breed Newsl 44:28–30

Chapter 4

TILLING and Ecotilling for Rice

Jennifer L. Cooper, Steven Henikoff, Luca Comai, and Bradley J. Till

Abstract

Mutagenesis is frequently used to test gene function and to aid in crop improvement. Targeting Induced Local Lesions in Genomes (TILLING) is a reverse genetic strategy first developed to identify induced point mutations in *Arabidopsis*. This general strategy has since been applied to many plant and animal species. Here, we describe a protocol for high-throughput TILLING in rice. Gene segments are amplified using fluorescently tagged primers, and products are denatured and reannealed to form heteroduplexes between the mutated and wild-type sequences. These heteroduplexes are substrates for cleavage by single-strand-specific nucleases. Following cleavage, products are analyzed on denaturing polyacrylamide gels using the LI-COR DNA analyzer system. Several rice TILLING populations have been described, and a public mutation screening service is now available. The basic methods used for TILLING can be adapted for the discovery and cataloguing of natural nucleotide variation in populations, a strategy known as Ecotilling, which was first used to study genetic diversity among *Arabidopsis* ecotypes, and has since been applied to crop plants.

Key words: Enzymatic mismatch cleavage, Mutation discovery, Reverse genetics, Nucleotide polymorphism, Celery juice extract

1. Introduction

Full genomic sequences are now available for many organisms; however, the task of assigning function to each gene remains a sizeable effort for many species. Sequence homologies and transcriptional profiles provide important clues to gene function, but proof of *in vivo* activities requires further work. A powerful strategy for assigning *in vivo* gene function is the phenotypic characterization of gene disruptions caused by induced and natural nucleotide variation. Indeed, this has been a major source of our understanding of fundamental biological principles, and has provided a means for crop domestication and breeding (1).

Reverse genetic methods utilize knowledge of gene function and accumulation of genome sequences to target specific genes for

Yinong Yang (ed.), *Rice Protocols*, Methods in Molecular Biology, vol. 956,
DOI 10.1007/978-1-62703-194-3_4, © Springer Science+Business Media, LLC 2013

disruption without the *a priori* need of phenotypic characterization (2). Large mutant populations are stored in early generations and only desired mutants are chosen for phenotypic characterization and genetic analysis. Additionally, reverse genetic strategies allow for the identification of alleles that confer no phenotypic consequences on their own, but do when combined with other alleles. This provides a mechanism for the development of traits that are either unlikely or impossible with traditional forward methods.

Targeting Induced Local Lesions in Genomes (TILLING) is a reverse genetic method that allows the rapid screening of thousands of mutagenized lines for mutations in a gene of interest (3, 4). In TILLING, DNA from mutagenized lines is pooled and arrayed in microtiter plates (Fig. 1). A target of ~1.5 kb in a gene of interest is amplified from the pooled DNA using fluorescently labeled PCR primers. The PCR products are denatured and reannealed so that heteroduplexes can form if a mutation is present. Heteroduplexes are cleaved using a crude protein extract from celery containing the nuclease CEL I, and the products are electrophoresed on denaturing polyacrylamide gels and visualized using fluorescence imaging. Mutations are identified by the presence of cleaved PCR products in a given pool. This enzymatic mismatch cleavage method is suitable for the recovery of single-nucleotide mutations and small indels, and has been used for the discovery of thousands of induced mutations (5, 6).

The success of TILLING is dependent on two factors: mutagenesis and mutation discovery. Mutagenesis can be performed with any mutagen that yields single-nucleotide changes or small indels. Chemical mutagens have been widely used to develop mutagenized TILLING populations in a variety of species, and much has been learned with regard to the spectrum and density of mutations from different chemicals (6). Ethylmethanesulfonate (EMS) and methylnitrosourea (MNU) have both been successfully used as mutagens in rice, and populations with densities between one mutation per 135 kb and one per 290 kb have been reported (7, 8). Success with rice has led to the development of a public TILLING service (http://tilling.ucdavis.edu/index.php/Rice_Tilling). Less is known about the applicability of ionizing radiation for reverse genetics using enzymatic mismatch cleavage. Gamma irradiation

Fig. 1. (continued) twice on the 96-well plate, but in two unique pools. The desired target is PCR amplified from the pooled DNA with two labeled primers, one label for each LI-COR channel (filled star = IRDye 700, open star = IRDye 800). The PCR products are heated to denature the DNA, and then slowly cooled to allow the formation of heteroduplexes if a mutation is present in the pool. The PCR products are then digested with Celery Juice Extract (CJE) to cut 3′ of the heteroduplex. Following sample cleanup, the products are denatured and electrophoresed on polyacrylamide gels in a LI-COR DNA analyzer. Putative mutations are visualized as above background bands (*boxes in gel image*) in the 700 and 800 channels that add up to the size of the full-length PCR product. With 2-dimensional pooling, a putative mutation will be observed twice on the same gel and will provide the identity of the individual line harboring the putative mutation. This figure is modified from ref. (7).

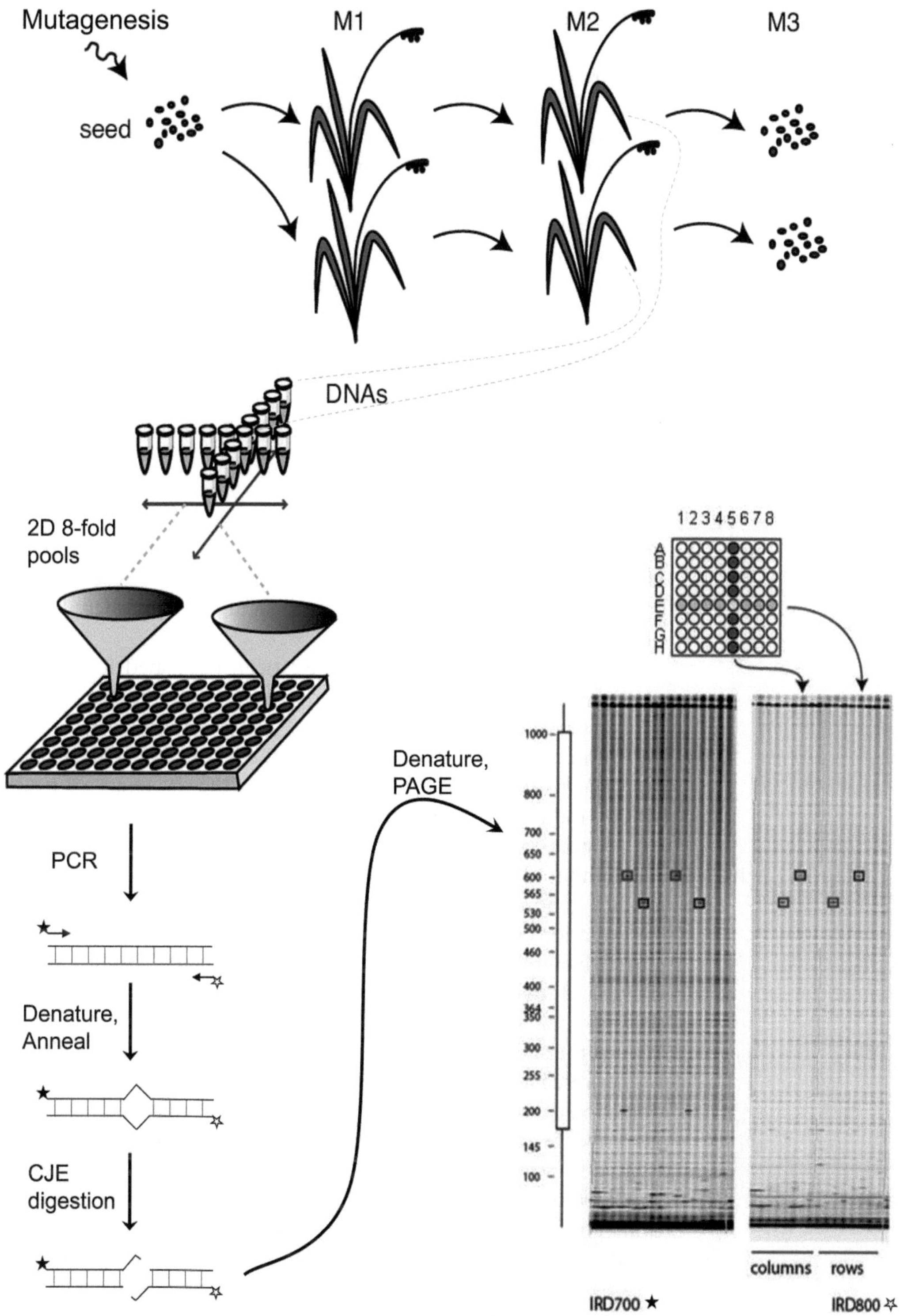

Fig. 1. Rice TILLING overview. Seeds are mutagenized and grown into M_1 plants, which are allowed to self-fertilize. Tissue is harvested from the M_2 plants for DNA isolation. The M_2 self-fertilizes and the resulting M_3 seed is saved for mutant analysis. DNA from eight M_2 plants is pooled into a well on a 96-well plate. In 2-dimensional pooling, each DNA is represented

has been used to mutagenize rice and a low density of SNP and indel mutations was reported (9). Fast neutron mutagenesis and other treatments that cause primarily large deletions can be considered useful for reverse genetics, but alternative mutation discovery methods such as PCR amplicon length polymorphism screening or array comparative genome hybridization must be used for deletion discovery (10–13). Because a high frequency of deletions may be deleterious to gene function, such mutagens also require larger populations to ensure a high probability of recovering mutations in every gene. Since mutagens causing point mutations can be used to achieve a high density of mutations randomly throughout the genome, they offer a major advantage because efficient recovery of alleles will be obtained with relatively small populations. For example, the *Arabidopsis* TILLING project averaged the delivery of ~14 mutations for every 1–1.5 kb gene target when screening between 3,000 and 6,000 mutant lines. This high throughput allows for a rapid and low-cost screening service (http://tilling.fhcrc.org/arab/status.html). Another major advantage is that point mutations can produce allelic series including knockout and missense. Hence, TILLING provides a rich source for phenotypic diversity and allows for the characterization of essential genes where knockouts would be lethal.

Several TILLING methods have been published for the discovery of point mutants. We describe here a modification for rice of the previously described method using fluorescently labeled PCR primers and a LI-COR DNA analyzer (14, 15). This method has been used since 2001 and resulted in the discovery of nearly 10,000 induced mutations by public TILLING services (6). These experiences have been used to estimate screening costs and sample throughput (16). For example, using one LI-COR DNA analyzer, ~2,300 genomic DNA samples pooled eightfold can be screened for mutations in one gene target per day. Assays are highly scalable, and with eight LI-COR analyzers, semi-automated 384 liquid handling and two dedicated technicians, 12,000 samples can be screened per day.

Further advantages of enzymatic mismatch cleavage followed by fluorescence detection using PAGE gels include simple visual detection of natural nucleotide polymorphisms that might be erroneously scored as true mutations using other methods. Additionally, multiple nucleotide polymorphisms can be discovered on a single amplicon because enzymatic cleavage does not proceed to completion. Thus, the same methods used for TILLING can be used for the discovery and characterization of natural nucleotide polymorphisms, a strategy called Ecotilling because it was first used to assay ecotypes of *Arabidopsis* (5). The method has been used for SNP discovery in organisms including rice, maize, switchgrass, humans, poplar, and melon (16–20). Major differences between TILLING and Ecotilling lie in the design of pooling and sample arraying and in data analysis as described in this chapter.

2. Materials

2.1. PCR

1. Ex-*Taq* DNA polymerase, Hot Start Version (Takara). Store at -20°C (see Note 1).
2. 10× Ex-*Taq* PCR buffer (Takara) supplied with Ex-*Taq*. Store at -20°C.
3. 2.5 mM (each) dNTPs (Takara) supplied with Ex-*Taq*. Store at -20°C.
4. 25 mM $MgCl_2$.
5. TE: 10 mM Tris–HCl, 1 mM ethylene diamine tetraacetic acid (EDTA), pH 7.4.
6. Left primer (T_m 67–73°C) unlabeled (MWG) 100 μM in TE. Store at -80°C (see Note 2).
7. Left primer (melting temperature (T_m) 67–73°C) labeled 5' with IRDye 700 (MWG) 100 μM in TE. Store at -80°C (see Note 3).
8. Right primer (T_m 67–73°C) unlabeled (MWG) 100 μM in TE. Store at -80°C.
9. Right primer (T_m 67–73°C) labeled 5′ with IRDye 800 (MWG) 100 μM in TE. Store at -80°C.
10. Primer mixture: 3 μL IRDye 700 left primer, 2 μL unlabeled right primer, 4 μL IRDye 800 right primer, 1 μL unlabeled right primer (see Note 4).
11. PCR mixture for a 96-well plate: 360 μL water, 57 μL 10x Ex-*Taq* buffer, 68 μL, 25 mM $MgCl_2$, 92 μL 2.5 mM dNTP mixture, 4 μL primer mixture, 6 μL Ex-*Taq*. Mix, adding polymerase last. Use immediately and discard the remainder after use (see Note 5).

2.2. CJE Digestion

1. Crude celery juice extract (CJE) prepared as previously described (14, 21) (see Note 6).
2. 10× CJE buffer: 5 mL 1 M $MgSO_4$, 5 mL 1 M 4-(2-hydroxyethyl)-1-piperazineethanesulfonic acid (HEPES), pH 7.5, 2.5 mL 2 M KCl, 100 μL 10% Triton® X-100, 5 μL 20 mg/mL bovine serum albumin, 37.5 mL water. Store aliquots at –20°C.
3. CJE reaction mixture for a 96-well plate: 2.4 mL water, 420 μL 10× CJE buffer, 36 μL CJE. The amount of CJE may vary based on the prep. Mix on ice, use immediately, and discard the remainder after use.
4. Stop solution: 0.225 M EDTA, pH 8.0.

2.3. Spin Plate Cleanup

1. 96-Well membrane plates (Millipore).
2. Sephadex® G50 medium (Amersham Pharmacia Biotech).

3. Sephadex spin plates: Use a Sephadex loading device (MultiScreen 45-μL column loader; Millipore; cat. no. MACL 096 45) to fill all wells of a 96-well membrane plate with approx 0.03 g of G50. Hydrate with 300 μL water. Allow the plate to stand for 1 h at room temperature. Plates may be stored at 4°C in a sealed container to prevent evaporation for up to 1 week.
4. Deionized formamide (Sigma).
5. Formamide load buffer: 38.5 mL deionized formamide, 1.5 mL 0.5 M EDTA, pH 8.0, 2.3 mg bromophenol blue. Store 1.8 mL aliquots at −20°C.
6. 200-bp markers: Perform PCR (Subheading 3.1) and spin plate cleanup (Subheading 3.3) with primers designed to yield a 200-bp fragment.

2.4. Electrophoresis

1. Ammonium persulfate (APS): Dissolve APS 10% (w/v) in water. Store at 4°C in small aliquots.
2. Polyacrylamide gel: Combine 20 mL gel mixture (6.5% acrylamide, 7 M urea; LI-COR), 15 μL *N,N,N′,N′*-tetramethyl ethylenediamine (TEMED), and 150 μL APS. Use immediately.
3. 0.8× TBE buffer: Dissolve 89.2 g Tris-base and 45.8 g boric acid in water, add 68 mL 0.25 M EDTA, pH 8.0, bring vol to 10 L.
4. IR^2 gel analyzer, 25-cm glass plates, and 25-mm spacers (LI-COR).
5. 100-tooth membrane combs (The Gel Company).
6. Gel loading tray (The Gel Company).
7. Size Standard IRDye 700 and IRDye 800 molecular weight markers (LI-COR), or create molecular weight standards by combining PCR amplicons of desired sizes prepared as described (Subheading 2.3, item 6).

3. Methods

3.1. PCR and Heteroduplex Formation

PCR amplification should be performed using equipment and consumables that are segregated from PCR products to avoid contamination that may result in a failure to detect novel nucleotide polymorphisms.

1. Deposit 5 μL of pooled genomic DNA per well in a 96-well microtiter plate. The final concentration of genomic DNA and the maximal allowable pooling may vary depending on the organism and DNA extraction method used (see Notes 7–9).
2. Add 5 μL freshly made PCR mixture to each well. Centrifuge for 2 min at 1,400 × *g* and place in thermal cycler.

3. Run the following thermal cycler program: 95°C for 2 min; loop 1 for 8 cycles (94°C for 20 s, 73°C for 30 s, reduce temperature 1°C per cycle, 72°C for 1 min); loop 2 for 45 cycles (94°C for 20 s, 65°C for 30 s, 72°C for 1 min); 72°C for 5 min; 99°C for 10 min; loop 3 for 70 cycles (70°C for 20 s, reduce temperature 0.3°C per cycle); hold at 8°C. After thermal cycling store samples in the dark at –20°C for use within 1 week. Extended storage may result in the loss of fluorescent signal.

3.2. CJE Digestion

1. Place PCR samples on ice and add 20 μL CJE reaction mixture to each sample. Centrifuge for 2 min at 1,000 × *g*. Mixing is not required but suggested when using 96 or 384 channel semi-automated pipettors.
2. Incubate at 45°C for 15 min.
3. Place samples on ice and stop reaction by adding 5 μL 0.225 M EDTA. Store samples in the dark at 4°C for use within 2 days or freeze at –20 C for up to 1 week.

3.3. Sample Purification

1. Assemble hydrated spin plate and an empty 96-well catch plate for centrifugation (see Note 10).
2. Spin hydrated plate containing Sephadex G50 for 2 min at 440 × *g*.
3. Remove the catch plate and insert a sample receptacle plate containing 1.5 μL formamide load dye in each well. Load the CJE digestion products onto the Sephadex plate within 10 min (see Note 11).
4. Spin for 2 min at 440 × *g*.
5. Add 200 base pair marker into sample wells representing every eighth gel lane starting with lane 4, 12, 20, and so on prior to reducing the volume. This marker facilitates lane identification and is compatible with the GelBuddy gel analysis software (Subheading 3.7, step 2). How the wells of the 96-well assay plate correspond to specific gel lanes depends on the method of sample loading (see Note 12).
6. In a fume hood, reduce the volume at 90°C to approx 1.5 μL (this takes approx 45 min), leaving formamide–bromophenol blue solution ready for loading. Transfer to ice until ready to load. Samples can be stored in the dark at –20°C for up to 1 week prior to use.

3.4. Denaturing Polyacrylamide Gel Preparation

1. Assemble 25-cm glass plates, 25-mm spacers, and casting rails. Plates can be preassembled and stored in a dust-free environment for weeks in advance.
2. Pour gels. For each 25-cm plate assembly, fill a 20-mL syringe with freshly prepared acrylamide mixture, and then dispense along the top avoiding bubbles by rapping continuously on

the plate just above the liquid edge. If any bubbles appear, remove them quickly after the gel is poured with a thin wire tool.

3. Leaving a little excess acrylamide at the well, insert the top spacer all the way into the glass, making sure that spacer is centered horizontally.
4. Insert the Plexiglass pressure plate between the glass plate and casting rails. Tighten the top screws as soon as the spacer is inserted, slightly compressing the rubber pads on the pressure plate.
5. Add acrylamide to the top glass edge where the comb is inserted and on the edges to assure that polymerization is not inhibited within the gel.
6. Let the gel set at least 90 min before placing in gel box. Gels can be stored in plastic wrap at 4°C for up to 24 h prior to use (see Note 13).
7. Prior to placing the gel in the gel box, wash the plates with distilled water, removing the comb spacer and excess polyacrylamide at the top edge. Dry the plates and wipe with isopropanol, making sure that the back plate is spotless where the laser shines through.
8. Insert the top buffer reservoir between the glass plate and the casting rails. If this is difficult, moisten the gasket with buffer and remove one casting rail in order to fit the top reservoir. Fill the lower buffer reservoir to the fill line with 0.8xTBE (approx 500 mL) and insert the gel.
9. Tighten the screws to seal the upper reservoir and fill with buffer. Rinse the slot vigorously using a large syringe without a needle (see Note 14).

3.5. Loading Samples onto Membrane Combs

1. Load 0.25 μL of sample onto the gel loading tray such that position A1 on the plate represents gel lane 3, position B1 equals lane 4, A2 equals 9, etc. Sample order will vary depending on the method of deposition, and a specific order is desirable when using eightfold pooling and two-dimensional arraying with the GelBuddy analysis program (see Note 15).
2. When the tray is filled, insert a membrane comb. Samples will absorb onto comb teeth.
3. Use a pipettor to add 0.25–0.5 μL IRDye 700 marker to comb tooth 1 and 2, and IRDye 800 marker to comb 99 and 100. This provides a means to confirm that the comb was properly inserted into the gel with tooth #1 on the left side of the gel.

3.6. Electrophoresis

1. Access the user controls (LI-COR) using an Internet browser.
2. Provide a unique gel run name, and hit Create Run (see Notes 16 and 17).

3. Start the prerun (20 min), waiting for the all ready signal from the scanner before proceeding. The prerun can be started while samples are being applied to the comb.
4. After the prerun, clean the slot out with a syringe and drain the top buffer reservoir until the level is below the glass edge. Wick out the remaining buffer between the glass plates, first with a paper towel and then with a 6-in-wide strip of Whatman 1 paper, sliding it into the slot left by the spacer.
5. Using a 1-mL pipettor, fill the slot with 1% Ficoll® leaving just a thin bead, approx 1 mm above the slot.
6. Hold the comb at a 45° vertical angle with lane 1 on the left, aim for the slot, and insert rapidly by pushing gently (see Note 18). Push the comb down until it just touches the gel surface.
7. Gently fill the reservoir to the fill line, insert the electrode cover, close the top, and then click on Collect image. The time the comb touches the slot until the time the current is applied should be no more than about 20 s or so to prevent diffusion.
8. After 10 min, open the lid (be sure that you hear the "pling" signal and the high-voltage light turns off), remove the comb, and gently rinse the slot with buffer (see Note 19). Replace the top electrode and close the lid. You should hear the pling and see the laser and high-voltage lights turn on. The gel can be monitored from a browser.

3.7. Gel Image Analysis for TILLING

1. Download GelBuddy from http://www.gelbuddy.org (see Note 20). Install and open the GelBuddy program. Under "File," select "Open 700 and 800 Channel Images." Select the corresponding 700 and 800 TIFF images (downloaded from the LI-COR). The two images can be alternated by clicking the 700/800 button.
2. Use the sliding bars at the top right to adjust the 700 and 800 images so that bands can be seen. Click the "Find Lanes" button (box with blue and red vertical lines). In the pop-up window select 96 lanes, segmented tracks, and both channels for finding lanes. Click "OK." 96 colored lines will appear over the gel lanes. Lane 1 is the first sample lane and lane 96 is the last sample lane. Blue lines indicate the lane containing the 200 bp marker (every 8 lanes starting with lane 4). Lane position can be corrected by clicking the "Lane Editing" button (box with blue and red vertical lines with arrow). Lanes can be dragged into position using the mouse. Lanes can be added or deleted by using functions in the "Edit" menu. Mark failed lanes by selecting the lane and choosing "Mark Current Lane as Failed" in the edit drop-down menu. Click the "Find Lanes" button again to eliminate the lane lines. The assigned lane number will remain.

3. Calibrate the fragment length markers (click the box with blue and green horizontal lines). Using the mouse, drag the red 100% line to the bottom of the gel. Drag the red 0% line to the full-length PCR product. Drag the blue 200 bp line to the position of the 200 bp lane marker, and the blue 700 bp line to the 700 bp marker position. Click the "Calibrate" button again to remove the calibration lines.
4. Select the appropriate mode of analysis. Click the "Find Mutations" button (box with rectangle and eye inside). In the pop-up window, fill in the full-length product size, pool size, and grouping information. Selecting "TILLING 2D" results in default settings for the grouping of row and column pool mutations.
5. Using the mouse, drag the rectangle over a putative mutation. Click the mouse to mark the mutation. Change to the other image channel and mark the corresponding band in the other wavelength image. Continue until all putative mutations are marked.
6. Under the "Report" menu, select "View Concise Report." This provides a text report summarizing the putative mutations sorted by lane or by molecular weight, as well as which lanes have equivalent molecular weight mutations. This report can be saved for future reference. More advanced features such as exporting data directly to a database are described on the GelBuddy Web site.
7. Under the "File" menu, select "Save Gel Markup" to save the marked-up gel image. This saved image can be opened by selecting "Open Gel Markup" under the "File" menu.

3.8. Gel Image Analysis for Ecotilling

1. Follow steps 1–3 of Subheading 3.7 (see Note 21).
2. Click on the "Find Mutations" button (box with rectangle and eye inside). In the pop-up window, fill in the full-length product size, pool size, and select "Ecotilling/Generic" in the presets section.
3. Using the mouse, drag the rectangle over a putative mutation. Click the mouse to mark the mutation. Change to the other image channel and mark the corresponding band in the other wavelength image. Continue until all putative mutations of the same molecular weight are marked.
4. Select the "Edit Signals Group" mode and check that all bands of the same molecular weight are grouped together. If not, alter the placement of marked bands in "Record Signals" mode. Alternatively, select all bands to be grouped in the "Edit Signals Group" mode by clicking boxed bands while holding the shift key. When all bands are selected, choose "Group Selected Signals" under the edit drop-down menu.

5. Under the report drop-down menu select "View Concise Report" and evaluate the output. Lanes grouped according to banding pattern (haplotype) are listed in the "Lane Classes" section of the report (see Note 22).
6. When complete, save the concise report and the gel markup as described in Subheading 3.7.

4. Notes

1. Alternative DNA polymerases can be considered for PCR. Hot start enzymes are advantageous because assays can be prepared at room temperature and artifactual gel banding due to primer mispriming is potentially reduced. All steps of the TILLING protocol were optimized for use with Ex-Taq polymerase and its reaction buffer. PCR and CJE optimization, therefore, may be required if using a different polymerase. It is suggested that side-by-side comparisons be performed with Ex-Taq when evaluating alternative polymerases. In addition, polymerase errors may increase false positive signals. This is not an issue when screening samples pooled two dimensionally as polymerase errors are random and two-dimensional pooling requires the presence of a replicating signal in precise gel coordinates (see Note 9).
2. Due to the high cost of gene-specific fluorescently end labeled primers, it is advisable to perform trial reactions with unlabeled primers to test for amplification efficiency. A yield of 7–10 ng/μL of final PCR product is required for robust and consistent identification of mutations when pooling samples eightfold. Primer trials can be discontinued when consistent amplification is achieved and failure rates are suitably low to render trials cost ineffective. Universal primer strategies aimed at using IRDye-labeled universal primers to reduce costs have been described (16, 22, 23). Such strategies may require additional optimization such as side-by-side comparisons with gene-specific end-labeled primers.
3. Primers should be designed to produce an amplicon size between 725 and 1,600 bp. Larger amplicons are possible, but different acrylamide mixtures and electrophoresis conditions are needed to improve band resolution at higher molecular weights. Primers are designed using the Primer 3 program with an optimal primer length of 24 bp and maximal primer length of 30 bp to accommodate a higher melting temperature primer that results in less assay background noise. Primer design and target selection can be performed in a single browser using the CODDLe input utility followed by CODDLe (24)

and http://www.proweb.org/input/. This system has been extensively used for public TILLING services and >90% primer success rate has been described when used for *Arabidopsis thaliana* gene targets (25). PCR amplification may be inhibited with amplicons of high GC (>50%). In this case, a final concentration of 3% TMSO or 0.5 M betaine has been used in some cases to improve amplification (B.J. Till, J.L. Cooper, C. Codomo, M. Darlow, C. Burtner, K. Young, L. Comai and S. Henikoff, unpublished). It has also been observed that the ramp time between primer annealing and Taq extension can be important for robust amplification of high GC templates, with faster ramp times producing higher yields (S. Henikoff and C. Codomo, unpublished).

4. Special care should be taken when using primers labeled with IRDye 700 and IRDye 800. When possible, avoid prolonged exposure of labeled primers and PCR products to fluorescent lights. Primer stocks should be diluted to no more than 100 μM, aliquoted, and stored at −80°C. The smallest amount of primer mix is made and then discarded after use. Fluorescent signal can dramatically decrease when using primer stocks that have undergone repeated freeze–thaw cycles. IRDye-labeled primers often produce lower yield in PCR amplifications than unlabeled versions containing the same oligonucleotide sequence. To obtain consistently high PCR product yield, a mixture of both labeled and unlabeled primers is used in the PCR mix. Presumably, unlabeled primers anneal better to genomic templates and are important to increase target molecules in the early steps of amplification to a critical mass for labeled amplification. An alternative two-step PCR method has been described for the amplification of partially degraded genomic samples from *Drosophila melanogaster* (26).
5. Reaction volumes for this protocol are scaled for 96-well assays. Pipetting is aided by the use of multichannel pipettors, and assays are compatible with liquid handling robotics such as Hydra and Apricot microdispensers. Assay volumes are suitable for standard 384-well microtiter plates, master mix volumes can be scaled, and the protocol followed as described. A single person can perform assays using six 96-well plates or four 384-well plates at one time.
6. A crude CJE containing the CEL I enzyme has been used for the discovery of thousands of mutations by the TILLING method. Use of a crude extract from *Brassica* petioles has also been described, as well as the enzyme ENDO1 from *Arabidopsis* (27). The commercially available Surveyor nuclease can be used for TILLING and Ecotilling in addition to S1 nuclease and crude and purified versions of mung bean nuclease (21).

Care should be taken when exploring alternative enzymes for mismatch cleavage as extensive optimization may be required.

7. For rice, we use a concentration of 0.06 ng/μl with samples pooled eightfold in a two-dimensional format (7). The optimal concentration for high-quality TILLING and Ecotilling images depends on many factors including PCR amplification, CJE digestion, and sample purification and therefore can vary from lab to lab. A genomic DNA titration can be performed in a single 96-well assay to determine the optimal amount of genomic DNA for TILLING or Ecotilling (28).
8. The quality and quantification of genomic DNA starting material are crucial for TILLING and Ecotilling applications. A variety of different kits and methods of DNA extraction may be suitable, including the FastDNA kit previously described for *Arabidopsis*, *Drosophila*, maize, rice, and soybean TILLING (7, 25, 26, 29, 30). Extraction and sample quantification methods should be thoroughly tested before deciding on a strategy and scaling-up operations. Accurate DNA quantification is essential when pooling samples prior to screening. Large-scale TILLING projects have relied on simple agarose gel assays for genomic DNA quantification, and this can be considered a standard by which to compare other methods (14, 25, 26).
9. DNA pooling provides higher throughput by allowing less machine time per sample for mutation discovery. To date, the highest pooling used for large-scale TILLING screens of diploid organisms has been eightfold. Studies in *Arabidopsis* have shown that heterozygous mutations representing 1 in 16 in an eightfold pool are recovered at the expected 2:1 ratio compared to homozygous mutations, suggesting that mutation discovery is robust in eightfold pools (31). Both one-dimensional and two-dimensional sample pooling and arraying strategies have been described (14). In one-dimensional pooling, 768 unique plants can be screened in a single mutation screen. Pools where mutations have been identified then need to be deconvolved and DNA from individual plants screened to determine the plant in the original pool harboring the mutation (15). In the two-dimensional strategy, samples are first arrayed in an 8 × 8 grid and then pooled by row and by column. 384 unique samples are screened in a single 96-well assay and a true mutation will produce bands on two lanes of the gel (Fig. 1). Lane coordinates provide a key to determine the exact plant harboring the mutation. While the one-dimensional strategy can offer a higher screening throughput, the two-dimensional strategy provides an independent confirmation of mutations in the same assay, potentially reducing discovery errors. The two-dimensional strategy

also reduces the number of liquid handling steps that can be a source of human error. For these reasons, it is suggested to start TILLING projects with two-dimensional pooling of samples. A one-dimensional strategy can be considered when TILLING or Ecotilling have been thoroughly established and increased throughput is desired.

10. Sample purification removes salts and detergents prior to gel loading and allows for sample volume reduction via evaporation. Sephadex columns provide a rapid and robust method for sample purification, but in theory any method that provides salt removal and allows for volume reduction can be considered. Alcohol precipitation has been described as an alternative method (32).
11. It is important to deposit sample directly over the center of the Sephadex column, thus avoiding any disturbance of the column such as touching the column with pipet tips.
12. When using a multichannel pipettor and the gel loading tray to load membrane combs, gel lane to sample well correspondence can differ depending on how samples are deposited into the tray (e.g. column or row priority, 8 or 12 channel pipettor). Decide on a method and then determine the sample wells that correspond to lanes 4, 12, 20, and so on. A simple guide for the deposition of the 200 bp marker into the proper wells can be created using and old 96 tip box insert. Color the rim of the holes corresponding to the appropriate wells. Prior to deposition of the 200 bp marker, place the guide above or below sample plate, and deposit 200 bp marker following colored circles.
13. Gel plates may be pre-poured and stored for up to several days at 4°C covered in plastic wrap. Each 25-cm plate requires approx 20 mL. To assemble new plates, clean the plates with dilute liquid detergent (i.e., 2% Tween® 20) and a soft scrub brush. Rinse plates with distilled water, wipe down with 0.2 N HCl, rinse with distilled water again, and wipe with isopropyl alcohol. Spacers are cleaned by wiping with a wet tissue. Assemble the pieces with the screws backed off and align the pieces by standing the assembly vertically while tightening the screws. Tighten just beyond where you begin to feel resistance (overtightening will crack plates). Place on a horizontal support. When handling acrylamide or polyacrylamide gels, always wear gloves, as unpolymerized acrylamide is a nerve toxin.
14. It is important that the slot is clean, as any loose acrylamide will inhibit insertion of the comb. For increased visibility of the loading well, insert a background card wedged behind the back plate such that it is centered between the ears of the front plate. Two vertical marks can be made on the card that are a comb width apart and will provide a guide for later inserting the

comb precisely in the middle of the gel, which is necessary so that all of the lanes are scanned.

15. For analysis of two dimensionally pooled samples with the GelBuddy program, row pools and column pools from an 8 × 8 grid should be in adjacent lanes on the gel (Fig. 1). This can be accomplished by several methods. Samples can be pooled into the sample plate in the described order and then with the use of an adjustable width multichannel pipettor sample can be deposited onto the loading tray such that position A1 on the plate represents lane 3, position B1 equals lane 4, A2 equals lane 9, etc. This can also be achieved without the loading tray by using a specialized comb loading robot (Aviso USA). Alternatively, large volume stock plates of pooled DNA can be prepared and arrayed in such a way that row pools and column pools are adjacent after loading into the sample tray by the method of choice (row or column deposition, single, 8 or 12 channel pipettor). This can be easily standardized and is currently the method of choice for TILLING at the Plant Breeding Unit of the FAO/IAEA Joint Programme. Combs can be preloaded and stored at 4°C; however samples are more stable in formamide loading buffer. It is suggested that combs be used within 2 h of application of sample onto teeth. Membrane combs can be washed and reused by soaking used combs in deionized water for a minimum of 40 min. Allow combs to air-dry before reuse.

16. Gels can be run twice, even after a day. The prerun is necessary if the plate has been moved because it is needed for focusing the lasers. After runs are complete, and a new gel is ready, remove the old gel, pour out the buffer from both reservoirs, and clean plates as described. Be sure that the current is off before touching a buffer chamber.

17. For a 1.5-kb fragment, enter the following LI-COR settings: collect time 4 h 15 min, run at 1,500 V, 40 mA, 40 W, 50°C.

18. Practice this step with used combs. Take special care to avoid bending any teeth. If a tooth sticks to the plate, it may not be possible to save it without moving the comb around and thus losing sample from all teeth.

19. Occasionally, LI COR gel images look fuzzy with diffuse and ill-defined lanes, for any of the following reasons: (1) urea is not thoroughly rinsed from the well before adding Ficoll; (2) the comb is pushed into acrylamide when loaded; (3) the comb moves laterally once inserted into the well; (4) the upper buffer chamber is filled too rapidly after the comb is inserted; (5) Ficoll is not rinsed out of well after the comb is removed; and (6) Ficoll is rinsed too thoroughly from the well after the comb is removed.

20. Alternative methods for data analysis are available. Prior to the advent of specialized software, TILLING data analysis was performed using PhotoShop (33). Commercial software is available through Softgenetics® (http://www.softgenetics.com/jelMarker_tilling.html).
21. The same bench protocols are used for TILLING and Ecotilling. The main modifications for Ecotilling are in experimental design of sample arraying and in data analysis. For sample arraying, non-pooled samples have been screened for many Ecotilling projects. When screened alone, only heterozygous mutations are discovered. To identify homozygous mutations, an equal amount of wild-type or reference DNA, preferably from a sequenced variety, is added to the test sample prior to PCR and mutation discovery. When the discovery of rare polymorphisms is desired, samples can be pooled and arrayed as in traditional TILLING. Pooling, however, may obscure the proper assignment of common polymorphisms (16). Data analysis for Ecotilling is distinct in that all lanes have inherent information. For example, two samples having no bands in a 1.5 kb gene fragment will be scored as genetically similar for that region. In TILLING, only those lanes with bands (putative induced mutations) are valuable and all other lanes can be disregarded. For Ecotilling, therefore, the proper assignment of failed lanes is crucial. It is also important to properly group bands by molecular weight to ensure accurate genotypic associations. Because of this, data analysis for Ecotilling can be more time consuming than for TILLING. Placement of genetically similar samples in adjacent lanes and the inclusion of technical or biological replicates can facilitate gel analysis. It is therefore advised to perform small-scale tests to evaluate different sample arraying strategies before engaging in a large-scale Ecotilling effort.
22. In gene targets with a high percentage of nucleotide polymorphisms, GelBuddy may erroneously link polymorphisms between the IRDye 700 and IRDye 800 channel images causing errors in banding pattern groupings. Because linkages between the two channels cannot be edited, proper banding patterns can be catalogued by marking bands in only one of the two image channels (B.J. Till, J. Jankowicz-Cieslak, and C. Mba, unpublished).

Acknowledgments

The methods described here were developed with support from the National Science Foundation Plant Genome Research Program.

References

1. Ahloowalia BS, Maluszynski M, Nichterlein K (2004) Global impact of mutation-derived varieties. Euphytica 135:187–204
2. Parry MAJ, Madgwick PJ, Bayon C, Tearall K, Hernandez-Lopez A, Baudo M, Rakszegi M, Hamada W, Al-Yassin A, Ouabbou H, Labhilili M, Phillips AL (2009) Mutation discovery for crop improvement. J Exp Bot 60:2817–2825
3. McCallum CM, Comai L, Greene EA, Henikoff S (2000) Targeted screening for induced mutations. Nat Biotechnol 18:455–457
4. Colbert T, Till BJ, Tompa R, Reynolds S, Steine MN, Yeung AT, McCallum CM, Comai L, Henikoff S (2001) High-throughput screening for induced point mutations. Plant Physiol 126:480–484
5. Comai L, Young K, Till BJ, Reynolds SH, Greene EA, Codomo CA, Enns LC, Johnson JE, Burtner C, Odden AR, Henikoff S (2004) Efficient discovery of DNA polymorphisms in natural populations by Ecotilling. Plant J 37:778–786
6. Till BJ, Afza R, Bado S, Huynh OA, Jankowicz-Cieslak J, Matijevic M, Mba C (2009) In: Shu QY (ed) Induced plant mutations in the genomics era. Food and Agriculture Organization of the United Nations, Rome, pp 237–239
7. Till BJ, Cooper J, Tai TH, Colowit P, Greene EA, Henikoff S, Comai L (2007) Discovery of chemically induced mutations in rice by TILLING. BMC Plant Biol 7:19
8. Suzuki T, Eiguchi M, Kumamaru T, Satoh H, Matsusaka H, Moriguchi K, Nagato Y, Kurata N (2008) MNU-induced mutant pools and high performance TILLING enable finding of any gene mutation in rice. Mol Genet Genomics 279:213–223
9. Sato Y, Shirasawa K, Takahashi Y, Nishimura M, Nishio T (2006) Mutant selection from progeny of gamma-ray-irradiated rice by DNA heteroduplex cleavage using Brassica petiole extract. Breeding Science 56:179–183
10. Bruce M, Hess A, Bai J, Mauleon R, Diaz MG, Sugiyama N, Bordeos A, Wang GL, Leung H, Leach JE (2009) Detection of genomic deletions in rice using oligonucleotide microarrays. BMC Genomics 10:129
11. Rios G, Naranjo MA, Iglesias DJ, Ruiz-Rivero O, Geraud M, Usach A, Talon M (2008) Characterization of hemizygous deletions in citrus using array-comparative genomic hybridization and microsynteny comparisons with the poplar genome. BMC Genomics 9:381
12. Li X, Lassner M, Zhang YL (2002) Deleteagene: a fast neutron deletion mutagenesis-based gene knockout system for plants. Comp Funct Genomics 3:158–160
13. Rogers C, Wen J, Chen R, Oldroyd G (2009) Deletion based reverse genetics in *Medicago truncatula*. Plant Physiol.
14. Till BJ, Zerr T, Comai L, Henikoff S (2006) A protocol for TILLING and Ecotilling in plants and animals. Nat Protoc 1:2465–2477
15. Till BJ, Colbert T, Codomo C, Enns L, Johnson J, Reynolds SH, Henikoff JG, Greene EA, Steine MN, Comai L, Henikoff S (2006) High-throughput TILLING for Arabidopsis. Methods Mol Biol 323:127–135
16. Till BJ, Zerr T, Bowers E, Greene EA, Comai L, Henikoff S (2006) High-throughput discovery of rare human nucleotide polymorphisms by Ecotilling. Nucleic Acids Res 34:e99
17. Gilchrist EJ, Haughn GW, Ying CC, Otto SP, Zhuang J, Cheung D, Hamberger B, Aboutorabi F, Kalynyak T, Johnson L, Bohlmann J, Ellis BE, Douglas CJ, Cronk QC (2006) Use of Ecotilling as an efficient SNP discovery tool to survey genetic variation in wild populations of *Populus trichocarpa*. Mol Ecol 15:1367–1378
18. Nieto C, Piron F, Dalmais M, Marco CF, Moriones E, Gomez-Guillamon ML, Truniger V, Gomez P, Garcia-Mas J, Aranda MA, Bendahmane A (2007) EcoTILLING for the identification of allelic variants of melon eIF4E, a factor that controls virus susceptibility. BMC Plant Biol 7:34
19. Weil CF (2009) TILLING in grass species. Plant Physiol 149:158–164
20. Raghavan C, Naredo MEB, Wang HH, Atienza G, Liu B, Qiu FL, McNally KL, Leung H (2007) Rapid method for detecting SNPs on agarose gels and its application in candidate gene mapping. Mol Breeding 19:87–101
21. Till BJ, Burtner C, Comai L, Henikoff S (2004) Mismatch cleavage by single-strand specific nucleases. Nucleic Acids Res 32:2632–2641
22. Wienholds E, van Eeden F, Kosters M, Mudde J, Plasterk RH, Cuppen E (2003) Efficient target-selected mutagenesis in zebrafish. Genome Res 13:2700–2707
23. Winkler S, Schwabedissen A, Backasch D, Bokel C, Seidel C, Bonisch S, Furthauer M, Kuhrs A, Cobreros L, Brand M, Gonzalez-Gaitan M (2005) Target-selected mutant screen by TILLING in Drosophila. Genome Res 15:718–723
24. CODDLe and PARSESNP Input Form. (Accessed September 16, 2009, at http://www.proweb.org/input/)

25. Till BJ, Reynolds SH, Greene EA, Codomo CA, Enns LC, Johnson JE, Burtner C, Odden AR, Young K, Taylor NE, Henikoff JG, Comai L, Henikoff S (2003) Large-scale discovery of induced point mutations with high-throughput TILLING. Genome Res 13:524–530
26. Cooper JL, Greene EA, Till BJ, Codomo CA, Wakimoto BT, Henikoff S (2008) Retention of induced mutations in a Drosophila reverse-genetic resource. Genetics 180:661–667
27. Triques K, Piednoir E, Dalmais M, Schmidt J, Le Signor C, Sharkey M, Caboche M, Sturbois B, Bendahmane A (2008) Mutation detection using ENDO1: application to disease diagnostics in humans and TILLING and Eco-TILLING in plants. BMC Mol Biol 9:42
28. Elias R, Till BJ, Mba C, Al-Safadi B (2009) Optimizing TILLING and Ecotilling techniques for potato (*Solanum tuberosum* L). BMC Res Notes 2:141
29. Till BJ, Reynolds SH, Weil C, Springer N, Burtner C, Young K, Bowers E, Codomo CA, Enns LC, Odden AR, Greene EA, Comai L, Henikoff S (2004) Discovery of induced point mutations in maize genes by TILLING. BMC Plant Biol 4:12
30. Cooper JL, Till BJ, Laport RG, Darlow MC, Kleffner JM, Jamai A, El-Mellouki T, Liu S, Ritchie R, Nielsen N, Bilyeu KD, Meksem K, Comai L, Henikoff S (2008) TILLING to detect induced mutations in soybean. BMC Plant Biol 8:9
31. Greene EA, Codomo CA, Taylor NE, Henikoff JG, Till BJ, Reynolds SH, Enns LC, Burtner C, Johnson JE, Odden AR, Comai L, Henikoff S (2003) Spectrum of chemically induced mutations from a large-scale reverse-genetic screen in Arabidopsis. Genetics 164:731–740
32. Draper BW, McCallum CM, Stout JL, Slade AJ, Moens CB (2004) A high-throughput method for identifying N-ethyl-N-nitrosourea (ENU)-induced point mutations in zebrafish. Methods Cell Biol 77:91–112
33. Till BJ, Colbert T, Tompa R, Enns LC, Codomo CA, Johnson JE, Reynolds SH, Henikoff JG, Greene EA, Steine MN, Comai L, Henikoff S (2003) High-throughput TILLING for functional genomics. Methods Mol Biol 236:205–220

Chapter 5

Functional Characterization of Rice Genes Using a Gene-Indexed T-DNA Insertional Mutant Population

Ki-Hong Jung and Gynheung An

Abstract

Despite the availability of the finished genome sequence and tools for its analyses, few rice genes have been characterized. Because *Agrobacterium*-mediated transformation causes random T-DNA insertions across the genome, T-DNA can be a good mutagen for functional genomics. Gene-indexed mutants with flanking sequences around inserted T-DNA are valuable resources for accelerating functional characterizations of rice genes. Such mutant lines, covering at least half the rice genome, have been generated through international efforts. Here, we describe approaches that use functional genomics with T-DNA insertional mutagenesis.

Key words: Activation tagging, Gene-indexed, Insertional mutagenesis, Reporter, Rice, T-DNA

1. Introduction

Completion of rice genome sequencing has established a valuable foundation for rice researchers. Availability of genome sequences and functional annotation of individual genes have facilitated further research of this crop. The rice genome, with a compact 389 Mb, is one-sixth the size of the maize genome and 40 times smaller than the wheat genome, making it an excellent model for the study of cereal genomes (1). A total of 32,000–42,000 non-transposable-element-related protein-coding sequences have been detected, and approximately 70% of the predicted rice proteins have homologues in the *Arabidopsis* proteome (2). Although powerful bioinformatics tools are becoming readily available for the annotation of genome sequences, in silico analysis by itself is only indicative and is generally not sufficient to determine functions. Instead, experimental evidence is required for functional validation (3). Out of ~26,000 genes identified in *Arabidopsis*, fewer than

Yinong Yang (ed.), *Rice Protocols*, Methods in Molecular Biology, vol. 956,
DOI 10.1007/978-1-62703-194-3_5, © Springer Science+Business Media, LLC 2013

10% have been functionally defined with great confidence (4); the number from rice is even lower (1). Furthermore, the prevalence of gene families poses major challenges in functional genomic analysis. TIGR rice genome pseudomolecules release 5.0 (http://rice.plantbiology.msu.edu/) has recently identified 3,842 rice paralogous gene families consisting of 20,729 protein sequences (5, 6). This poses a particular obstacle to functional studies because it is often the case that a gene, when mutated, will display no detectable phenotype (7, 8).

Agrobacterium-mediated transformation has been used as an effective mutagen and as a tool for functional genomics in higher plants such as rice and *Arabidopsis* (1, 9, 10). Approximately 172,500 flanking sequence tags (FSTs) have been reported in rice. Using the FST database in the Rice Functional Genomic Express Database (RiceGE, http://signal.salk.edu/cgi-bin/RiceGE), we have found that 27,551 (48%) of the 57,142 loci (given by the International Rice Genome Sequencing Project and the TIGR rice genome pseudomolecules release 5.0) contain insertions in their genic regions or 5′-untranslated regions (1). These gene-indexed insertional populations are very useful for facilitating the functional analysis of a gene of interest.

Another type of insertional population consists of lines that bear gain-of-function phenotypes. Gene cassettes of such activation-tagged lines have a strong enhancer element near one border of the T-DNA, which can boost the expression of genes located up to 10 kb away from the enhancer sequence, independent of the direction of transcription. This results in ectopic or increased expression of the targeted gene (11, 12). We have produced more than 47,900 lines carrying tetramerized cauliflower mosaic virus 35S enhancer sequences, and have generated 27,621 FSTs from those lines (12). This activation-tagging approach is one means to deal with the problem of gene redundancy that is coupled with conventional screens for loss-of-function mutations. A main benefit of this approach is that high-throughput gain-of-function studies can be conducted by utilizing both genome-wide expression profiles and a gene-indexed activation-tagging population. However, phenotypes identified by this system must be corroborated by general gain-of-function analyses because more than one gene at the insertion site can be simultaneously activated by the enhancer.

Singletons or predominant gene family members in association with a specific condition can be identified from genome-wide expression profiles. Analyzing insertional mutants in these genes increases the chance of finding mutant phenotypes. By screening T-DNA insertional knockout lines of 25 singletons or gene family members that are significantly up-regulated by light, we have obtained ten mutant lines in which the phenotypes co-segregate with T-DNA (13). Multi-silencing of redundant genes can be a solution for elucidating functions when a single mutant does not lead to any phenotypic changes (13, 14). In addition, genome-wide transcriptomes provide a list of significant genes associated with several biotic/abiotic stresses or at various developmental stages. More than 3,000 hybridizations are currently available from the NCBI public microarray database Gene Expression Omnibus (GEO, http://www.ncbi.nlm.nih.gov/geo/) (15). This information enables us to take systemic functional approaches related to agriculturally valuable traits, including disease resistance and tolerance to drought or submergence.

Another useful tool is the entrapment-tagging system, which is engineered to carry a promoterless reporter gene, e.g., *beta-glucuronidase* (*GUS*) or *green fluorescent protein* (*GFP*), next to a border of T-DNA (16–18). The *GUS* gene has frequently been used for entrapment tagging because its gene product can be accurately detected by a simple assay. At least 5% of the lines generated with a GUS-entrapment vector are GUS positive in various tissues, such as roots, leaves, flowers, and seeds (9, 11). Utilization of reporter genes can facilitate the functional identification of tagged genes by screening defective phenotypes linked to expression patterns. Using this approach, we have characterized several mutants that show defective phenotypes in various organs. One problem with the GUS reporter, however, is its destructive nature during staining and destaining procedures (16). Therefore, nondestructive reporter genes, such as *GFP* or *luciferase*, are also useful for entrapment (17, 18).

Here, we detail three approaches for characterizing rice genes using a T-DNA gene-indexed mutant population.

2. Materials

2.1. Isolation of Flanking Sequence Tags via Inverse PCR

1. Rice (*Oryza sativa ssp. japonica* cv. Dongjin or Hawyoung), which is insertionally mutagenized by using T-DNA of *pGA2707*, *pGA2715*, *pGA2717*, and *pGA2772* vectors (9, 11, 12, 17, 19) (Fig. 1a).
2. Genomic DNAs of pooled seedlings.
3. Extraction buffer: 2% (w/v) cetyl trimethylammonium bromide (CTAB), 1.42 M NaCl, 20 mM ethylenediaminetetraacetic acid (EDTA, pH 8.0), 2% polyvinylpyrrolidone 40, 5.5 mM ascorbic acid, and 4 mM diethylthiocarbamic acid; store at 4°C and incubate at 65°C before use.
4. Restriction enzymes for inverse PCR: *Pst*I, *Cla*I, *Xho*I, and *Sac*I.
5. T4 DNA ligase.
6. ExTaq polymerase, 10× buffer, 0.2 mM dNTP.

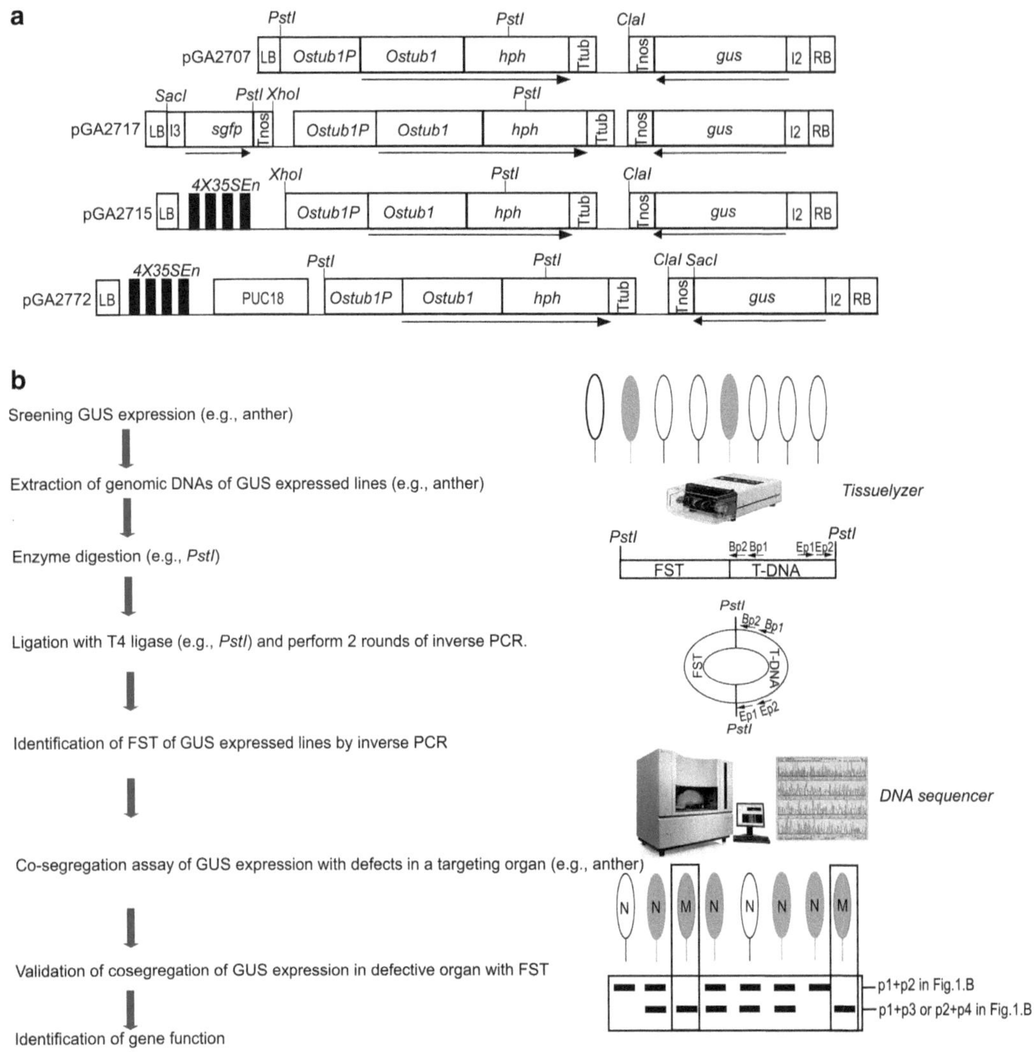

Fig. 1. Functional identification of gene-entrapment lines showing tissue/organ-preferential GUS expression. (**a**) T-DNA vectors with *GUS* reporter gene for gene entrapment. pGA2707, pGA2715, pGA2717, and pGA2772 are T-DNA vectors developed by our group. *Arrows* indicate direction of coding transcripts; *I2* and *I3,* intron 2 and intron 3 of *OsTubA1*, respectively, carrying three putative splicing acceptor and donor sites; *OsTub1P*, promoter of *OsTubA1*; *OsTub1*, first intron of *OsTubA1*; *Ttub*, *OsTubA1* terminator; *Tnos*, nos terminator; *Cla*I, *Pst*I, *Sac*I, and *Xho*I, restriction enzyme sites; *sgfp*, *synthetic green fluorescent protein* (*gfp*) gene; RB and LB, right and left borders of T-DNA, respectively; *hph*, *hygromycin phosphotransferase* gene; *gus, beta-glucuronidase* gene; *4X35SEn*, a tetramer of *35S* promoter enhancer; *PUC18*, pUC18 sequence for plasmid rescue. (**b**) Procedure for functional identification via gene-entrapment system. The first step is to screen GUS-positive lines and then identify flanking sequence tags by inverse PCR. Next, screen the phenotypes in next generation associated with GUS expression and select lines in which all defective mutants are GUS positive. The last step is to confirm cosegregation of the phenotype and T-DNA insertion by genotyping analysis. Gray color indicates GUS expression; Bp1, reverse primer near border of T-DNA for 1st round of inverse PCR; Bp2, for 2nd round; Ep1 is forward primer near restriction enzyme site (e.g., *Pst*I) for 1st round of inverse PCR; Ep2, for 2nd round; FST, flanking sequence tags obtained by inverse PCR of T-DNA tagging lines; M, progeny showing defective phenotypes associated with GUS expression; N, progeny with normal phenotype; p1, forward primer of target gene for genotyping; p2, reverse primer; p3, reverse primer near RB for detection of T-DNA; p4, forward primer near LB.

7. Primers for inverse PCR (17, 19).
8. 96-well PCR machine.
9. 1% agarose.
10. Gel electrophoresis.
11. ABI Prism 3700 DNA analyzer.

2.2. GUS Assay

1. 1 mM EDTA, pH 8.0.
2. 5 mM potassium ferricyanide; store at 4°C.
3. 5 mM potassium ferrocyanide; store at 4°C.
4. 100 mM sodium phosphate, pH 7.0; store at 4°C.
5. 1% Triton-X-100.
6. 5-bromo-4-chloro-3-indolyl-beta-D-glucuronic acid cyclohexylammonium salt (X-Gluc); store at –20°C (for Gus solution preparation refer to step 9).
7. 20% methanol.
8. Dimethyl sulfoxide (DMSO).
9. Prepare GUS solution containing 1 mM X-Gluc, 0.5 mM potassium ferricyanide, 0.5 mM potassium ferrocyanide, 0.1% (v/v) Triton X-100, 100 mM sodium phosphate buffer (pH 7.0), several drops of DMSO, and 20% methanol (see Note 2).

2.3. Genotyping of Gene-Indexed T-DNA Insertional Lines

1. Web browser useful for identification of T-DNA insertional gene-indexed knockout or activation-tagging lines: RiceGE (http://signal.salk.edu/cgi-bin/RiceGE) or the Rice Annotation Project database (RAP-DB, http://rapdb.dna.affrc.go.jp/) (1, 20).
2. Seeds of T-DNA insertional mutants at a target gene.
3. Primers for genotyping: Two gene-specific primers and one primer complementary to T-DNA near the left border (21) or the right border (GUS reverse primer, 5'-CATCACTTCCTGATTATTGACC-3' or 5'-AACGCTGATCAATTCCACAG-3', which are common for all T-DNA insertional tagging vectors developed by POSTECH) (see Note 1). Gene-specific primers can be deduced through the iSect Primer Tool at RiceGE (http://signal.salk.edu/cgi-bin/RiceGE?JOB=TEXT&TYPE=TDNA&QUERY=PFG_2C-30054.L).
4. 96-well PCR Thermal Cycler.
5. 10× TBE buffer: 1 M Tris base, 0.9 M boric acid, and 0.01 M EDTA (pH 8.0).
6. Agarose gel, 1.0–1.2% in 1× TBE buffer.
7. Standard gel electrophoresis equipment.

2.4. Validation of Enhanced Expression of Target Genes in Activation-Tagging System

1. TRIzol reagent (Invitrogen); store at 4°C.
2. 70% and 100% ethanol.
3. TE buffer: 10 mM Tris–HCl and 1 mM EDTA (pH 8.0).
4. 3 M sodium acetate (pH 5.2); store at −20°C.
5. Primers to 3′UTR of the transcript.
6. SuperScript III reverse transcriptase (Invitrogen).
7. RQ1 RNase-free DNaseI (Promega).
8. RNeasy MinElute Cleanup Kit (Qiagen).
9. Other materials are the same as those in Subheading 2.1.

3. Methods

3.1. Identification of Organ/Tissue-Preferential Expression Patterns Through GUS Assay

1. Harvest samples.
2. Soak samples in a reaction tube containing GUS solution and incubate at 37°C for 24 h.
3. Remove X-Gluc solution and incubate samples at 65°C for 1 h after adding 75% ethanol.
4. Repeat step 4 until chlorophylls are removed.
5. Observe GUS staining by microscopy.
6. Identify T-DNA insertional lines showing tissue-preferential GUS activity.

3.2. Identification of FSTs for T-DNA Insertional Lines Showing Organ/Tissue-Preferential Expression Patterns

1. Digest 10 μl of genomic DNAs with a restriction enzyme (e.g., *Pst*I) in a 30-μL reaction at 37°C overnight.
2. Deactivate enzyme by incubating reaction solution at 75°C for 10 min.
3. Self-ligate genomic DNA with 1 μl of T4 DNA ligase in a ligation buffer, and deactivate ligase at 75°C for 10 min.
4. Remove 1 μL of reaction solution and perform two rounds of inverse PCR with nested primer pairs in a 20-μL reaction.
5. Separate bands for amplified fragments on 1.0–1.2% agarose gel in 1× TBE buffer.
6. Elute DNA from bands with a purification kit (Promega) (see Note 3).
7. Sequence 1–10 ng of purified DNAs (Fig. 1b).

3.3. Genotyping of Gene-Indexed T-DNA Insertional Lines

1. Collect 25–60 mg of fresh young rice leaf tissue in a 1.5-mL microcentrifuge tube, and cool on ice.
2. Add a 5- or a 3-mm steel ball to each tube, close their lids, and dip tubes in a liquid nitrogen bath.

3. Homogenize frozen tissue immediately using a Mixer mill MM200 (Retsch) with a 5-mm ball at 13–15 frequencies per second for 30 s, or with a 3-mm ball at 17–20 frequencies per second for 1 min, until tissue turns into fine powder.
4. Add 700 μl of prewarmed extraction buffer (65°C) and 7 μl of RNase A (20 mg/mL), mix well by vortexing briefly, and incubate at 65°C for at least 5 min.
5. Add 570 μl of chloroform:isoamyl alcohol (24:1) to tube and invert several times to mix; do not vortex.
6. Spin tubes in a microcentrifuge at full speed (15,700×*g*) for 10 min and transfer supernatants to clean tubes.
7. Precipitate DNA by adding 0.7 volumes of isopropanol. Mix and centrifuge at full speed for 10 min.
8. Wash pellet in 300 μl of 70% ethanol, spin down in centrifuge for 2 min, and remove all ethanol carefully. Air-dry and resuspend pellet in 20 μl of TE.
9. Remove 1 μl of solution and measure its concentration with a Nanodrop ND-1000 (Nanodrop) or by spectrophotometer after dilution in distilled water.
10. Perform PCR with 10–50 ng of genomic DNA as described in Figs. 1b and 2b.

3.4. Functional Identification of Organ/Tissue-Preferential Genes with T-DNA Insertional Lines Carrying GUS Activity

1. Identify lines showing preferential GUS activity in a desired organ or tissue.
2. Isolate flanking sequences of these lines with inverse PCR.
3. Grow and genotype 15–20 progeny.
4. Observe phenotypic defects in organs or tissues where GUS was expressed.
5. Confirm co-segregation of genotype and phenotype (Fig. 1b).

3.5. Functional Identification of Gene Family Members Associated with Specific Expression Pattern Using Related Genome-Wide Expression Profiles

1. Generate a candidate gene list based on interesting genome-wide expression data (5, 13).
2. Identify candidate genes that have at least two alleles in the RiceGE Web browser.
3. Query the candidate gene list to the rice paralogous gene family database (6, 13).
4. Identify singletons or predominantly expressed gene family members associated with specific treatments (Fig. 2a) (see Notes 4 and 5).
5. Grow 15–20 progenies.
6. Observe phenotype showing defects.
7. Perform two sets of PCRs with genomic DNAs from normal and defective segregants (Fig. 2b).

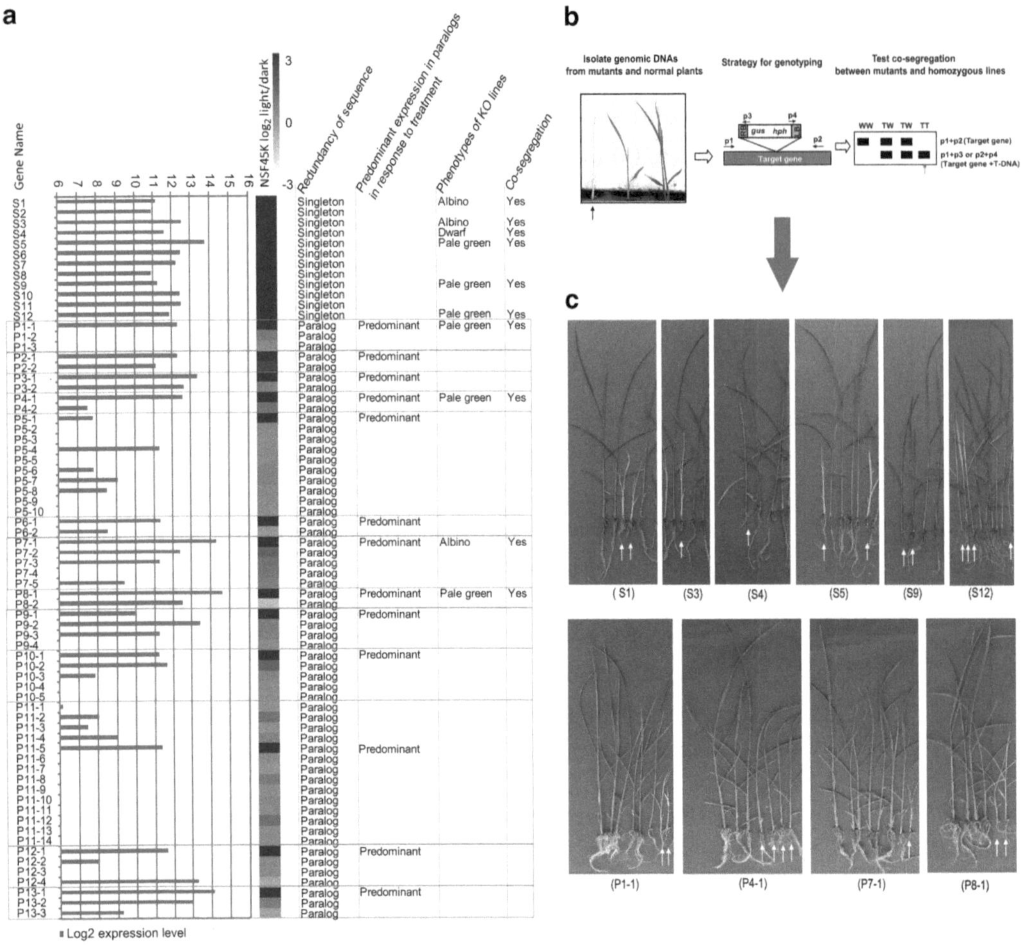

Fig. 2. (**a**) Functional identification of singletons or predominantly expressed gene family members associated with light response using gene-indexed T-DNA insertional lines. White and black colors in heat map indicate down-regulation and up-regulation, respectively. (**b**) Strategy for genotyping T-DNA insertional lines. *WW* wild-type progeny, *TW* heterozygous progeny, *TT* homozygous progeny. (**c**) Examples of T-DNA insertional lines defective in light response showing cosegregation of T-DNA with observed phenotypes. Where T-DNA is inserted into singleton, lines are marked as S. Where T-DNA is inserted into a gene family member that is predominantly expressed in light, lines are marked as P. *Black arrows* in (**b**) and white ones in (**c**) indicate progenies showing mutant phenotypes. For other legends, refer to Fig. 1.

8. Examine co-segregation of genotype and phenotype (Fig. 2c).
9. Confirm phenotype with another allele.

3.6. Identification of Activation-Tagging Lines Showing Enhanced Expression of Target Genes

1. Collect 50–100 mg of fresh tissues from putative activation-tagging lines for target genes in a 1.5-mL RNA-free microcentrifuge tube; hold on ice.
2. Perform steps 2 and 3 outlined in Subheading 3.3.
3. Add 1 mL of TRIzol reagent to tubes containing fine powders.
4. Extract total RNA according to manufacturer's instructions (Invitrogen).

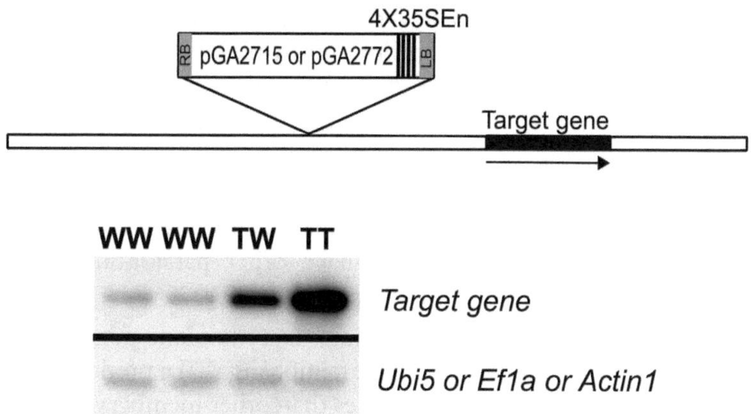

Fig. 3. Procedure for evaluating T-DNA-mediated activation-tagging lines. Refer to Figs. 1 and 2 for the procedure and abbreviations.

5. Add DNase I to tubes and incubate for 30 min at 37°C or RT.
6. Purify total RNAs according to manufacturer's instructions (Qiagen).
7. Synthesize cDNAs with 1–5 μg of total RNAs in a 20-μL reaction at 50°C for 1 h according to manufacturer's instructions (Invitrogen).
8. Normalize cDNAs with a set of control primers designed from housekeeping genes such as *Ubi5* (Os03g08020), *Ef1a* (Os01g22490), or *Act1* (*Os03g50885*): 20–22 cycles of PCR are required for *Ubi5* primer pairs and 22–24 cycles for *Ef1a* and *Act1* (22).
9. Execute quantitative reverse transcriptase-PCR with primer pairs for a target gene.
10. Confirm genotypes after gel electrophoresis (Fig. 3).

4. Notes

1. Left-border primers are 5′-TTGGGGATCCTCTAGAGTCGAG-3′ (for pGA2715), 5′-ACAAGCCGTAAGTGCAAGTG-3′ (for pGA2707), 5′-ACCTCGTCGAGAATTCAGTAC-3′ (for pGA2717), and 5′-TAGCTAGAGTCGAGAATTCAGT-3′ (for pGA2772). These are applicable with gene-specific reverse primers when the combination of the GUS reverse primer and the gene-specific primer cannot generate a PCR product.
2. Methanol is included to reduce endogenous *GUS* activity, which is quite high in some tissues.

3. It is often more efficient to sequence the PCR product directly without purifying DNA fragments through gel electrophoresis if PCR produces a single product.
4. If no match is found for a queried gene in the rice paralogous gene family list provided by Jung et al. (5), the gene is likely a singleton and a good candidate for screening mutant phenotypes using T-DNA insertional lines. Otherwise, the candidate genes have other paralogous gene family members in the genome.
5. Mutations in redundant genes often cause clear phenotypic changes, implying that a predominant member plays more crucial roles than other members. The predominant ones could be selected by their expression level and differential expression patterns.

References

1. Jung KH, An G, Ronald PC (2008) Towards a better bowl of rice: assigning function to tens of thousands of rice genes. Nat Rev Genet 9:91–101
2. IRGSP (2005) The map-based sequence of the rice genome. Nature 436:793–800
3. Radhamony RN, Prasad AM, Srinivasan R (2005) T-DNA insertional mutagenesis in Arabidopsis: a tool for functional genomics. Electron J Biotechnol 8:82–106
4. Brown DM, Zeef LA, Ellis J, Goodacre R, Turner SR (2005) Identification of novel genes in Arabidopsis involved in secondary cell wall formation using expression profiling and reverse genetics. Plant Cell 17:2281–2295
5. Jung KH, Dardick C, Bartley LE, Cao P, Phetsom J, Canlas P, Seo YS, Shultz M, Ouyang S, Yuan Q, Frank BC, Ly E, Zheng L, Jia Y, Hsia AP, An K, Chou HH, Rocke D, Lee GC, Schnable PS, An G, Buell CR, Ronald PC (2008) Refinement of light-responsive transcript lists using rice oligonucleotide arrays: evaluation of gene-redundancy. PLoS One 3:e3337
6. Lin H, Ouyang S, Egan A, Nobuta K, Haas BJ, Zhu W, Gu X, Silva JC, Meyers BC, Buell CR (2008) Characterization of paralogous protein families in rice. BMC Plant Biol 8:18
7. Harrison R, Papp B, Pal C, Oliver SG, Delneri D (2007) Plasticity of genetic interactions in metabolic networks of yeast. Proc Natl Acad Sci USA 104:2307–2312
8. Laeremans T, Coolsaet N, Verreth C, Snoeck C, Hellings N, Vanderleyden J, Martinez-Romero E (1997) Functional redundancy of genes for sulphate activation enzymes in Rhizobium sp. BR816. Microbiology 143(Pt 12):3933–3942
9. Jeon JS, Lee S, Jung KH, Jun SH, Jeong DH, Lee J, Kim C, Jang S, Yang K, Nam J, An K, Han MJ, Sung RJ, Choi HS, Yu JH, Choi JH, Cho SY, Cha SS, Kim SI, An G (2000) T-DNA insertional mutagenesis for functional genomics in rice. Plant J 22:561–570
10. Errampalli D, Patton D, Castle L, Mickelson L, Hansen K, Schnall J, Feldmann K, Meinke D (1991) Embryonic lethals and T-DNA insertional mutagenesis in Arabidopsis. Plant Cell 3:149–157
11. Jeong DH, An S, Kang HG, Moon S, Han JJ, Park S, Lee HS, An K, An G (2002) T-DNA insertional mutagenesis for activation tagging in rice. Plant Physiol 130:1636–1644
12. Jeong DH, An S, Park S, Kang HG, Park GG, Kim SR, Sim J, Kim YO, Kim MK, Kim J, Shin M, Jung M, An G (2006) Generation of a flanking sequence-tag database for activation-tagging lines in japonica rice. Plant J 45: 123–132
13. Jung KH, Lee J, Dardick C, Seo YS, Cao P, Canlas P, Phetsom J, Xu X, Ouyang S, An K, Cho YJ, Lee GC, Lee Y, An G, Ronald PC (2008) Identification and functional analysis of light-responsive unique genes and gene family members in rice. PLoS Genet 4:e1000164
14. Miki D, Shimamoto K (2004) Simple RNAi vectors for stable and transient suppression of gene function in rice. Plant Cell Physiol 45:490–495
15. Barrett T, Troup DB, Wilhite SE, Ledoux P, Rudnev D, Evangelista C, Kim IF, Soboleva A, Tomashevsky M, Marshall KA, Phillippy KH,

Sherman PM, Muertter RN, Edgar R (2009) NCBI GEO: archive for high-throughput functional genomic data. Nucleic Acids Res 37:D885–D890

16. Jefferson RA, Kavanagh TA, Bevan MW (1987) GUS fusions: beta-glucuronidase as a sensitive and versatile gene fusion marker in higher plants. EMBO J 6:3901–3907
17. Ryu CH, You JH, Kang HG, Hur J, Kim YH, Han MJ, An K, Chung BC, Lee CH, An G (2004) Generation of T-DNA tagging lines with a bidirectional gene trap vector and the establishment of an insertion-site database. Plant Mol Biol 54:489–502
18. Haseloff J, Siemering KR, Prasher DC, Hodge S (1997) Removal of a cryptic intron and subcellular localization of green fluorescent protein are required to mark transgenic Arabidopsis plants brightly. Proc Natl Acad Sci USA 94:2122–2127
19. An S, Park S, Jeong DH, Lee DY, Kang HG, Yu JH, Hur J, Kim SR, Kim YH, Lee M, Han S, Kim SJ, Yang J, Kim E, Wi SJ, Chung HS, Hong JP, Choe V, Lee HK, Choi JH, Nam J, Park PB, Park KY, Kim WT, Choe S, Lee CB, An G (2003) Generation and analysis of end sequence database for T-DNA tagging lines in rice. Plant Physiol 133:2040–2047
20. Tanaka T, Antonio BA, Kikuchi S, Matsumoto T, Nagamura Y, Numa H, Sakai H, Wu J, Itoh T, Sasaki T, Aono R, Fujii Y, Habara T, Harada E, Kanno M, Kawahara Y, Kawashima H, Kubooka H, Matsuya A, Nakaoka H, Saichi N, Sanbonmatsu R, Sato Y, Shinso Y, Suzuki M, Takeda J, Tanino M, Todokoro F, Yamaguchi K, Yamamoto N, Yamasaki C, Imanishi T, Okido T, Tada M, Ikeo K, Tateno Y, Gojobori T, Lin YC, Wei FJ, Hsing YI, Zhao Q, Han B, Kramer MR, McCombie RW, Lonsdale D, O'Donovan CC, Whitfield EJ, Apweiler R, Koyanagi KO, Khurana JP, Raghuvanshi S, Singh NK, Tyagi AK, Haberer G, Fujisawa M, Hosokawa S, Ito Y, Ikawa H, Shibata M, Yamamoto M, Bruskiewich RM, Hoen DR, Bureau TE, Namiki N, Ohyanagi H, Sakai Y, Nobushima S, Sakata K, Barrero RA, Sato Y, Souvorov A, Smith-White B, Tatusova T, An S, An G, Oota S, Fuks G, Messing J, Christie KR, Lieberherr D, Kim H, Zuccolo A, Wing RA, Nobuta K, Green PJ, Lu C, Meyers BC, Chaparro C, Piegu B, Panaud O, Echeverria M (2008) The Rice Annotation Project Database (RAP-DB): 2008 update. Nucleic Acids Res 36:D1028–D1033
21. Jung KH, Hur J, Ryu CH, Choi Y, Chung YY, Miyao A, Hirochika H, An G (2003) Characterization of a rice chlorophyll-deficient mutant using the T-DNA gene-trap system. Plant Cell Physiol 44:463–472
22. Jain M, Nijhawan A, Tyagi AK, Khurana JP (2006) Validation of housekeeping genes as internal control for studying gene expression in rice by quantitative real-time PCR. Biochem Biophys Res Commun 345:646–651

Chapter 6

QTL Analysis and Map-Based Cloning of Salt Tolerance Gene in Rice

Ji-Ping Gao and Hong-Xuan Lin

Abstract

Most agronomic traits are governed by quantitative trait loci (QTLs) and exhibit continuous distribution in a segregating population. The hereditary characteristics of these traits are more complicated than those of monogenic traits. Detection and isolation of these QTLs can greatly improve crop production throughout the world. In recent times, significant progress has been made toward understanding the molecular basis underlying quantitative traits. Herein, we describe a QTL-mapping protocol for detecting and cloning a major QTL regulating rice shoot K^+ concentration under salt stress conditions. This QTL-mapping approach combined with the marker-assisted selection technique can be applied for the elucidation of complex traits in rice and other cereal crops.

Key words: Rice, QTL analysis, Map-based cloning, Salt tolerance

1. Introduction

Many important agricultural traits such as plant height, crop yield, and abiotic stress tolerance are controlled by multiple genes. These traits are called continuous traits because they always exhibit a continuous distribution in a segregating population. For example, if two inbred lines (or varieties) are crossed, the phenotypic characteristics of the F_2 population are mostly intermediate to those of the two parents and the distribution of the F_2-generation phenotypes resembles a normal or Gaussian distribution. These traits cannot be classified into several discrete phenotypic classes that are usually used for predicting the genotypes of the discontinuous traits. Because phenotypic evaluations of continuous traits often yield quantitative data, these traits are also called quantitative traits, and the loci controlling these traits are referred to as quantitative trait loci (QTLs).

Yinong Yang (ed.), *Rice Protocols*, Methods in Molecular Biology, vol. 956,
DOI 10.1007/978-1-62703-194-3_6, © Springer Science+Business Media, LLC 2013

Unlike discontinuous traits, quantitative traits do not follow the typical patterns of dominance and recessiveness of Mendelian inheritance. The continuous distribution of traits is attributed to the independent segregation of many small-effect Mendelian factors and the additive effect of each involved locus. Furthermore, the phenotypes of these genes are complicated by distinct inter-gene interactions and the interactions between genes and their environment.

There has been a long-standing interest in attempts to understand the genetic basis of quantitative traits. QTL analysis is a statistical study that links genotypes with phenotypes, aiming to define the number and chromosomal locations of QTLs related to a trait of interest, to assess the effect of each QTL on the phenotype, and to demonstrate whether these QTLs interact with one another (1, 2). A genetic marker, which is an identifiable region of variable DNA, has a statistical association with a given trait. The basis of QTL analysis is the identification of associations between genetic markers and quantitative phenotypes. In order to facilitate this, a set of genetic markers must be developed from the parents of a mapping population. A mapping population is then used for genome-wide linkage map construction and phenotype evaluation. Subsequently, statistical methods are applied to determine whether an association exists between any of the markers and the quantitative trait. The simplest approach in QTL analysis is the analysis of variance (ANOVA) at the marker loci, i.e., the use of a t-test or an F-test to determine whether the differences between genotype group averages are significant. However, this single-marker ANOVA approach has several weaknesses, including the inability to provide separate estimates of QTL location and QTL effect. Furthermore, the ability to detect QTLs will decrease when the markers are widely spaced. In contrast to single-marker analysis, interval mapping is used to provide evidence for the presence of a QTL between two markers. This approach is based on maximum likelihood and calculates the likelihood of a QTL across all intervals; this will locate the most likely QTL position and estimate its effect with improved precision and reliability. Thus, interval mapping is currently the most commonly used analytical approach for detecting QTLs.

Rice is one of the most important staple cereal crops in the world. The elucidation of complex agronomic traits by QTL mapping can have a great impact on rice breeding. During the past two decades, developments in molecular biology have dramatically advanced the application of QTL analyses in rice (3, 4). The ultimate objective of QTL mapping is to identify candidate genes, or even nucleotide polymorphisms, underlying the natural variations associated with the trait of interest. While achieving this objective is still a major challenge, a few traits, e.g., heading date (5–8), yield (9–11), and stress tolerance (12–14), have been analyzed by QTL mapping.

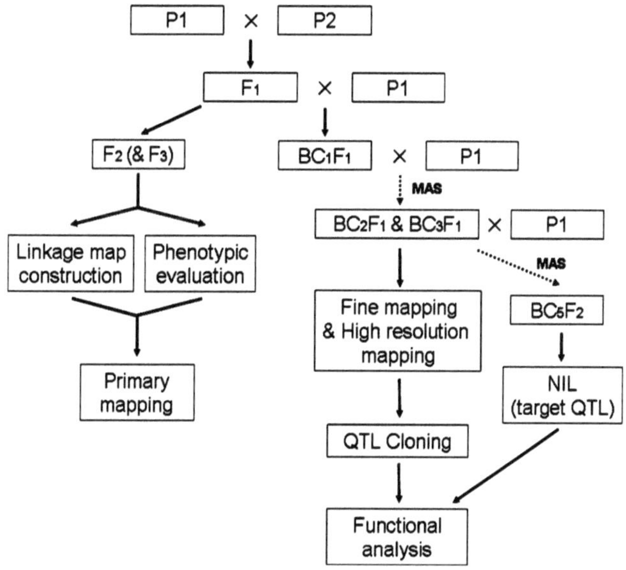

Fig. 1. Flowchart depicting the strategy for mapping and cloning of QTL. *MAS* marker-assisted selection, *NIL* near-isogenic line.

Soil salinity is a key abiotic stress affecting crop productivity worldwide. Under salt stress, excess Na^+ ions are transported from the root to the shoot and accumulate in photosynthetic tissues, resulting in growth inhibition and yield reduction. Salt tolerance in plants is the ultimate manifestation of several physiological processes, including Na^+ uptake and exclusion, Na^+ compartmentation, and ion balance (especially K^+/Na^+ ratio) and distribution. The capacity of plants to maintain a high K^+/Na^+ ratio is a key determinant of achieving salt tolerance. Some physiological traits related to rice salt tolerance have been investigated using QTL analyses, such as days of seedling survival, Na^+ and K^+ concentration, Na^+ and K^+ quantity, and K^+/Na^+ ratio (15–19). Researchers at our laboratory have identified *SKC1*, an important QTL that maintains shoot K^+ concentration (SKC) in a salt-tolerant variety under salt stress, and found that it encodes a high-affinity K^+ transporter (HKT)-type sodium transporter (12, 19). Herein, we describe a detailed procedure for mapping and cloning of *SKC1* by QTL analysis and a map-based cloning approach (Fig. 1).

2. Materials

2.1. Generation of QTL Mapping Populations

1. Seeds of rice varieties: *Oryza sativa* ssp. *indica* cv. Nona Bokra (a variety with high salt tolerance) and *Oryza sativa* ssp. *japonica* cv. Koshihikari (a susceptible elite variety). Though we use rice varieties that differ with respect to salt tolerance as an example,

the approach outlined here can be applied to identify QTLs for other traits. In general select varieties that have a large difference for your trait of interest.

2. Equipment for crossing: Scissors, tweezers, and glassine paper bags.

2.2. Construction of Restriction Fragment Length Polymorphism Linkage Map

1. Cetyl trimethylammonium bromide (CTAB).
2. Restriction enzymes: *Apa*I, *Bam*HI, *Bgl*II, *Dra*I, *Eco*RI, *Eco*RV, *Hind*III, and *Kpn*I.
3. Agarose.
4. 0.4 N NaOH.
5. Nylon membrane.
6. 2× standard saline citrate (SSC) buffer: Dissolve 17.53 g of NaCl and 8.82 g of sodium citrate in dH_2O, adjust pH to 7.0, and bring the final volume to 1 L using dH_2O.
7. Enhanced chemiluminescence (ECL) direct nucleic acid labeling and detection system (RPN3001; Amersham).
8. X-ray film and X-ray film cassettes.
9. Software for linkage mapping analysis: MAPMAKER/EXP 3.0.

2.3. Evaluation of Shoot K^+ Content Under Salt Stress

1. Yoshida's cultural solution (20): Add 1 mL aliquots of each stock solution (see below) to 800 mL water. Adjust pH between 5.8 and 6.2 and add water to achieve a final volume of 1 L.

   1,000× Stock solution A: NH_4NO_3 80 g/L, $NaH_2\ PO_4{\cdot}2H_2O$ 93 g/L, K_2SO_4 52.4 g/L.

   1,000× Stock solution B: $CaCl_2{\cdot}2H_2O$ 44.2 g/L.

   1,000× Stock solution C: $MgCl_2{\cdot}6H_2O$ 122 g/L.

   1,000× Stock solution D: FeNaEDTA 19 g/L.

   1,000× Stock solution E: H_3BO_3 3.01 g/L, $MnSO_4{\cdot}5H_2O$ 2.17 g/L, $CuSO_4{\cdot}5H_2O$ 0.075 g/L, $ZnSO_4{\cdot}7H_2O$ 0.2008 g/L, $Na_2MoO_4{\cdot}2H_2O$ 0.024 g/L.
2. NaCl.
3. 100 mM acetic acid.
4. Atomic absorption spectrophotometer (AA-680; Shimadzu, Japan).

2.4. QTL Analysis

Software for QTL analysis: MAPMAKER/QTL.

2.5. Fine Mapping and High-Resolution Mapping

1. TPS buffer: 100 mM Tris–HCl, 1 M KCl, 10 mM EDTA; pH 8.0.
2. Isopropanol.
3. 70% ethanol.

4. TE buffer: 10 mM Tris–HCl, 1 mM EDTA; pH 8.0.
5. 2 mM dNTP.
6. *Taq* DNA polymerase.

2.6. Validation of the Candidate Gene

1. *Pfu* DNA polymerase.
2. Restriction enzymes: *Bam*HI, *Eco*RI, *Kpn*I, and *Xba*I.
3. Plasmids: pUC18, pBluescript SK+, and pHB.
4. *Escherichia coli* strain DH5α and *Agrobacterium tumefaciens* strain EHA105.
5. Antibiotics: Ampicillin, kanamycin, and hygromycin B.
6. *Oryza sativa* ssp. *japonica* cv. Zhonghua 11.

3. Methods

3.1. Generation of A Primary QTL-Mapping Population

1. The individuals of the two parental lines, namely, Nona Bokra, a highly salt-tolerant *indica* variety, and Koshihikari, a salt-susceptible *japonica* variety, are crossed by hand pollination (see Note 1) and the F_1-generation seeds are harvested.
2. F_1 plants are grown and selfed to obtain F_2-generation seeds (see Note 2). Several F_1 plants are backcrossed with Koshihikari plants (recurrent parent) to generate a BC_1F_1 population, and individuals from this BC_1F_1 population are used to perform fine mapping and high-resolution mapping (as described in Subheading 3.5).
3. F_2 plants ($n = 133$) are grown (see Note 3), selfed, and F_3-generation seeds are harvested from each F_2 parent plant.

3.2. Construction of a Genome-Wide RFLP Linkage Map

1. Genomic DNAs are extracted from green leaves of the two parent plants and the 133 F_2 plants by the CTAB method (21).
2. To identify polymorphic restriction fragment length polymorphism (RFLP) markers for linkage map construction, Southern blotting is performed using genomic DNAs isolated from the leaves of Nona Bokra and Koshihikari plants. Each restriction digestion is performed using 2 μg of genomic DNA and one of the following enzymes: *Apa*I, *Bam*HI, *Bgl*II, *Dra*I, *Eco*RI, *Eco*RV, *Hind*III, and *Kpn*I. The resulting fragments are separated by electrophoresis on a 0.6% agarose gel and transferred onto a positively charged nylon membrane in 0.4 N NaOH for 4 h. Then, the membrane is washed using 2×SSC solution, dried, and baked at 120°C for 20 min for DNA fixation. All probes (genomic DNAs or cDNAs) are amplified by polymerase chain reaction (PCR) and labeled with horseradish peroxidase (HRP) according to the protocol provided in the ECL

direct nucleic acid labeling and detection system. HRP-labeled probes are hybridized to the membrane; ECL reagents are added to the membrane and the membrane is then exposed to an X-ray film for 1–3 h.

3. The signals on the X-ray film revealed a total of 161 RFLP markers spanning the whole rice genome indicating polymorphisms between the two parent varieties; these markers are selected for conducting RFLP analysis of the F_2 population.
4. All 161 markers are used to determine RFLP genotypes of the F_2 plants by Southern blotting using the abovementioned procedure. The genotype scores of each of the analyzed markers for all the individuals of the mapping population are recorded, and the genotype of each plant in the F_2 population is then determined.
5. The genotype data is entered in an Excel file according to the instructions of the MAPMAKER/EXP 3.0 program (22, 23). The genotype scores are represented as follows: A (or 1), homozygous for Koshihikari; H (or 2), heterozygous; and B (or 3), homozygous for Nona Bokra.
6. The MAPMAKER/EXP 3.0 software is used to construct a genome-wide RFLP linkage map, and the map distances are presented in terms of Kosambi centiMorgans (cM). A logarithm of odds (LOD) score of 3.0 is used to determine both the linkage groups and the order of markers. A linkage map of rice covering 12 chromosomes and spanning 1,704.4 cM of the rice genome with an average distance of 12.3 cM between two marker loci is shown in Fig. 2.

3.3. Evaluation of Shoot K^+ Content Under Salt Stress

1. The seeds of 133 F_3 lines originating from different F_2 plants and their parental lines, Nona Bokra and Koshihikari, are exposed to 45°C for 1 week to break any possible dormancy (see Note 4).
2. The seeds are exposed to 35°C for 48 h for germination and 16 uniform germinated seeds of each line are sown in the holes of a thin Styrofoam board with a nylon net bottom. The seedlings are then incubated in water for 3 days at 30°C.
3. The seedlings are transferred to Yoshida's cultural solution and grown under a 13-h light/11-h dark photoperiod at 26°C. The culture solution is renewed after every 2 days.
4. At 8 days after sowing, the seedlings are transferred into a fresh Yoshida's cultural solution containing 140 mM NaCl. This culture solution is changed after every 2 days.
5. At 10 days after NaCl treatment, the aerial parts of the 16 plants are harvested and pooled as one sample. Each sample is dried at 90°C for 4 h, weighed, and extracted in 15 mL of 100 mM acetic acid for 2 h at 90°C.

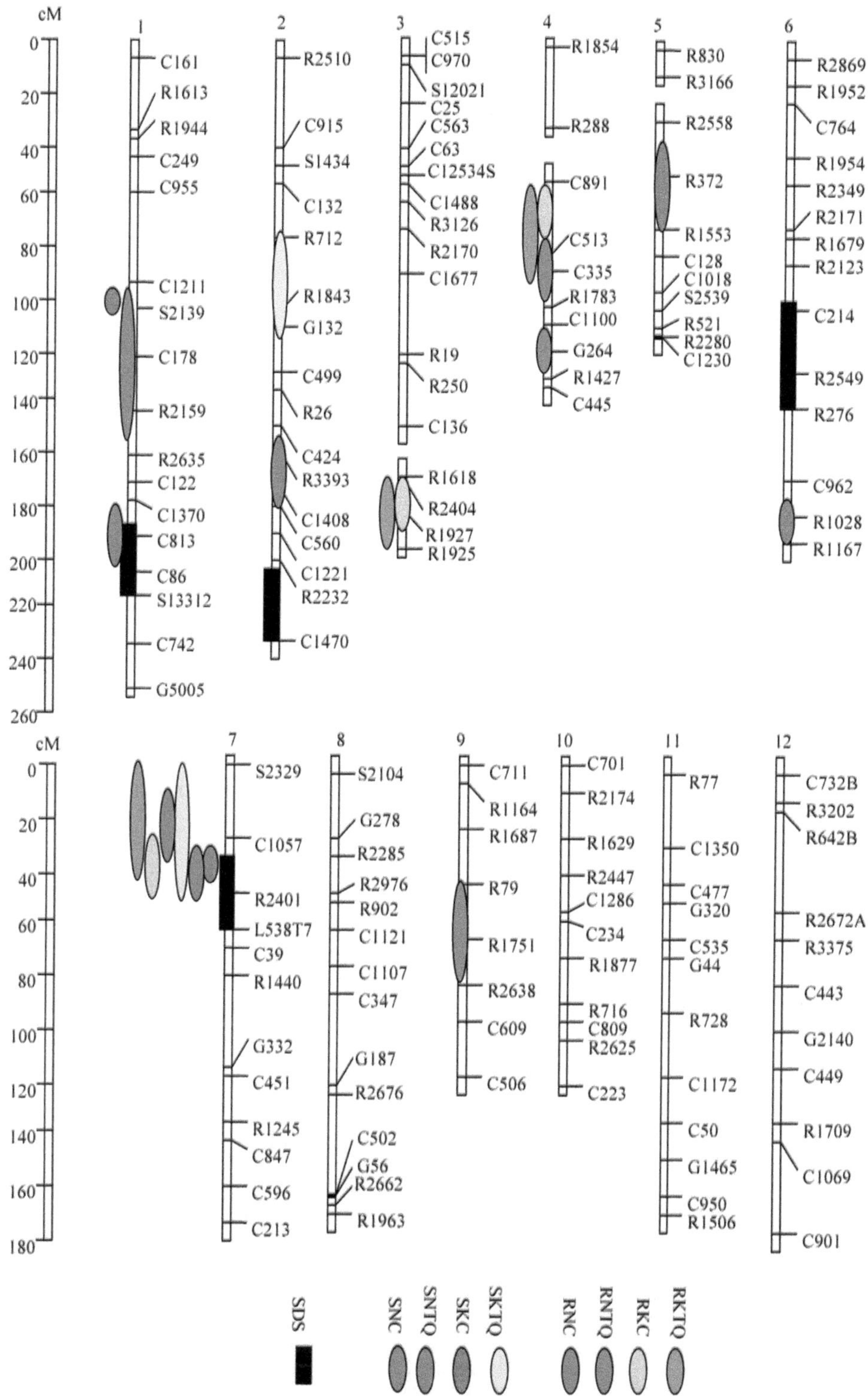

Fig. 2. Genetic linkage map showing the location of QTLs for *SKC1* and other traits related to salt tolerance in the Nona Bokra/Koshihikari F_2 population. Distances are in Kosambi centiMorgans. The *boxes* or the *ellipses* on the chromosomes represent putative regions of QTLs. A reduction of a one LOD value from LOD peaks was used to define left and right borders of the confidence interval for QTLs. *SDS* survival days of seedling, *SNC* shoot Na^+ concentration, *SNTQ* shoot Na^+ total quantity, *SKC* shoot K^+ concentration, *SKTQ* shoot K^+ total quantity, *RNC* root Na^+ concentration, *RNTQ* root Na^+ total quantity, *RKC* root K^+ concentration, *RKTQ* root K^+ total quantity (reproduced from ref. 19 with permission from Springer).

6. The K^+ content of the extract is determined three times for each sample by using atomic absorption spectrophotometer. The K^+ in the shoot is expressed in terms of dry weight (μmol/mg dry weight).

3.4. QTL Analysis

1. The K^+ concentration values are entered in an Excel file, and the mean and standard deviation are estimated for each F_3 line.
2. The phenotype data (K^+ concentration) is combined with the genotype data (RFLP scores of each line), according to the instructions of the MAPMAKER/QTL program (24, 25).
3. Interval mapping analysis is performed by using the MAPMAKER/QTL program to identify the QTLs affecting SKC under salt stress conditions. A LOD score of 3.0 is used to identify the presence of putative QTLs in a given genomic region. The phenotypic variation explained by each QTL is calculated in terms of percentage and the additive effect of different QTLs is estimated by using the MAPMAKER/QTL software. One major QTL, namely, *SKC1*, accounted for 40.1% of the total phenotypic variance for SKC. This *SKC1* locus was mapped to the C1211-S2139 region on chromosome 1 (Fig. 2), and the additive effect of this QTL in the Nona Bokra plants (salt tolerant variety) resulted in an increase in the SKC by 0.11 μmol/mg dry weight (19).

3.5. Fine Mapping and High-Resolution Mapping

1. BC_1F_1 plants are grown and their genotype is analyzed using the 161 RFLP markers. These BC_1F_1 plants are backcrossed with the Koshihikari plants (recurrent parent); the C1211-S2139 region is heterozygous and almost all other regions are homozygous in the former with respect to the corresponding regions in the latter. Finally, BC_2F_1-generation seeds are harvested.
2. To obtain a near-isogenic line, the previously described backcrossing and marker-assisted selection procedure is repeated, and BC_5F_1-generation seeds are harvested.
3. PCR-based markers, such as cleaved amplified polymorphic sequence (CAPS), sequence-tagged site (STS), and simple sequence repeats (SSR, or microsatellites), are used instead of RFLP markers in the subsequent experiments (see Note 5). Genomic DNA is extracted and PCR is performed according to the following procedure:

   A small piece of rice leaf (about 2 cm) is crushed in a 1.5-mL microfuge tube containing 300 μl of TPS buffer and the tube is incubated at 75°C for 20 min. Then, the solution is centrifuged at 12,000 × *g* for 10 min and the DNA obtained in the resulting supernatant is precipitated by adding the same volume of isopropanol. The pellet thus obtained is washed with 70% ethanol and dissolved in 50 μl 0.1 × TE. A 1-μl aliquot of this DNA extract is used as the template for PCR amplification. The 10-μl reaction volume contains 1 μl of template DNA, 1 μl 10×PCR buffer, 1 μl 25 mM $MgCl_2$, 1 μl of each dNTP (2 mM), 2 μl of 50% glycerol, 0.1 μl *Taq* DNA polymerase (5 U/μl), 0.2 μl of a 20 μM solution of each primer, and 3.7 μl dH_2O. The PCR program is as follows: 94°C for 5 min; 30 cycles of 94°C for 30 s, 60°C for 1 min, and 72°C for 1 min; and 72°C for 10 min.

4. The BC_2F_1 plants are selfed, and 192 BC_2F_2 plants are grown and used for fine mapping. The CAPS or STS markers in the C1211-S2139 region are identified and the genotype of each BC_2F_2 plant is analyzed using these markers.
5. The SKC of the BC_2F_3 lines under salt stress is measured as described in Subheading 3.3.
6. QTL analysis is performed as described in Subheading 3.4. Fine mapping revealed that the *SKC1* locus is delimited to the region between markers K159 and K061 (Fig. 3a).
7. BC_3F_1 plants are selfed, and 2,973 BC_3F_2 plants are grown for high-resolution mapping. The K159 and K061 markers are used to detect recombination events in the BC_3F_2 population. The genotypes of the recombinants are determined using the newly developed markers within the K159-K601 region (Fig. 3b).
8. BC_3F_3 population is derived from the recombinant plants of BC_3F_2 generation. The BC_3F_3 plants are screened for the homozygous recombination products with the markers in the K159-K601 region.
9. The fixed recombinant plants (BC_3F_4) are used to measure the SKC under salt stress by the procedure described in Subheading 3.3. Such progeny-testing revealed that the *SKC1* locus is delimited to a 7.4-kb region between the markers K036 and Pr (Fig. 3b, c).
10. The genotype of the BC_5F_2 population is analyzed to identify a near-isogenic line, i.e., NIL (*SKC1*), in which only the K159-K061 region is homozygous for the donor parent, while the other regions are homozygous for the recurrent parent. This result is confirmed by measuring the SKC in the NIL (*SKC1*) and Koshihikari plants by the method described in Subheading 3.3 (Fig. 3d).

3.6. Validation of Candidate Gene

1. Between the two markers K036 and Pr, only one gene encodes the HKT-type transporter. Two pairs of primers are designed as follows: one containing *Eco*RI/*Kpn*I sites to amplify *SKC1* genomic DNA and the other pair containing *Bam*HI/*Xba*I sites to amplify *SKC1* cDNA (see Note 6). A 2,829-bp *Eco*RI/*Kpn*I fragment (containing a 2,554-bp region upstream of the initial codon and a 275-bp coding region) is amplified

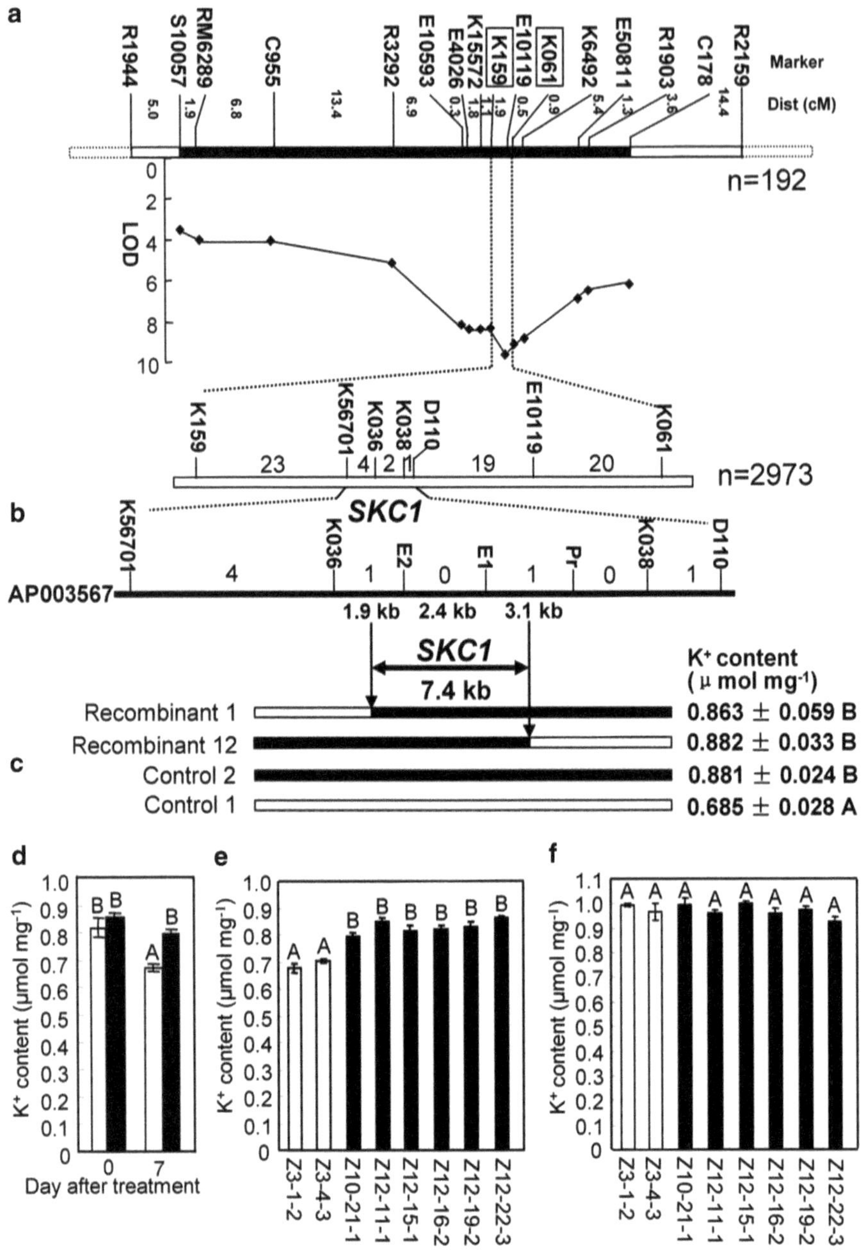

Fig. 3. Identification of the *SKC1* QTL and genetic complementation test. (**a**) The location of *SKC1* on rice chromosome 1 in 192 BC_2F_2 plants in which a small region (*filled bar*) is segregating. (**b**) High-resolution linkage map of the *SKC1* region produced with 2,973 BC_3F_2 plants. The number of recombinants between the adjacent markers is indicated on the linkage map. The *filled bar* shows a part of the PAC clone AP003567. Cleaved amplified polymorphic sequence markers developed from the PAC sequences uncovered a single recombination event between markers K036 and E2 and markers E1 and Pr. (**c**) Progeny testing of fixed recombinant plants (BC_3F_4) delimited the *SKC1* locus to the region between markers K036 and Pr. Shoot K^+ contents (means ± s.e., $n = 4$) of two recombinant lines and control 2 (homozygous for the Nona Bokra in the target region) were substantially higher than those of control 1 (homozygous for the Koshihikari in the target region) under 140 mM NaCl for 7 days. *Filled and open bars* show homozygous chromosomal segments for Nona Bokra and Koshihikari, respectively. (**d**) Shoot K^+ contents of NIL(*SKC1*) (*filled bars*) and Koshihikari (*open bars*) under non-salt stress and 140 mM NaCl. Data are shown as means ± s.e. ($n = 4$). (**e**, **f**) Shoot K^+ contents (means ± s.e., $n = 5$) in T_1 rice transformants under 140 mM NaCl for 7 days (**e**) or under normal condition (**f**). *Open bars* represent empty vector lines and *filled bars* denote *SKC1* lines. Columns with different letters were significantly different ($P < 0.01$, least significant difference test; **c**–**f**). (Reproduced from ref. 12 with permission from Nature publishing group).

from Nona Bokra genomic DNA by using the first pair of primers and cloned into pUC18. A full-length *SKC1* open reading frame (ORF; length, 1,665 bp; containing a *Bst*XI site at the position of 184 bp downstream of the initial codon) is amplified from a cDNA library derived from Nona Bokra by using the second pair of primers and cloned into pBluescript SK+. For each construct, namely, the 2,829-bp *Eco*RI/*Kpn*I fragment and the full-length *SKC1* ORF, 5 independent plasmids are sequenced.

2. To design an *SKC1* promoter-ORF (4,219 bp) construct, the plasmid carrying *SKC1* ORF is digested using *Bst*XI and *Xba*I and the *Bst*XI/*Xba*I-ORF fragment is purified and ligated to the *Bst*XI/*Xba*I sites of pUC18 carrying the *SKC1* genomic DNA.
3. After digestion with *Eco*RI and *Xba*I, the 4,219-bp *SKC1* promoter-ORF fragment is purified and ligated to the *Eco*RI/*Xba*I sites of pHB, a plant binary vector modified from pCAMBIA 3301 (26).
4. This *SKC1* construct is transformed into *Agrobacterium tumefaciens* strain EHA105 and subsequently introduced into rice calli derived from immature embryos of a *japonica* variety, namely, Zhonghua 11, by *Agrobacterium*-mediated transformation (27). The empty pHB vector is also introduced into Zhonghua 11 as a control.
5. Hygromycin B-resistant plantlets (T_0 generation) are transferred into the soil in pots and grown to maturity in a greenhouse. The seeds (T_1 generation) are harvested following selfing of the transformants.
6. After germination, the T_1-generation seeds of six independent *SKC1* transgenic lines and two control lines are grown in water for 3 days at 30°C; subsequently, the seedlings are transferred to Yoshida's cultural solution containing 50 mg/L hygromycin B under a 13-h light/11-h dark photoperiod at 26°C for 4 days.
7. The hygromycin B-resistant seedlings are transferred to a normal Yoshida's cultural solution. After 4 days, the seedlings are divided into two groups (16 plants for each line and five replicates per group). The plants in the salt-stress group are treated with 140 mM NaCl for 7 days, while plants in the other group are grown under normal conditions.
8. The shoots are harvested and the K^+ concentration is measured as described in Subheading 3.3. In comparison with the empty vector control, the SKC in all 6 *SKC1* transgenic lines are substantially high under salt stress but not under normal conditions (Fig. 3e, f). This result confirms that the candidate gene is *SKC1* (12).

4. Notes

1. Most QTL analyses are conducted on individual biparental populations. Therefore, the probability to detect QTL essentially depends on intrinsic trait difference between the two parental lines. In this study, we used a salt-tolerant variety, Nona Bokra, as the donor parent and a salt-susceptible variety, Koshihikari, as the recurrent parent. In addition, a large amount of polymorphism existing in the mapping populations derived from interspecific or intersubspecies crosses facilitates the development of molecular markers.
2. Both immortalized and segregative populations have been used as primary mapping populations in rice. Because of their fixed genotypes, immortalized populations such as recombinant inbred lines (RILs), backcross inbred lines (BILs), doubled-haploid lines (DHLs), and introgression lines (ILs) can be used at different locations and for multiple studies over years; particularly, these populations can be used for evaluating the interaction between QTLs and the environment. When studying a new parent plant, a segregative population of F_2 plants or an advanced backcross population (BC_2F_2) is preferred because these populations can be generated faster than immortalized populations.
3. The precision of QTL analysis depends on the population size; the probability of detection of QTLs in large populations is higher than that in small populations. Previous studies have revealed that QTLs with an explained variance of 10% have an 80% chance of being detected in a population of 200 plants, the probability of detection reducing more or less linearly with population sizes (2). Thus, if the expected explained variance per QTL is less than 10%, one should resort to large-sized populations (2).
4. Phenotype differences can often be detected in F_2 population. However, in some conditions, it is necessary to destroy the F_2 population to detect the phenotype. In this study, we used an equivalent F_3 population to analyze K^+ concentration in plant shoots.
5. Southern blotting analysis is a labor-intensive and time-consuming procedure. This disadvantage of using RFLP markers can be circumvented by using PCR-based markers. Accordingly, we have converted RFLP markers used in linkage map construction into PCR-based markers and developed new PCR-based markers for the subsequent experiments. Meanwhile, we applied a different DNA extraction method (28) to isolate a small amount of genomic DNA because a single PCR reaction required only about 50 ng of DNA.

6. Genomic DNA is amplified by using the first pair of primers (SKC1-*Eco*RI-F1, 5′-GGG GAA TTC CGA CGT GAC AGA TTT GGA TTA C-3′; and SKC1-*Kpn*I-R1, 5′-GGG GGT ACC AGC TGG GAG TTG GAG AAG GAC T-3′). The full-length *SKC1* ORF is amplified by using the second pair of primers (SKC1-*Bam*HI-F2, 5′-GGG GGA TCC ATG AGT TCT CTG GAT GCC AC-3′; and SKC1-*Xba*I-R2, 5′-GGG TCT AGA TTA TTC TAT CTT CCA TGC CTG ACC-3′).

References

1. Xu SZ (2002) QTL analysis in plants. Methods Mol Biol 195:283–310
2. Alonso-Blanco C, Koornneef M, van Ooijen JW (2006) QTL analysis. Methods Mol Biol 323:79–99
3. McCouch SR, Doerge RW (1995) QTL mapping in rice. Trends Genet 11:482–487
4. Ashikari M, Matsuoka M (2006) Identification, isolation and pyramiding of quantitative trait loci for rice breeding. Trends Plant Sci 11:344–350
5. Yano M, Katayose Y, Ashikari M, Yamanouchi U, Monna L, Fuse T, Baba T, Yamamoto K, Umehara Y, Nagamura Y, Sasaki T (2000) Hd1, a major photoperiod sensitivity quantitative trait locus in rice, is closely related to the Arabidopsis flowering time gene CONSTANS. Plant Cell 12:2473–2484
6. Takahashi Y, Shomura A, Sasaki T, Yano M (2001) Hd6, a rice quantitative trait locus involved in photoperiod sensitivity, encodes the alpha subunit of protein kinase CK2. Proc Natl Acad Sci USA 98:7922–7927
7. Kojima S, Takahashi Y, Kobayashi Y, Monna L, Sasaki T, Araki T, Yano M (2002) Hd3a, a rice ortholog of the *Arabidopsis* FT gene, promotes transition to flowering downstream of Hd1 under short-day conditions. Plant Cell Physiol 43:1096–1105
8. Doi K, Izawa T, Fuse T, Yamanouchi U, Kubo T, Shimatani Z, Yano M, Yoshimura A (2004) Ehd1, a B-type response regulator in rice, confers short-day promotion of flowering and controls FT-like gene expression independently of Hd1. Genes Dev 18:926–936
9. Ashikari M, Sakakibara H, Lin S, Yamamoto T, Takashi T, Nishimura A, Angeles ER, Qian Q, Kitano H, Matsuoka M (2005) Cytokinin oxidase regulates rice grain production. Science 309:741–745
10. Song XJ, Huang W, Shi M, Zhu MZ, Lin HX (2007) A QTL for rice grain width and weight encodes a previously unknown RING-type E3 ubiquitin ligase. Nat Genet 39:623–630
11. Fan C, Xing Y, Mao H, Lu T, Han B, Xu C, Li X, Zhang Q (2006) GS3, a major QTL for grain length and weight and minor QTL for grain width and thickness in rice, encodes a putative transmembrane protein. Theor Appl Genet 112:1164–1171
12. Ren ZH, Gao JP, Li LG, Cai XL, Huang W, Chao DY, Zhu MZ, Wang ZY, Luan S, Lin HX (2005) A rice quantitative trait locus for salt tolerance encodes a sodium transporter. Nat Genet 37:1141–1146
13. Ueda T, Sato T, Hidema J, Hirouchi T, Yamamoto K, Kumagai T, Yano M (2005) qUVR-10, a major quantitative trait locus for ultraviolet-B resistance in rice, encodes cyclobutane pyrimidine dimer photolyase. Genetics 171:1941–1950
14. Xu K, Xu X, Fukao T, Canlas P, Maghirang-Rodriguez R, Heuer S, Ismail AM, Bailey-Serres J, Ronald PC, Mackill DJ (2006) Sub1A is an ethylene-response-factor-like gene that confers submergence tolerance to rice. Nature 442:705–708
15. Lin HX, Yanagihara S, Zhuang JY, Senboku T, Zheng KL, Yashima S (1998) Identification of QTLs for salt tolerance in rice via molecular markers. Chinese J Rice Sci 12:72–78
16. Gong JM, He P, Qian Q, Shen LS, Zhu LH, Chen SY (1999) Identification of salt-tolerance QTL in rice (Oryza sativa L.). China Sci Bull 44:68–71
17. Prasad SR, Bagali PG, Hittalmani S, Shashidhar HE (2000) Molecular mapping of quantitative trait loci associated with seedling tolerance to salt stress in rice (*Oryza sativa* L.). Curr Sci 78:162–164
18. Koyama ML, Levesley A, Koebner RMD, Flowers TJ, Yeo AR (2001) Quantitative trait loci for component physiological traits determining salt tolerance in rice. Plant Physiol 125:406–422
19. Lin HX, Zhu MZ, Yano M, Gao JP, Liang ZW, Su WA, Hu XH, Ren ZH, Chao DY (2004)

QTLs for Na^+ and K^+ uptake of the shoots and roots controlling rice salt tolerance. Theor Appl Genet 108:253–260

20. Yoshida S, Forno DA, Cock JH, Gomez KA (1976) Laboratory manual for physiological studies of rice, 3rd edn. International Rice Research Institute, Manila, Philippines
21. Murray MG, Thompson WF (1980) Rapid isolation of high molecular weight plant DNA. Nucleic Acids Res 8:4321–4326
22. Lander ES, Green P, Abrahamson J, Barlow A, Daly MJ, Lincoln SE, Newberg LA (1987) MAPMAKER: an interactive computer package for constructing primary genetic linkage maps of experimental and natural populations. Genomics 1:174–181
23. Stephen EL, Lincoln SE, Lander ES (1990) Constructing genetic maps with MAPMAKER: a tutorial and reference manual. A Whitehead Institute for Biomedical Research, Cambridge, USA
24. Lander ES, Botstein D (1989) Mapping mendelian factors underlying quantitative traits using RFLP linkage maps. Genetics 121:185–199
25. Lincoln S, Dalyand M, Lander E (1993) Mapping genes controlling quantitative traits with Mapmaker/QTL1.1: a tutorial and reference manual, 2nd edn. Whitehead Institute Technical Report, Cambridge, Massachusetts
26. Mao J, Zhang YC, Sang Y, Li QH, Yang HQ (2005) A role for Arabidopsis cryptochromes and COP1 in the regulation of stomatal opening. Proc Natl Acad Sci USA 102:12270–12275
27. Hiei Y, Ohta S, Komari T, Kumashiro T (1994) Efficient transformation of rice (*Oryza sativa* L.) mediated by Agrobacterium and sequence analysis of the boundaries of the T-DNA. Plant J 6:271–282
28. Lin HX, Liang ZW, Sasaki T, Yano M (2003) Fine mapping and characterization of quantitative trait loci *Hd4* and *Hd5* controlling heading date in rice. Breed Sci 53:51–59

Chapter 7

Site-Specific Gene Integration in Rice

Vibha Srivastava

Abstract

Site-specific gene integration is a powerful technique for ensuring stable transgene expression. Transgenic plants produced by conventional transformation techniques often display highly variable transgene expression, which is mostly attributed to integration patterns consisting of multiple copies of transgene constructs. Therefore, it is desirable to generate single-copy integrations, preferably in a characterized genomic position. Precise integration of foreign genes into a selected genomic position can be obtained by employing site-specific recombination systems derived from bacteria or yeast. P1 bacteriophage Cre-*lox* system has been particularly successful in directing precise integration of foreign genes into "previously engineered" genomic sites. The resulting transgenic plants display stable expression through successive generations. Therefore, site-specific integration approach is useful for streamlining production of transgenic plants.

Key words: Transgenic rice, Transgene expression, Site-specific recombination, Site-specific integration, Cre-*lox*

1. Introduction

Since the production of first transgenic plant in 1983, significant progress has been made in extending plant transformation technology from model species to crop plants. Tremendous efforts have also been made in understanding transgene integration and expression processes, which have been crucial in developing strategies for avoiding gene silencing and maximizing gene expression. A survey of transgene expression studies indicates that transgene loci consisting of complex multi-copy integrations express at suboptimal levels and may become suppressed in subsequent generations (1–4). Single-copy loci, on the other hand, mostly remain stable through generations. Although single-copy loci are not completely immune to silencing, it is still considered prudent to select them for long-term propagation (4). While different factors may be involved in silencing single-copy loci, use of strong viral promoters is common to many cases (5). Therefore, use of plant promoters in

Yinong Yang (ed.), *Rice Protocols*, Methods in Molecular Biology, vol. 956,
DOI 10.1007/978-1-62703-194-3_7, © Springer Science+Business Media, LLC 2013

transgene constructs may alleviate the problems associated with silencing of 35S promoters. Transgene integration studies found that DNA repair and nonhomologous recombination, involving formation of concatamers, serve as the predominant mechanisms of foreign gene integration (6–9). As a result, multiple copies of the introduced DNA are integrated; and single-copy integrations are rarely found. *Agrobacterium*-mediated transformation tends to produce a higher percentage of single-copy loci compared to particle-bombardment-mediated transformation (10). Therefore, *Agrobacterium*-mediated transformation methods were developed for a range of crop plants including monocot species (11). However, production of precise single-copy plants is still inefficient (6, 7). In addition, binary vector backbone or even *Agrobacterium* chromosomal DNA may be transferred along with the gene of interest (12–14). Therefore, molecular strategies for engineering single-copy loci have been developed, including the strategies for enhancing homologous recombination (HR) between the introduced DNA and plant nuclear genome (15). Meantime, site-specific recombination (SSR)-mediated transgene integration has shown promise for crop biotechnology (16). In the SSR-mediated transgene integration approach, the gene construct (without the vector backbone) becomes integrated into the "marked" or "previously engineered" genomic site by the action of an SSR system such as Cre-*lox* (17, 18). In the Cre-*lox* system, Cre represents the recombinase enzyme, which catalyzes a recombination reaction between two *lox* sites. A reaction between two *lox* sites, one located on a circular DNA fragment and the other located on a chromosome, results in integration of the circular DNA into the chromosome. This mechanism has been utilized to develop plant transformation procedure designed to generate precise single-copy lines at high rate (19, 20). The present chapter describes a protocol for generating site-specific transgene integrations in rice.

The molecular strategy of site-specific gene integration in rice is based on Cre-mediated recombination between a circular DNA molecule and the plant chromosome, each containing a *lox* site (Fig. 1a). As recombination between wild-type *lox* sites (*loxP* X *loxP*) is freely reversible, mutant *lox* sites are employed to stabilize the integration structure (Fig. 1b). A 34 bp *loxP* sequence consists of a core 8 bp flanked by 13 bp arms (palindromic sequences). Mutant *lox* sites containing 7 bp mutations in either left (*lox75*) or right arm (*lox76*) are utilized for site-specific integrations (Fig. 1c). Recombination between these mutants (*lox75* X *lox76*) results in the formation of a *loxP* and a double-mutant (dm) *lox* site. Since dm *lox* site is a poor substrate of Cre enzyme, a reverse reaction is mostly prevented (Fig.1b). The detailed strategy, shown in Fig. 2, involves the use of two components: (1) "donor" vector containing the gene of interest (GOI) and (2) a rice genotype (target line) containing a single copy of pVS52 (Fig. 3) T-DNA. GOI (complete

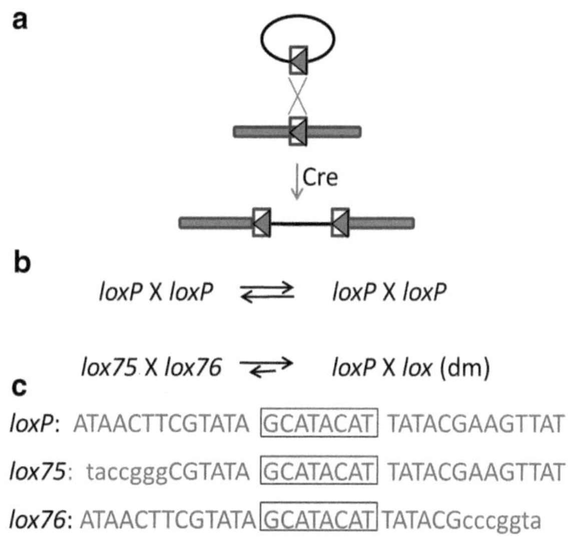

Fig. 1. Mechanism of site-specific integration. (**a**) Co-integration strategy. Cre-mediated recombination between two *lox* sites (*boxed arrowheads*), one resident to the plasmid and another to the chromosome, results in the formation of a co-integrate structure. (**b**) A reaction between *loxP* X *loxP* is reversible. Therefore, mutant *lox* sites are employed, which recombine to produce a double-mutant (dm) site. Since *dm lox* is a poor substrate of Cre recombinase, the reverse reaction is minimized. (**c**) Each *lox* site consists of a 34 bp sequence: 8 bp (boxed) spacer and 13 bp flanking elements. DNA sequences of wild-type (*loxP*) and mutant *lox* sites (*lox75* and *lox76*) are shown.

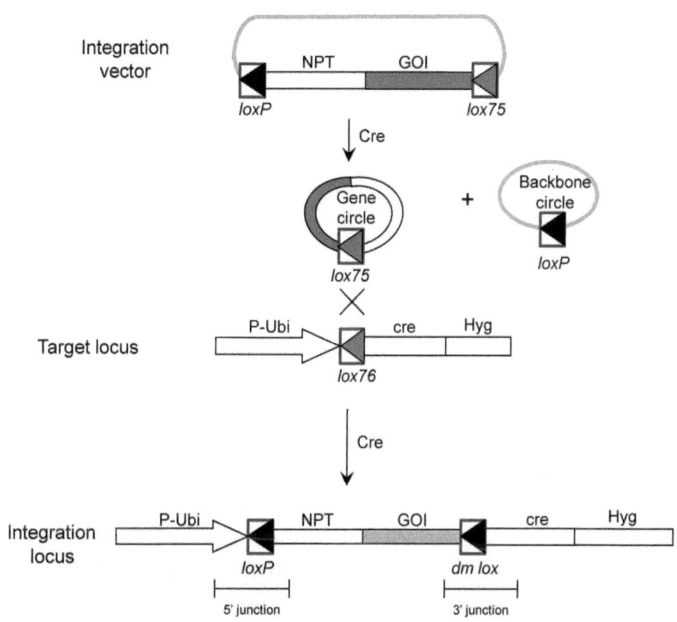

Fig. 2. Detailed strategy of site-specific integration. Introduction of integration vector into target cells by particle bombardment results in separation of gene construct from vector backbone, followed by integration into the target locus. Cre activity provided by the target locus catalyzes recombination between lox75 X lox76 to generate a defined integration locus consisting of distinct 5′ and 3′ junctions. Site-specific placement of promoterless marker gene (NPT) facilitates its expression and makes the event selectable. NPT, promoterless neomycin phosphotransferase gene; GOI, gene of interest; P-Ubi, maize Ubiquitin promoter; Hyg, hygromycin resistance gene.

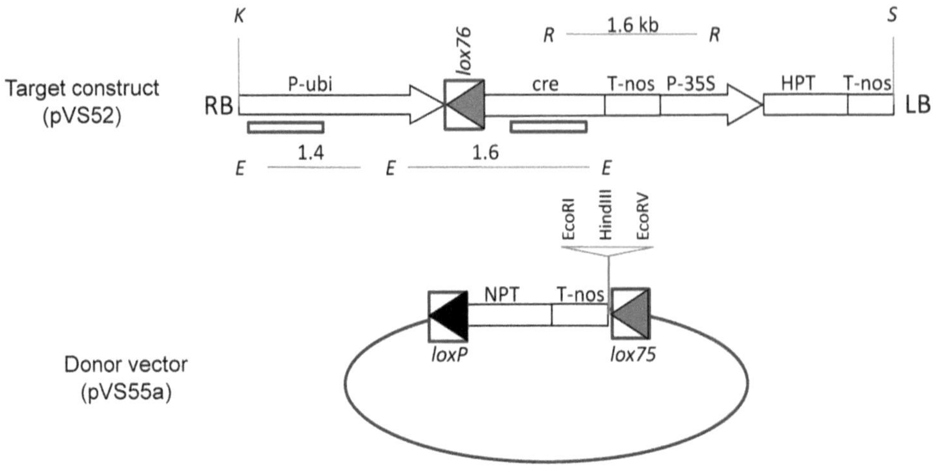

Fig. 3. Detailed map (not to scale) of the target construct (pVS52) in binary vector pPZP200 backbone and donor construct (pVS55a) in pBluescriptSK backbone. Restriction map of pVS52 is indicated. *Open box* below the construct indicates DNA probes used in Fig. 4. RB and LB, right and left T-DNA borders; P-ubi, maize Ubiquitin promoter; T-nos, transcription terminator from nopaline synthase gene; P-35S, CaMV 35S promoter; HPT, hygromycin phosphotransferase gene; NPT, promoterless neomycin phosphotransferase gene; *K*, *Kpn*I; *R*, *Eco*RV; *S*, *Sac*I. Sequence of *lox* sites are shown in Fig. 1.

transcription unit) is incorporated into the multi-cloning site (*Eco*RI, *Hind*III, or *Eco*RV) of the donor vector to generate the final transformation construct. Introduction of this construct into target cells results in Cre-mediated intramolecular recombination separating vector backbone from the "gene circle" and integration of "gene circle" into the chromosomal target site. The resulting site-specific integration locus is stabilized by the formation of a dm *lox* site and selected by the expression of the previously promoterless marker gene (NPT) (see Note 1).

2. Materials

2.1. Design of "Target" and "Donor" Vectors

Target (pVS52) and donor (pVS55a) vectors can be obtained from V. Srivastava, University of Arkansas. The target vector described below was generated by assembling different gene fragments in pPZP200 vector (21). pPZP200 is described at the Addgene vector database (http://www.addgene.org). pVS52 (Fig. 3) contains the following DNA fragments between the right and left border of pPZP200 backbone (in the order given below):

1. Promoter of maize polyubiquitinin gene: 2 kb *Pst*I fragment (NCBI accession no. S94464).
2. A 34 bp *lox76* sequence ((19); Fig. 1).
3. Coding sequence of *cre* gene (1.03 kb) (NCBI reference no. NC_005856).

4. Transcription termination sequence derived from nopaline synthase (*nos*) gene of *Agrobacterium* C58 strain (NCBI accession no. NC_003065).
5. Cauliflower Mosaic Virus 35S promoter (22).
6. Coding region of hygromycin phosphotransferase gene (HPT).
7. Transcription termination sequence from *nos* gene.

The donor vector (pVS55a) can also be obtained from V. Srivastava, University of Arkansas, or assembled using standard cloning techniques. pVS55a was generated by assembling the following DNA fragments in the order described below in a pBluescript SK backbone (Stratagene Inc.):

1. A 34 bp *loxP* sequence (Fig. 1).
2. A promoterless neomycin phosphotransferase II gene (*NPT*) consisting of only the coding sequence (0.8 kb).
3. Transcription termination sequence of *nos* gene.
4. Cloning sites: Such as *Eco*RI, *Hind*III, and *Eco*RV.
5. A 34 bp *lox75* sequence ((19); Fig. 1).

2.2. Agrobacterium-Mediated Transformation of Rice

Rice target lines in Nipponbare or Taipei309 background can be obtained from V. Srivastava, University of Arkansas. Each target line contains a single copy of pVS52 T-DNA. Alternately, new target lines can be developed by transforming rice using *Agrobacterium* harboring pVS52. *Agrobacterium*-mediated rice transformation is described by Mei et al. (23). For selection of rice transformants, use hygromycin B (50 mg/L).

2.3. Generation of Integration Vector

A suitable integration vector (Fig. 2) can be generated by cloning a GOI in the *Eco*RI, *Hind*III, or *Eco*RV site of pVS552 (Fig. 3). After proper verification by multiple restriction digestions, isolate large quantity of integration vector DNA using Endo-free® Plasmid Mega kit (Qiagen). Dissolve in water at 1 μg/μL concentration.

2.4. Biolistic Transformation of Rice

1. Mature seeds of target line.
2. Gene gun: PDS1000/He system (Bio-Rad, Inc.).
3. 1 μg/μL integration vector DNA.
4. 1 μm gold particles: Suspend 60 mg particles in 95% ethanol, vortex vigorously, centrifuge, remove ethanol, rinse with 1 mL sterile water, and add 1 mL of water.
5. Bombardment accessories (Bio-Rad Inc.): Macrocarriers, rupture disks (1,100 psi), screens.
6. Spermidine: 0.1 M in water stored in −20°C.
7. 2.5 M Calcium chloride ($CaCl_2 \cdot 2H_2O$): Make in water and store in −20°C.
8. 200 proof Absolute alcohol.

9. Geneticin (BRL Inc.): 100 mg/mL (store at 4°C).
10. Callus induction medium (N6D): 3.98 g Chu (N6) basal salt, 0.1 g of MS vitamin powder, 2 mg of 2,4-D, 0.5 g casamino acids, 2.5 g proline, 30 g of sucrose, and 2 g of Gelrite. Adjust pH to 5.7, add water to 1 L, and autoclave.
11. Bombardment medium: Callus induction medium with the addition of 63 g sorbitol per liter.
12. Selection medium (N6D-S medium): Callus induction medium containing geneticin (100 mg/L). Geneticin is added after autoclaving.
13. Regeneration medium: 4.4 g of MS basal salt, 0.1 g of MS vitamins, 1 g casamino acids, 2 mg of kinetin, 0.1 mg of α-naphthaleneacetic acid, 30 g of sucrose, 30 g of sorbitol, 3 g Gelrite. Adjust pH to 5.7; add water to 1 L, and autoclave.
14. Rooting medium: 4.4 g of MS basal salt, 0.1 g of MS vitamin, 30 g of sucrose, and 2 g of Gelrite. Adjust pH to 5.7, add water to 1 L, and autoclave. Add 100 mg/L geneticin after autoclave.

2.5. Analysis of Transgenic Plants

1. Cetyltrimethyl ammonium bromide (CTAB) Buffer: 100 mL of 1 M Tris–HCl pH 8.0, 280 mL of 5 M NaCl, 40 mL of 0.5 M EDTA, 20 g of CTAB. Dissolve, and bring total volume to 1 L with ddH_2O.
2. TE buffer: 10 mL of 1 M Tris–HCl pH 8.0, 2 mL of 0.5 M EDTA. Bring total volume to 1 L with ddH_2O.
3. 7.5 M Ammonium acetate pH 5.2.
4. Chemicals for polymerase chain reaction (PCR) and Southern hybridizations.

3. Methods

3.1. Cloning of the Gene of Interest in the "Donor" Vector

In the following section, GUS gene was used as the GOI. Other GOI can be cloned into the pVS55a multi-cloning site using standard cloning techniques (see Note 2). A *Hind*III GUS gene fragment consisting of a maize ubiquitin promoter transcriptionally fused to GUS gene, followed by *nos3′* transcription terminator (Ubi-GUS-*nos3′*), was obtained from pAHC25 (24).

3.2. Generation of Target Lines by Agrobacterium-Mediated Transformation

3.2.1. Agrobacterium Preparation (Also See Ref. 23)

1. Transform pVS52 into *Agrobacterium tumefaciens* strain EHA105 using electroporator (Bio-Rad micropulser) according to the manufacturer's protocol.
2. After electroporation, add 1 mL of YEP medium to the *Agrobacterium* cells and incubate for 3 h at 28°C.
3. Plate about 100 μL of the cells on a YEP solid medium containing 100 μg/mL spectinomycin and 60 μg/mL rifampicin. Incubate the plates at 28°C for 2 days.

4. Pick a single colony and verify the presence of pVS52 by isolating plasmid and performing restriction digestion: *Eco*RV digestion will generate 1.6 and 10.9 kb fragments. Store the verified *Agrobacterium* suspension in a glycerol solution at −70°C for future use.

3.2.2. Transformation of Rice

Use *Agrobacterium strain* EHA105 harboring pVS52 and follow a standard rice transformation protocol (23). Use hygromycin B (50 mg/L) for selection of rice transformants.

3.3. Generation of Site-Specific Integration Lines

3.3.1. Preparation of Callus

1. Dehusk about 100 mature seeds of "target line" and surface sterilize with 70% ethanol for 1 min and then with 30% Clorox® containing 0.1% SDS solution for 30 min with vigorous shaking.
2. Remove Clorox and rinse seeds in sterile water five times. Dry them on sterile filter paper.
3. Using sterile forceps place seeds on callus induction medium (N6D) in Petri dishes.
4. Seal the plates, and incubate them in light at 28°C.
5. After 2–3 weeks, collect the scutellar callus emerging from seeds for transformation.
6. Select ten different pieces of embryogenic callus and place them as a cluster in the middle of bombardment medium on 60 × 15 mm Perti plates.

3.3.2. Preparation of Gold Particles

1. In a 1.5 mL microcentrifuge tube add 50 μL suspension of gold particles (60 mg/mL). Then add 5–10 μg of Integration Vector DNA. Mix by tapping the tube.
2. Lay the tube on its side and place a 50 μL drop of 2.5 M $CaCl_2 \cdot 2H_2O$ solution on the wall of the tube followed by a 20 μL drop of 0.1 M spermidine. Note: The two drops should remain separate.
3. Carefully close the cap of the tube, quickly lift it, and hold it on vortex for vigorous mixing for 3 min.
4. Centrifuge at full speed, remove all liquids, and rinse with 1 mL of absolute ethanol.
5. Remove ethanol, and add 100 μL of absolute ethanol.

3.3.3. Bombardment of Callus Plates

1. Rub the tube (containing DNA-coated gold particles) on an empty box of pipette tips several times to break the gold particle pellet and develop even suspension.
2. Quickly pipette 5–10 μL of gold particles on macrocarriers.
3. Let dry, and bombard on callus plates using 1,100 psi rupture disks.

4. Leave bombarded callus on the bombardment medium overnight at 28°C.
5. Transfer clusters of callus to callus induction medium (N6D) the next day, and incubate plates in light at 28°C.

3.3.4. Tissue Culture

1. After 1 week of incubation on N6D medium, transfer callus to selection medium (N6D containing 100 mg/L geneticin) (see Note 3).
2. Pick geneticin-resistant clones and transfer to a fresh selection plate.
3. Transfer geneticin-resistant callus to regeneration medium, and incubate in light at 28°C (see Note 4).
4. Transfer regenerated shoots to rooting medium containing 100 mg/L geneticin.
5. After 2–3 weeks, rice plantlets can be transplanted to soil in pots.

3.3.5. Analysis of Target- and Site-Specific Integration Lines

1. Extract genomic DNA from leaves of target- and site-specific integration plants using CTAB method (25).
2. Digest 10 μg of target line DNA with *Eco*RV in 200 μL reaction at 37°C overnight. Add 5 μL of RNase (10 mg/mL) to the reaction; incubate at room temperature for an additional 30 min.
3. Re-precipitate the digested DNA and suspend in 30 μL of water.
4. Separate the digested DNA on a 0.8% agarose gel and transfer to a nylon membrane using standard Southern blotting procedure.
5. Hybridize Southern blot to radioisotope-labeled DNA probes overnight (see Note 5). See Fig. 3 for examples of DNA probes.
6. Identify single-copy lines (displaying a 1.6 kb band internal to the construct and an additional band representing transgene–host DNA junction). See Fig. 3 for examples of single-copy lines.
7. Use single-copy target lines for site-specific integration experiments.
8. Once putative site-specific integration lines are obtained, carry out PCR and Southern analysis for confirmation.
9. As described above, digest 10 μg of site-specific integration plant DNA with *Eco*RI or other suitable enzyme, prepare Southern blot, and hybridize sequentially with three different

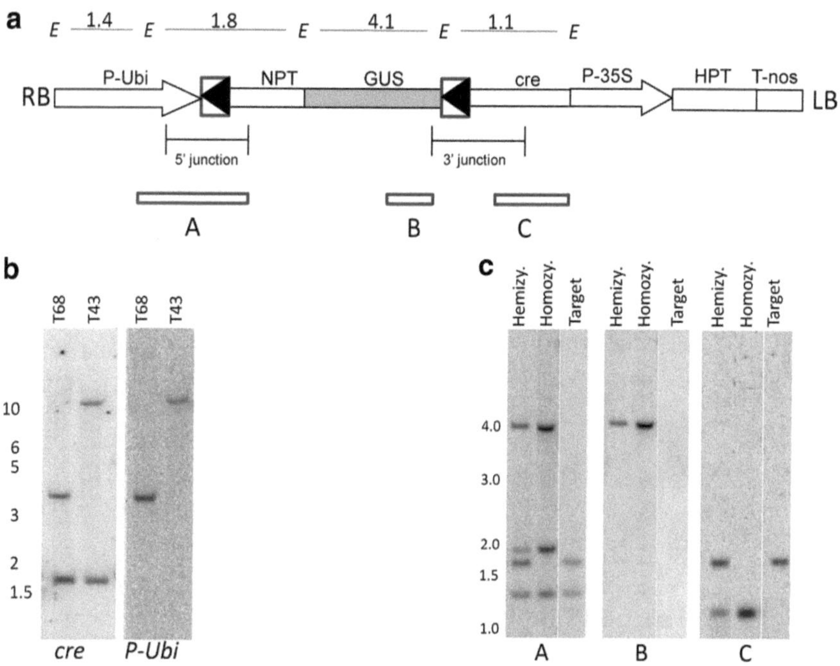

Fig. 4. Analysis of target- and site-specific integration lines. (**a**) Map of site-specific integration locus containing β-glucuronidase gene (GUS) as the gene of interest. *Eco*RI (*E*) map of the locus and location of DNA probes (*open boxes below the map*) is shown. (**b**) *Eco*RV (*R*) digested genomic DNA of target lines, T43 and T68, hybridized with cre and P-Ubi probes (see Fig. 3). Each line represents a single-copy integration of pVS52 construct. (**c**) *Eco*RI-digested genomic DNA of site-specific integration plants hybridized with A, B, and C probes. Target line DNA is present as a control. Hybridization with A probe will generate distinct patterns for hemizygous or homozygous integrations (provided the target locus was homozygous).

DNA probes: 5′ junction probe (PUbi-NPT) (see Note 6), 3′ junction probe (cre), and GOI (see Note 7 and Fig. 4).

10. Generation of 1.8 kb band with PUbi-NPT and 1.1 kb band with cre probes indicates the presence of site-specific integration locus. Absence of additional hybridizing bands with PUbi-NPT and GOI probes suggests the absence of random background integrations. If site-specific integration is derived from a homozygous target locus, presence of target locus bands in addition to the site-specific integration bands is expected unless integration occurs in both alleles of the target locus (see Fig. 4 for examples).
11. Select single-copy lines for additional analysis using PCR. Carry out PCR with primers specific to 5′ and 3′ junctions (see Note 8 and Fig. 2), and determine the presence of PCR products on an agarose gel. Amplification of expected size products indicates the presence of a precise site-specific integration structure.

4. Notes

1. This protocol could also be adapted to other plant species such as maize or wheat. However, maize transformations may require the use of phosphinothricin transferase gene (*bar*) as selectable marker. Therefore, an alternative donor vector, pVS48, containing *bar* gene instead of NPT gene is also available.
2. If the GOI is less than 5 kb, recovery of site-specific integration lines is expected to be at the same rate as that of random integration lines (4–5 lines per bombarded plate). The majority of lines displaying expected PCR products for 5′ and 3′ junctions contain precise junctions as per DNA sequencing. Rarely DNA sequence changes comprising short deletions in promoter or NPT sequences are found.
3. To avoid chimerism (mixture of non-transgenic and transgenic cells) in callus, it is important to pick the geneticin-resistant clones within 3–4 weeks of selection.
4. While no selection is utilized on the regeneration medium, plantlets are rooted in the presence of selection to remove any escapes.
5. Southern blots can be probed with other DNA fragments that can identify site-specific integration bands.
6. Probe A can be amplified from genomic DNA of site-specific integration line using *PUbi* forward and *NPT* reverse primers. Alternately, NPT probe can be used to determine the presence of predicted 5′ junction.
7. As promoter–NPT junction of site-specific integration locus is selectable, most of the lines that are found contain this junction. However a small percentage of lines (~5%) may lack the correct 3′ junction (see Fig. 3 and 4).
8. Primers for 5′ junction are the following: *PUbi* forward: 5′-TCTACTTCTGTTCATGTTTGTG-3′ and *NPT* reverse 5′-CTCGATGCGATGTTTCGCTT-3′. Primers for 3′ junctions are the following: *nos3′* forward: 5′-GATTAGAGTC CCGCAATTAT-3′, and *cre* reverse 5′-CTAATCGCCATC TTCCAGCA-3′. With these primers, a 1.4 kb 5′ junction and 1.1 kb 3′ junction products are expected.

References

1. Breyne P, Gheysen G, Jacobs A, Van Montagu M, Depicker A (1992) Effect of T-DNA configuration on transgene expression. Mol Gen Genet 235:389–396
2. Iglesias VA, Moscone EA, Papp I, Neuhuber F, Michalowski S, Phelan T, Spiker S, Matzke M, Matzke AJ (1997) Molecular and cytogenetic analyses of stably and unstably expressed transgene loci in tobacco. Plant Cell 9:1251–1264
3. Kumpatla SP, Hall TC (1998) Recurrent onset of epigenetic silencing in rice harboring a multicopy transgene. Plant J 14:129–135
4. De Wilde C, Podevin N, Windels P, Depicker A (2001) Silencing of antibody genes in plants with single-copy transgene inserts as a result of gene dosage effects. Mol Genet Genomics 265:647–653
5. Elmayan T, Vaucheret H (1996) Expression of single copies of a strongly expressed 35S transgene can be silenced post-transcriptionally. Plant J 9:787–797
6. Jorgensen R, Snyder C, Jones JDG (1987) T-DNA is organized predominantly in inverted repeat structures in plants transformed with *Agrobacterium tumefaciens* C58 derivatives. Mol Gen Genet 207:471–477
7. Mayerhofer R, Koncz-Kalman Z, Nawrath C, Bakkeren G, Crameri A et al (1991) T-DNA integration: a mode of illegitimate recombination in plants. EMBO J 10:697–704
8. Kohli A, Leech M, Vain P, Laurie DA, Christou P (1998) Transgene organization in rice engineered through direct DNA transfer supports a two-phase integration mechanism mediated by the establishment of integration hot spots. Proc Natl Acad Sci USA 95:7203–7208
9. Pawlowski WP, Somers DA (1998) Transgenic DNA integrated into the oat genome is frequently interspersed by host DNA. Proc Natl Acad Sci USA 95:12106–12110
10. Cheng ZQ, Huang XQ, Ray W (2001) Comparison of biolistic and Agrobacterium-mediated transformation methods on transgene copy number and rearrangement frequency in rice. Acta Botanica Sinica 43:826–833
11. Shrawat AK, Lörz H (2006) *Agrobacterium*-mediated transformation of cereals: a promising approach crossing barriers. Plant Biotechnol J 4:575–603
12. Kononov ME, Bassuner B, Gelvin SB (1997) Integration of T-DNA binary vector 'backbone' sequences into the tobacco genome: evidence for multiple complex patterns of integration. Plant J 11:945–957
13. De Buck S, De Wilde C, Van Montagu M, Depicker A (2000) T-DNA vector backbone sequences are frequently integrated into the genome of transgenic plants obtained by *Agrobacterium*-mediated transformation. Mol Breed 6:459–468
14. Ulker B, Li Y, Rosso MG, Logemann E, Somssich IE, Weisshaar B (2008) T-DNA-mediated transfer of Agrobacterium tumefaciens chromosomal DNA into plants. Nat Biotechnol 26:1015–1017
15. Wright DA, Townsend JA, Winfrey RJ Jr, Irwin PA, Rajagopal J, Lonosky PM, Hall BD, Jondle MD, Voytas DF (2005) High-frequency homologous recombination in plants mediated by zinc-finger nucleases. Plant J 44:693–705
16. Ow DW (2002) Recombinase-directed plant transformation for the post-genomic era. Plant Mol Biol 48:183–200
17. Srivastava V, Ariza-Nieto M, Wilson AJ (2004) Cre-mediated site-specific gene integration for consistent transgene expression in rice. Plant Biotechnol J 2:169–179
18. Srivastava V, Ow DW (2002) Biolistic-mediated site-specific integration in rice. Mol Breed 8:345–350
19. Albert H, Dale EC, Lee E, Ow DW (1995) Site-specific integration of DNA into wild-type and mutant lox sites placed in the plant genome. Plant J 7:649–659
20. Vergunst AC, Jansen LE, Hooykaas PJJ (1998) Site-specific integration of *Agrobacterium* T-DNA in Arabidopsis thaliana mediated by Cre recombinase. Nucleic Acids Res 26:2729–2734
21. Hajdukiewicz P, Svab Z, Maliga P (1994) The small, versatile pPZP family of Agrobacterium binary vectors for plant transformation. Plant Mol Biol 25:989–994
22. Odell JT, Nagy F, Chua NH (1985) Identification of DNA sequences required for activity of the cauliflower mosaic virus 35S promoter. Nature 313:810–812
23. Mei C, Zhou X, Yang Y (2007) Use of RNA interference to dissect defense-signaling pathways in rice. Methods Mol Biol 354:161–171
24. Christensen AH, Sharrock RA, Quail PH (1992) Maize polyubiquitin genes: structure, thermal perturbation of expression and transcript splicing, and promoter activity following transfer to protoplasts by electroporation. Plant Mol Biol 18:675–689
25. Murray MG, Thompson WF (1980) Rapid isolation of high molecular weight plant DNA. Nucleic Acids Res 8:4321–4325

Chapter 8

Transcriptome Profiling Analysis Using Rice Oligonucleotide Microarrays

Ki-Hong Jung and Pamela C. Ronald

Abstract

Transcriptome analysis using oligonucleotide microarrays is a powerful tool for detecting changes in genome-wide transcripts under a given biological condition. Although the rice genome sequence is available, the number of functionally characterized genes in rice is still very limited. Genome-wide transcriptome analysis is a useful tool for elucidating the functions of rice genes that have not yet been determined. Currently, more than 3,000 arrays are publicly available. Here, we introduce methods for genome-wide transcriptome analysis in rice.

Key words: Rice, Microarray, GEO, aRNA, Hybridization

1. Introduction

One way to understand gene function is to monitor gene expression in response to stimuli or throughout development. The introduction of DNA microarrays and other transcriptome profiling approaches now makes it possible to monitor gene expression on a genome-wide scale. Gene expression profiling using microarrays allows researchers to measure expression levels of thousands of genes in a single RNA sample through the hybridization of fluorescent dye-labeled nucleic acids to pre-designed oligonucleotides or cDNAs (1). Microarrays exploit the preferential binding of complementary single-stranded nucleic acid sequences. There are two types of microarrays, cDNAs and oligonucleotides, which differ according to the way probes are prepared. The first generation of rice microarrays employed 8,927 cDNA clones as probes. The Rice Gene Expression Database (RED, http://red.dna.affrc.go.

Yinong Yang (ed.), *Rice Protocols*, Methods in Molecular Biology, vol. 956,
DOI 10.1007/978-1-62703-194-3_8, © Springer Science+Business Media, LLC 2013

Table 1
Summary of publicly available rice oligonucleotide microarray data

Platform	No. of probes/ probe sets	Oligo length in nucleotides	No. of hybridizations available at NCBI GEO
Affymetrix	55,515	25	1,771
NSF/ISU45K	43,311	50–70	231
NSF20K	20,123	50–70	128
BGI/Yale 60K	60,727	70	400
Agilent 44K	40,901	60	670
Agilent 22K	22,175	60	295
Total			3,495

jp/RED/) was used to assemble and assess cDNA microarray data (2, 3). This cDNA microarray had limited capacity because of the size of the cDNA used as probes. In addition, conserved sequence regions among cDNA probes caused cross-hybridization among probes. The development of oligonucleotide microarrays has significantly overcome these two main obstacles in the cDNA microarray platform. In rice, at least six platforms have been developed for oligonucleotide microarrays. Four of them are genome-wide microarray platforms: Affymetrix; Beijing Genomics Institute (BGI)/Yale 60 K; NSF45K; and Agilent 44 K (4–9). Overall features of these genome-wide microarray platforms are summarized in Table 1. Oligonucleotide microarrays are classified as spotted arrays or in situ arrays, based on how slides are printed. The BGI/Yale and NSF45K array platforms are spotting arrays and have been used to compare two samples (i.e., treatment vs. reference) labeled with different dyes (i.e., Cy3 and Cy5) at a time. Affymetrix and Agilent 44 K are arrays synthesized in situ. These two platforms employ a single channel (dye) for detection of transcripts in a biological sample, which allows many samples to be compared in parallel after normalizing expression profiles from multiple samples. A set of 11 different 25-mer oligonucleotide probe pairs is used in the Affymetrix array platform to measure a specific transcript in an RNA sample. The other three oligonucleotide array platforms make use of oligonucleotides that are 50- to 70-mer long. Computational methods were employed to design the probes on these array platforms, and this provided the researchers a systematic way of avoiding redundancy in designed probe regions, thereby minimizing the possibility of cross-hybridization among probes

(6, 10). In addition to using smaller probes and reducing the risk of cross-hybridization, the NSF45K and Affymetrix array platforms have revealed a third advantage of oligonucleotide microarrays. Both platforms were designed based on The Institute for Genomic Research (TIGR) rice genome annotation, including about 5,000 loci having alternative splicing variants. Splicing variants in around 20% of these loci can be distinguished by multiple probe sets specific to gene models in these loci (6, 11).

Microarray experiments involve four steps: experimental design, technical performance, statistical analysis, and data mining. Generating a clearly designed set of experiments is the most important part of a microarray experiment. To obtain valuable results from microarray experiments, researchers must consider several points before starting the experiments: the number of biological replicates (generally at least 3) needed to support statistical analysis; sampling time (i.e., day or night); specific types of tissue to be harvested (e.g., root or root tip); growing conditions (e.g., drought or salinity); genetic background (e.g., *Oryza sativa ssp. japonica* or *indica*); and the stage of plant growth (e.g., 1 day vs. 7 days). The existence of marker genes associated with treatment and distinguishable phenotypic symptoms after treatment helps researchers to evaluate biological samples for microarray experiments. If harvested samples satisfy all requirements for experimental design, researchers proceed through the technical steps to produce the image data. Then, statistical analysis suggests candidate genes for further analysis. Tools for data mining such as Gene Ontology (GO) enrichment analysis and integration of metabolic pathways with gene expression data help refine the candidate gene list from the several hundred or thousand genes that were initially selected based on differential expression alone (5, 6).

The results of gene expression profiles developed by individual researchers are generally scattered and therefore not easily available for public users. To make the data more accessible, the National Center for Biotechnology Information (NCBI) prepared the Gene Expression Omnibus (GEO, http://www.ncbi.nlm.nih.gov/geo/). In addition, the NSF45K rice microarray project team has made efforts to map oligos from different microarray platforms to TIGR gene models. This enables researchers to acces gene expression data from different array platforms and augment the usefulness of genome-wide gene expression data in rice (5, 6). To date, more than 3,000 hybridizations of rice microarray data have been generated. Effective application of vast gene expression information from public data will contribute to the acceleration of functional genomics and the improvement of agronomically important traits. In this chapter, we describe overall procedures for microarray experiments using the two-channel rice oligonucleotide microarray platform (i.e., the NSF45K array).

2. Materials

2.1. Total RNA Isolation and Removal of DNA Contamination

1. Trizol Reagent and store at 4°C.
2. TissueLyser (Qiagen).
3. TissueLyser adapter set 2× 24 (Qiagen).
4. 3 mm tungsten carbide beads or 5 mm stainless steel beads (Qiagen).
5. 2.0 mL microcentrifuge tubes to extract genomic DNA or total RNA with TissueLyser.
6. Liquid nitrogen.
7. Refrigerated microcentrifuge.
8. 1.5-mL RNase-free microcentrifuge tubes.
9. RNAase-free pipette tips.
10. Chloroform.
11. Isopropanol.
12. 75% and 100% ethanol.
13. Diethyl pyrocarbonate (DEPC)-treated water.

2.2. Total RNA Cleanup and mRNA Isolation

1. RNeasy MinElute Cleanup Kit (Qiagen).
2. RNAase-free DNase set (Qiagen).
3. Microcentrifuge.
4. 75% and 100% ethanol.
5. Oligotex mRNA isolation Kit (Qiagen).

2.3. RNA Quality Control and Quantification

2.3.1. Total RNA Analysis Using Agarose Gel Electrophoresis

1. 10× TBE buffer: 1 M Tris base, 0.9 M boric acid, 0.01 M EDTA.
2. 1% Agarose gel in TBE buffer.
3. Nanodrop spectrophotometer (Thermo Scientific).

2.3.2. Total RNA Analysis Using Agilent Bioanalyzer

1. Agilent 2100 Bioanalyzer (Agilent).
2. Agilent RNA 6000 LabChip Kit (Agilent).

2.4. Hybridization of Two-Channel Array

2.4.1. Indirect Fluorescent Dye-Labeled RNA Amplification for Two-Channel Array

1. SuperScript indirect RNA amplification system (Invitrogen) and store at –20°C.
2. Cy3 Mono-Reactive Dye Pack (Amersham Biosciences) and store at –20°C.
3. Cy5 Mono-Reactive Dye Pack (Amersham Biosciences) and store at –20°C.

4. 3 M Sodium acetate, pH 5.2.
5. Thermal cycler set at 16°C, 46°C, and 70°C.
6. Heat block set at 16°C, 46°C, and 70°C.
7. Incubator set at 37°C.
8. Vortex mixer, microcentrifuge, and speed-vac concentrator.
9. 1.5-mL RNase-free microcentrifuge tubes.
10. RNase-free pipette tips.
11. Disposable glove.
12. Isopropanol.
13. 75% and 100% ethanol.
14. Ice.
15. 5× first-strand buffer (Invitrogen): 250 mM Tris–HCl (pH 8.3), 375 mM KCl, and 15 mM $MgCl_2$.
16. 5× second-strand buffer (Invitrogen): 100 mM Tris–HCl (pH 6.9), 450 mM KCl, 23 mM $MgCl_2$, and 0.75 mM β-NAD+, 50 mM $(NH_4)_2SO_4$.
17. Mixtures to synthesize first-strand cDNA (Invitrogen): T7-Oligo(dT) primer, 10 mM dNT_P mix, 5× first-strand buffer, 0.1 M $_D$TT, RNaseOUT™, SuperScript™ III RT.
18. Mixtures to synthesize second strand cDNA (Invitrogen): *E. coli* DNA polymerase I, *E. coli* DNA ligase, 5× second-strand buffer, 10 mM dNTP mix, DEPC-treated water, *E. coli* RNase H.
19. Mixtures to amplify amplified RNA (aRNA) containing amino-allyl UTP (aa-UTP) (Invitrogen): 100 mM ATP, 100 mM CTP, 100 mM GTP, 100 mM UTP, 50 mM amino-allyl-UTP (aa-UTP), 10× T7 reaction buffer, T7 enzyme mix, and DEPC-treated water.
20. Low-elution volume spin cartridges (Invitrogen).
21. cDNA loading buffer after adding 9 mL of 100% isopropanol (Invitrogen).
22. cDNA wash buffer after adding 4 mL of 100% ethanol (Invitrogen).
23. PureLink™ spin columns (Invitrogen).
24. aRNA binding buffer (Invitrogen).
25. aRNA wash buffer after adding 7 mL of 100% isopropanol (Invitrogen).
26. Dimethyl Sulfoxide (DMSO).
27. 2x Coupling buffer (Invitrogen).
28. RNase-free amber tube.

2.4.2. Sodium Borohydride Pretreatment Protocol for Two-Channel Array on Amino Slides

1. 2× Saline Sodium Citrate (SSC) solution: 300 mM sodium chloride, 30 mM sodium citrate (pH 7.0).
2. Blocking solution: 2× SSC, 0.05% SDS, 0.25% $NaBH_4$ (2.5 mg/mL).
3. 0.1% SDS.
4. Nanopure water.
5. 1× SSC.
6. 0.2× SSC.
7. Coplin jars.
8. Microplate centrifuge.
9. Ar_2 or N_2.

2.4.3. Hybridization of Two-Channel Array

1. Microscope slide holders.
2. 10 mL disposable pipettes and 50 mL tubes.
3. Sterile measuring cylinder.
4. Thermal cycler and incubator.
5. Liquid blocking reagent (Amersham).
6. 2% SDS20× SSC.
7. DEPC-treated water.
8. Geneframes hybridization solution (MWG Biotech).
9. 5× SSPE: 0.75 M sodium chloride, 50 mM NaH_2PO_4, 5 mM EDTA (pH 7.0).
10. 6 M Urea.
11. 0.5% Tween-20.
12. 10× Denhardt's solution.
13. Tecan HS4800 hybridization station (Tecan).

3. Methods

3.1. Preparation of Total RNA or mRNA for Microarray Experiments

1. Harvest 0.2–0.4 g of tissues from reference and treatment samples (Fig. 1).
2. Isolate 100–200 μg of total RNA by using Trizol Reagent (Invitrogen) and following the manufacturer's introduction.
3. Digest 50 μg of RNA with RNase-free DNase set (Qiagen), use 0.6 units per 1 μg RNA, and incubate at room temperature for 30 min.
4. Purify ~50 μg DNA-free total RNA with RNeasy MinElute Cleanup Kit (Qiagen) as introduced in the handbook of the manufacture (http://www.qiagen.com/HB/RNeasyMinEluteCleanup).

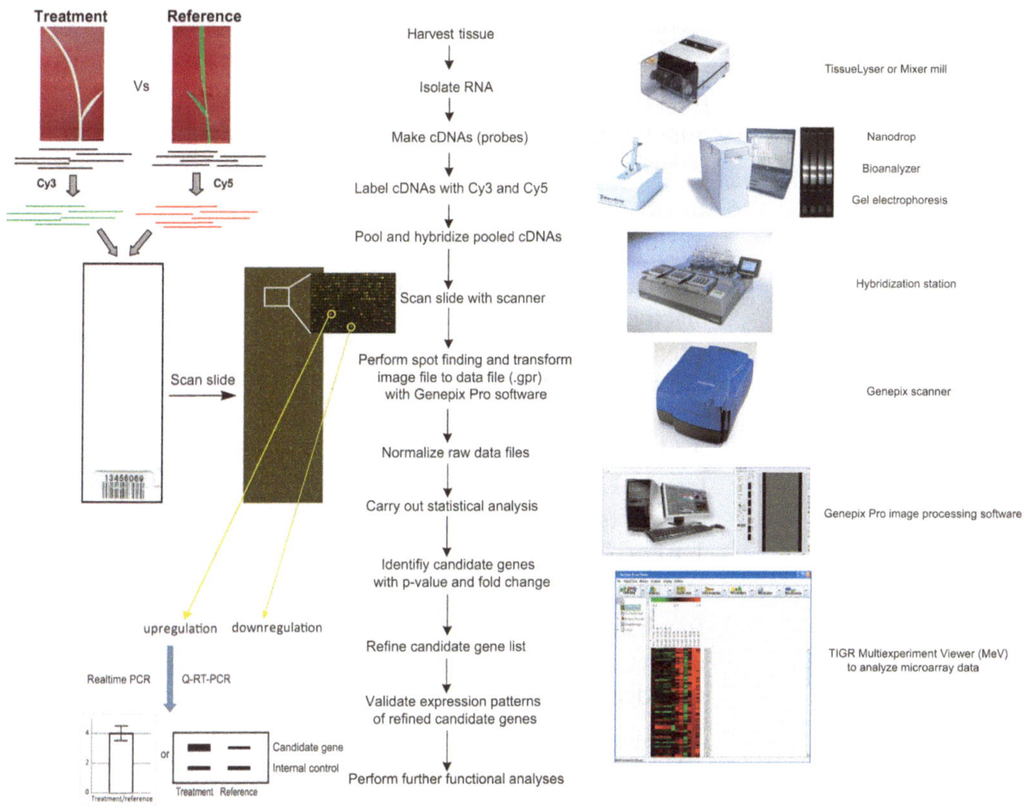

Fig. 1. Overall procedure for two-channel microarray experiments in rice.

5. Determine the concentration using a nanodrop with 1 μL of purified total RNA solution. Yields are generally around 25–40 μg of total RNA. Check the quality of the RNA by running the gel with 1 μg of the RNA.
6. Check DNA contamination by performing normal PCR reaction with internal control primers such as *EF1a* (*Os03g08020*) or *Ubi5* (*Os01g22490*) (12).
7. Purify mRNA from isolated total RNA by using Oliogtex mini kit (Qiagen) and following the manufacturer's introduction.
8. Determine the concentration of mRNA using a nanodrop. Between 100 and 500 pg of mRNA are generally needed for the further steps (Fig. 1) (see Note 1).

3.2. Preparation of Fluorescent Dye-Labeled Probes Using Amplification (See Note 2)

3.2.1. First-Strand cDNA Synthesis

1. Use 0.5–1 μg of total RNA or 50–100 ng of mRNA as template to make first-strand cDNA (Fig. 1).
2. Add template and T7-Oligo(dT) primer and make 10 μL with DEPC-treated water in a 0.2-mL RNase-free PCR tube.
3. Incubate the tube in thermal cycler set at 70°C for 10 min, and then place on ice for at least 5 min.

4. Centrifuge the tube briefly to collect the contents.
5. Add 4 μL 5× first-strand buffer, 2 μL 0.1 M DDT, 1 μL 10 mM dNTP mix, 1 μL RNaseOUT™ (40 U/μL), and 2 μL SuperScript™ III RT (200 U/μL) to the tube at room temperature (Fig. 1).
6. Incubate the tube in a thermal cycler set at 46°C for 3 h after mixing and centrifuging it.
7. Incubate the tube in a thermal cycler set at 70°C for 10 min to inactivate the reverse transcriptase and keep the tube on ice after mixing and centrifuging it (or the first-stranded cDNA can be stored at −20°C before using it).

3.2.2. Second-Strand cDNA Synthesis

1. Add 91 μL DEPC-treated water, 30 μL 5× second-strand buffer, 3 μL 10 mM dNTP mix, 4 μL *E. coli* DNA polymerase I (10 units/μL), 1 μL *E. coli* DNA ligase (10 units/μL), and 1 μL *E. coli* RNase H (2 units/μL) to the tube in step 7 of Subheading 3.2.1 to make the Second-Strand cDNA.
2. Incubate the tube in thermal cycler set at 16°C for 3 h after mixing and centrifuging it and keep it on ice (or the double-stranded cDNA can be stored at −20°C before using it).
3. Transfer solution in the reaction tube to 1.5 mL RNase-free microcentrifuge tube and label the tube.

3.2.3. Purification of the Second-Strand cDNA

1. Add 500 μL of cDNA loading buffer to the tube and mix it thoroughly by pipetting up and down.
2. Load the cDNA/buffer solution directly onto low-elution volume spin cartridge preinserted into a collection tube after clearly labeling the column.
3. Centrifuge at 6,000 × *g* at room temperature in a microcentrifuge for 1 min and remove the collection tube and discard the flow-through.
4. Place the spin cartridge in the same collection tube and add 700 μL of cDNA wash buffer to the column.
5. Centrifuge at 6,000 × *g* at room temperature for 1 min and remove the collection tube and discard the flow-through.
6. Place the spin cartridge in the same collection tube and centrifuge at 6,000 × *g* at room temperature for an additional 2 min to remove any residual wash buffer.
7. Discard the collection tube and place the spin cartridge into a new recovery tube.
8. Add 24 μL of DEPC-treated water to the center of the spin cartridge and incubate at room temperature for 5 min.
9. Centrifuge at 10,000 × *g* at room temperature for 1 min to collect the purified double-strand cDNA in the recovery tube.

3.2.4. Amplification of aRNA

1. At room temperature, add 1.5 μL 100 mM ATP, 100 mM 1.5 μL CTP, 1.5 μL 100 mM GTP, 0.75 μL 100 mM UTP, 2 μL 50 mM aa-UTP, 4 μL 10× T7 reaction buffer, and 7 μL T7 enzyme mix to the recovery tube and make final volume 40 μL with DEPC-treated water.
2. Incubate the tube at 37°C for 12–14 h after mixing them. This reaction tube contains aRNA (which has amino-allyl UTP to incorporate fluorescent dye).

3.2.5. Purification of aRNA

1. Add ~160 μL of aRNA binding buffer to the aRNA-containing solution and mix it thoroughly by pipetting up and down.
2. Add 100 μL of 100% ethanol to the reaction tube and mix it thoroughly by pipetting up and down.
3. Load the entire aRNA/buffer/ethanol solution directly onto a clearly labeled PureLink™ spin column preinserted into a collection tube.
4. Centrifuge at 12,000 × *g* in a microcentrifuge for 15 s at room temperature, remove the collection tube, and discard the flow-through.
5. Place the spin column in the same collection tube and add 500 μL of aRNA Wash Buffer to the column.
6. Centrifuge at 12,000 × *g* for 15 s at room temperature, remove the collection tube, and discard the flow-through.
7. Repeat steps 3 and 4.
8. Place the spin column in the same collection tube and centrifuge at 12,000 × *g* for an additional 2 min to remove any residual wash buffer.
9. Discard the collection tube and place the spin column into a new collection tube.
10. Add 100 μL of DEPC-treated water to the center of the spin cartridge and incubate at room temperature for 5 min.
11. Centrifuge at 12,000 × *g* for 2 min to collect the purified aRNA.
12. Determine the concentration of aRNA using a nanodrop. Yields are generally between 10 and 100 μg of aRNA (see Note 1).

3.2.6. Concentration of aRNA

1. Add 10 μL of 3 M sodium acetate (pH 5.2) to the purified aRNA-containing tube.
2. Add 300 μL of ice-cold 100% ethanol and mix the contents well.
3. Incubate the tube at −20°C for at least 1 h or at −80°C for at least 10 min.
4. Spin the tube at 14,000 × *g* for 10–20 min and carefully remove and discard the supernatant.

5. Add 250 μL of 75% ethanol and spin the tube at 14,000 × *g* for 2 min.
6. Carefully remove and discard the supernatant.
7. Spin the tube at 14,000 × *g* for 1 min and discard the supernatant with a 20 μL pipette.
8. Air-dry the sample for 10 min.

3.2.7. Coupling aRNA with Fluorescent Dyes and Purification of Dye-Labeled aRNA

1. Resuspend the samples in 5 μL of 2× Coupling buffer.
2. Aliquot aRNA/Coupling buffer solution containing 5 μg aRNA to a new RNase-free amber tube and make 5 μL with 2× Coupling buffer.
3. Add 5 μL of DMSO directly to each dye vial of Cy3 and Cy5 in an individual reaction size (RPN5661) or 45 μL of DMSO directly to each dye vial of Cy3 and Cy5 in large reaction size (PA23001 and PA25001).
4. Add 5 μL the dye solution to the amber tube after vortexing and centrifuging it.
5. Incubate the tube at room temperature in the dark for 1 h and purify the dye-coupled aRNA as described in Subheading 3.2.5 (the aRNA-containing solution purified in this step is dye-coupled unlike the aRNA initially purified in Subheading 3.2.5).
6. Determine the concentration of dye-coupled aRNAs with a nanodrop (see Note 1).
7. The quality of dye-coupled aRNAs can be measured by using the Agilent 2100 bioanalzyer with RNA 6000 Labchip Kit.
8. Use 100 p mole of dye-coupled aRNA for the hybridization of two-channel arrays such as the NSF45K array (6).

3.3. Sodium Borohydride Pretreatment Protocol for Two-Channel Array on Amino Slides

1. Preheat the block solution (minus $NaBH_4$) at 42°C until ready to use (see Note 3).
2. Add 0.5 g $NaBH_4$ to the preheated block solution, allowing the $NaBH_4$ to dissolve (it should bubble, take ~5 min to dissolve and reach a pH of ~9.0).
3. Transfer the block solution to appropriate Coplin jars at 42°C and place slides in the block solution.
4. Incubate at 42°C for 20 min and transfer slides to 1× SSC in a Coplin jar at room temperature; allow slides to cool for 5 min.
5. Transfer slides to slide holder for vigorous washing and do not allow slides to dry.
6. Submerge slides in 800 mL of 1× SSC with vigorous stirring for 5 min at room temperature, and repeat this procedure with fresh 1× SSC three more times (for a total of four washes).

7. Submerge slides in 800 mL of 0.2× SSC with vigorous stirring for 2 min at room temperature, and repeat this procedure with fresh 0.2× SSC three more times (for a total of four washes).
8. Submerge slides in 800 mL of Nanopure water with vigorous stirring for 2 min at room temperature and dry slides by spin-drying at 1,000 rpm for 10 min or using a stream of N_2.
9. If the slides are not used immediately for hybridization, seal them under Ar_2 or N_2 until use; this allows the slides to be safely stored for a number of hours (see Notes 4 and 5).

3.4. Microarray Hybridizations (See Note 6)

1. Pre-hybridize slides in a Tecan HS4800 Hybridization Station for 15 min at 50°C in 5× SSPE, 6 M Urea, 0.5% Tween-20, and 10× Denhardt's solution (Sigma) (Fig. 1) (see Notes 7 and 8).
2. Mix Cy3- or Cy5-labeled aRNA targets and dry by vacuum centrifugation.
3. Suspend the targets in 100-μL GeneFrames hybridization solution (MWG Biotech).
4. Incubate the targets in boiling water for 3 min and centrifuge at 14,000 × *g* for 5 min; inject into the hybridization station or keep at room temperature until the injection.
5. Hybridize samples for 16 h at 50°C with medium agitation (see Note 9).
6. Wash the slides in solution containing 2× SSC and 0.2% SDS for 1 min at 37°C, in 1× SSC solution for 1 min at 30°C, and in 0.5× SSC for 1 min at 30°C.
7. Keep slides under N_2 until they are scanned.

3.5. Scanning Slides

1. Scan the slides using a GenePix 4000B dual-laser microarray scanner (Axon Instruments) at 5-μm resolution using 100% laser power for both lasers (532 and 635 nm) (Fig. 1).
2. Scan the slides twice—once using a high photomultiplier tube (PMT) and again using low PMT settings.
3. Save the image as a TIFF file.

3.6. Generation and Normalization of Raw Data File

1. Quantify spot intensities using Axon GenePix Pro 4.0 image analysis software (Fig. 1).
2. Load the TIFF image file using high PMT and low PMT settings into the GenePix Pro program.
3. Generate the array layout in GenePix Array Layout (GAL) format mapped to Rice Genome Annotation Project rice gene models.
4. Load the gal file corresponding to the first slide you wish to analyze. Later, repeat this step for subsequent slides.
5. Adjust the features by looking for mismatches in the grid and manually flagging obvious problems such as probe spots that are saturated.

6. Extract the data and save as GenePix Result files (gpr) (Fig. 1).
7. Normalize replicated data to minimize the variations caused by technical procedures using the Lowess normalization method, such as the LMGene Package (http://cran.r-project.org/web/packages/LMGene/vignettes/LMGene.pdf), encoded in R language.

3.7. Generation of Candidate Gene List and Refinement of the List

1. Generate the candidate gene list by considering threshold criteria such as p-value and $\log_2$ fold change values of treatment over reference.
2. Perform clustering analysis of differential expression patterns of candidate genes selected in each comparison on the basis of the above criteria, using publicly available software for analyzing microarray data (e.g., Multiexperiment Viewer (MeV), http://www.tm4.org/mev.html) in the case of datasets with multiple comparisons (Fig. 1).
3. Integrate featured expression patterns resulting from clustering analysis to a pathway tool (e.g., RiceCyc, http://pathway.gramene.org/RICE/expression.html).
4. Identify pathways co-regulated by a treatment.
5. Validate the expression patterns of significant candidate genes and perform further functional analyses of the genes (Fig. 1).

4. Notes

1. Total RNA quality can be analyzed either by agarose gel electrophoresis in TBE buffer or using an Agilent Bioanalyzer. The Bioanalyzer requires very small quantities of RNA (50–150 ng) for analysis whereas agarose gel electrophoresis requires 1–2 μg.
2. The protocol for aRNA amplification and labeling with fluorescent dyes for two-channel arrays is a modification of the protocol provided by the manufacturer (Invitrogen).
3. Store slides at room temperature and protect them from light in heat-sealed pouches until use.
4. Reseal any unused slides again under argon until they are ready to use. This helps prevent oxidation of spotted material and will extend shelf life. Otherwise, store the slides in desiccators.
5. Read all safety information regarding sodium borohydride before using. This solution will bubble H_2; therefore take all precautions to avoid having an open flame or sparks near this solution.
6. For best results, pretreat arrays with sodium borohydride (after blocking) and just prior to hybridization. This step reduces the background fluorescence of the glass as well as any autofluorescence from the spotted material.

7. Hybridizations performed in-house by an automated hybridization station (Tecan HS4800) significantly reduce variation caused by handling slides during manual hybridization.
8. For manual hybridizations using a coverslip, hybridization buffers containing formamide have worked well and are recommended. Many commercially available hybridization buffers have worked well with these arrays.
9. Hybridizations are performed in a clean room environment (HEPA- and carbon-filtered) to minimize exposure of microarrays and labeled targets to dust and ozone.
10. Deep sequencing technology is now emerging as a new method for genome-wide transcriptome analysis; as costs for deep sequencing decrease microarrays may soon become obsolete.

References

1. Bowtell DD (1999) Options available–from start to finish–for obtaining expression data by microarray. Nat Genet 21:25–32
2. Yazaki J, Shimatani Z, Hashimoto A, Nagata Y, Fujii F, Kojima K, Suzuki K, Taya T, Tonouchi M, Nelson C, Nakagawa A, Otomo Y, Murakami K, Matsubara K, Kawai J, Carninci P, Hayashizaki Y, Kikuchi S (2004) Transcriptional profiling of genes responsive to abscisic acid and gibberellin in rice: phenotyping and comparative analysis between rice and Arabidopsis. Physiol Genomics 17:87–100
3. Yazaki J, Kojima K, Suzuki K, Kishimoto N, Kikuchi S (2004) The rice PIPELINE: a unification tool for plant functional genomics. Nucleic Acids Res 32:D383–D387
4. Walia H, Wilson C, Condamine P, Liu X, Ismail AM, Zeng L, Wanamaker SI, Mandal J, Xu J, Cui X, Close TJ (2005) Comparative transcriptional profiling of two contrasting rice genotypes under salinity stress during the vegetative growth stage. Plant Physiol 139:822–835
5. Jung KH, An G, Ronald PC (2008) Towards a better bowl of rice: assigning function to tens of thousands of rice genes. Nat Rev Genet 9:91–101
6. Jung KH, Dardick C, Bartley LE, Cao P, Phetsom J, Canlas P, Seo YS, Shultz M, Ouyang S, Yuan Q, Frank BC, Ly E, Zheng L, Jia Y, Hsia AP, An K, Chou HH, Rocke D, Lee GC, Schnable PS, An G, Buell CR, Ronald PC (2008) Refinement of light-responsive transcript lists using rice oligonucleotide arrays: evaluation of gene-redundancy. PLoS One 3:e3337
7. Ma L, Chen C, Liu X, Jiao Y, Su N, Li L, Wang X, Cao M, Sun N, Zhang X, Bao J, Li J, Pedersen S, Bolund L, Zhao H, Yuan L, Wong GK, Wang J, Deng XW (2005) A microarray analysis of the rice transcriptome and its comparison to Arabidopsis. Genome Res 15:1274–1283
8. Satoh K, Doi K, Nagata T, Kishimoto N, Suzuki K, Otomo Y, Kawai J, Nakamura M, Hirozane-Kishikawa T, Kanagawa S, Arakawa T, Takahashi-Iida J, Murata M, Ninomiya N, Sasaki D, Fukuda S, Tagami M, Yamagata H, Kurita K, Kamiya K, Yamamoto M, Kikuta A, Bito T, Fujitsuka N, Ito K, Kanamori H, Choi IR, Nagamura Y, Matsumoto T, Murakami K, Matsubara K, Carninci P, Hayashizaki Y, Kikuchi S (2007) Gene organization in rice revealed by full-length cDNA mapping and gene expression analysis through microarray. PLoS One 2:e1235
9. Suwabe K, Suzuki G, Takahashi H, Shiono K, Endo M, Yano K, Fujita M, Masuko H, Saito H, Fujioka T, Kaneko F, Kazama T, Mizuta Y, Kawagishi-Kobayashi M, Tsutsumi N, Kurata N, Nakazono M, Watanabe M (2008) Separated transcriptomes of male gametophyte and tapetum in rice: validity of a laser microdissection (LM) microarray. Plant Cell Physiol 49: 1407–1416
10. Chou HH, Hsia AP, Mooney DL, Schnable PS (2004) Picky: oligo microarray design for large genomes. Bioinformatics 20:2893–2902
11. Jung KH, Bartley LE, Cao P, Canlas PE, Ronald PC (2009) Analysis of alternatively spliced rice transcripts using microarray data. Rice 2:44–55
12. Jain M, Nijhawan A, Tyagi AK, Khurana JP (2006) Validation of housekeeping genes as internal control for studying gene expression in rice by quantitative real-time PCR. Biochem Biophys Res Commun 345:646–651

Chapter 9

Cloning of Small RNAs for the Discovery of Novel MicroRNAs in Plants

Guru Jagadeeswaran and Ramanjulu Sunkar

Abstract

Endogenous small RNAs can be grouped into several distinct classes of 21-nt-long microRNAs (miRNAs), short interfering RNAs (siRNAs), trans-acting siRNAs (tasiRNAs), and 24-nt long heterochromatic siRNAs. miRNAs are increasingly being recognized as significant effectors of gene regulation in a wide range of organisms. These molecules are typically ~21-nt long and are amenable for cloning by streamlined protocols. Here we detail the methodology for cloning small RNAs in rice to identify novel miRNAs and other important small RNAs. Briefly, small RNA molecules are size fractionated, attached to adaptors, and subsequently converted into cDNA and PCR amplified. Current high-throughput sequencing technologies allow sequencing of the PCR products directly.

Key words: MciroRNAs, Small RNAs, Posttranscriptional gene regulation, Rice

1. Introduction

MicroRNAs (miRNAs) are single-stranded RNA molecules of 21–22 nt length and act as negative regulators of gene expression in plants and animals (1). miRNA-guided posttranscriptional gene regulation constitutes one of the most conserved and well-characterized gene regulatory mechanisms, which is important for development, stress responses, and myriad of other biological processes in eukaryotes (1–4). Therefore, identification of miRNAs in diverse plant/animal species has been one of the most sought research topics in recent years. There are two broad approaches to identify miRNAs in eukaryotes which are (1) searching sequence databases and predicting potential miRNAs using computational tools and (2) cloning and sequencing small RNAs from target organisms. The cloning approach would be particularly effective for small RNAs that are specifically induced under certain conditions such as biotic or abiotic stress (5–8). With the advent of new sequencing technologies, cloning

Yinong Yang (ed.), *Rice Protocols*, Methods in Molecular Biology, vol. 956,
DOI 10.1007/978-1-62703-194-3_9, © Springer Science+Business Media, LLC 2013

of miRNAs has become the method of choice to identify/validate and profile novel miRNAs in different tissues or conditions.

The overall strategy in making small RNA libraries is to first isolate small RNAs of 21–25 nt in length from a polyacrylamide gel by size fractionation. Adaptors are then ligated to the 5′ and 3′ ends of these single-stranded RNA molecules to aid in subsequent reverse transcription and PCR amplification. In plants, small RNAs are generated by the activity of Dicer on miRNA hairpin precursors, which leaves 5′ monophosphate and 3′ hydroxyl ends (9). Cloning strategies make use of these biochemical features to preferentially clone miRNAs and other small RNAs that are derived as a result of Dicer activity. Also, the adaptors involved in this strategy are suitably modified to avoid self-ligation. After PCR amplification and purification, samples are sequenced using sequencing technologies such as sequencing by synthesis (SBS) technology. The protocol outlined here is suitable for Illumina, but by modifying the adapter sequences the libraries can be sequenced using diverse compatible sequencing technologies.

2. Materials

2.1. Isolation of Total RNA and Small RNA

1. Trizol Reagent (Invitrogen).
2. Chloroform.
3. Isopropanol.
4. 5 M NaCl.
5. 50% PEG8000.
6. Absolute ethanol.
7. Centrifuge.
8. 5 mg/mL glycogen (Ambion).
9. DEPC-treated water or RNase-free sterile water.

2.2. Denaturing PAGE Gel Electrophoresis

1. 19:1 Acrylamide:Bis acrylamide.
2. 8 M urea.
3. 0.5× TBE buffer: 50 mM Tris, 50 mM boric acid, 1 mM EDTA.
4. 10% Ammonium persulfate.
5. TEMED.
6. Vertical electrophoresis unit and combs.

2.3. Recovery of Small RNAs from Gel

1. Gel loading buffer: 98%w/v deionized formamide, 10 mM EDTA, pH 8.0.

   0.02%w/v xylene cyanol, 0.02%w/v bromophenol blue.

2. RNA Markers: 20 and 24 nt.
3. 6,000 ci/mmol ^{32}P-γ-ATP.
4. T-4 Polynucleotide kinase.
5. Phosphorimager or X-ray film.
6. Fluorescent labels.
7. 1.5 mL siliconized Eppendorf tubes.
8. 0.3 M NaCl.
9. Rotating shaker for microcentrifuge tubes.
10. 5 mg/mL glycogen (Ambion).
11. Absolute ethanol.

2.4. Adaptors for Ligation with Small RNAs

1. Solexa 5′ RNA adaptor (from Dharmacon): 5′ GUUCAGAG UUCUACAGUCCGACGAUC 3′.
2. Solexa 3′ RNA adaptor (from Dharmacon):
   5′ P-UCGUAUGCCGUCUUCUGCUUGUidT 3′
   (idT-inverted deoxy-thymidine).
3. 10× NEB Ligation Buffer.
4. T4 RNA ligase (NEB).
5. RNase OUT.
6. DEPC-treated water or RNase-free sterile water.

2.5. RT-PCR

2.5.1. Reverse transcription

1. 200 U/μL Superscript II Reverse Transcriptase (Invitrogen).
2. 5× First-strand buffer.
3. 0.1 M DTT solution.
4. 2.5 mM dNTP solution.
5. Solexa RT primer,
   5′-CAAGCAGAAGACGGCATACGA-3′.

2.5.2. PCR Amplification

1. PCR primer 1 (44 nt, DNA oligo):
   5′-AATGATACGGCGACCACCGACAGGTTCAGAGTT CTACAGTCCGA-3′.
2. PCR primer 2 (same as Solexa RT primer).
3. 10× PCR Buffer.
4. 2.5 mM dNTP solution.
5. Solexa RT primer (21 nt, DNA oligonucleotide):
   5′-CAAGCAGAAGACGGCATACGA-3′.
6. *KOD Taq* DNA polymerase (Novagen).
7. PCR machine.

2.6. Gel Purification of PCR Product

1. Low gelling temperature agarose (Agarose II™ obtained from Amresco) 3% gels in 1× TAE buffer.
2. 25 bp DNA ladder (Invitrogen).
3. 1× TAE buffer: 40 mM Tris–acetate pH 8.4, 1 mM EDTA pH 8.0.
4. 6× DNA loading buffer: 30% Glycerol, 0.02%w/v xylene cyanol, 0.02%w/v bromophenol blue.
5. Gel extraction kit (Qiagen).
6. Clean razor.

2.7. Cloning of PCR Product

1. pGEM-T Cloning vector system (Promega).
2. JM109 High Efficiency Competent Cells.
3. S.O.C medium (BD Diagnostics): To make the medium yourself, measure 20 g Bacto-tryptone, 5 g Bacto-yeast extract, 0.5 g NaCl, 2.5 mL 1 M KCl and bring to 1,000 mL with ddH$_2$O. Adjust pH to 7.0 with 10 N NaOH, autoclave to sterilize, and add 20 mL of sterile 1 M glucose (sterilized by passing through a 0.2 μm filter) immediately before use.
4. Ampicillin (Sigma).
5. 5-bromo-4-chloro-3-indolyl-beta-D-galactopyranoside (X-gal) (Promega).
6. Isopropyl-β-D-thio-galactoside (IPTG) (Promega).

3. Methods

3.1. Isolation and Preparation of RNA

1. Isolate total RNA from tissues with Trizol reagent following the manufacturer's directions. Do not exceed more than 200 mg of ground tissue per 1 mL of Trizol. For extracting RNA from up to 200 mg ground tissue, use 2 mL Eppendorf tubes, and for larger volumes use 15 mL Sorvall tubes. Wash the RNA pellet using 80% ethanol and not 70% ethanol.
2. Dissolve the RNA pellet in DEPC-treated or nuclease-free water, preferably at 3–4 μg/μL (see Note 1).
3. If polysaccharides and other materials co-precipitate with RNA, carry out one more round of chloroform extraction followed by sodium acetate precipitation.
4. Isolate small-RNA fraction from total RNA (see Note 2). If there is a small yield of total RNA, this step would cause further loss of RNA. If this is the case then proceed directly to step 11.

5. Add and mix the following in a 15 mL tube:

25% PEG 6000	1.0 mL
5 M NaCl	0.5 mL
Total RNA plus water to	5.0 mL

6. Incubate the tubes on ice for 2 h.
7. Spin at 16,000 ×*g* for 30 min at 4°C (see Note 3).
8. Transfer supernatant to new tubes as follows:

Supernatant (from step 7)	2.5 mL
Ice-cold ethanol (2.5 volumes)	6.25 mL

9. Incubate at −20°C for at least 2 h or overnight. Spin at 16,000 ×*g* for 30 min at 4°C. Air-dry the pellet briefly.
10. Dissolve the RNA pellet in an appropriate volume of DEPC-treated or nuclease-free water.
11. Determine RNA concentration and purity by UV absorbance using a spectrophotometer or a NanoDrop.
12. Run a small aliquot of total RNA in a 1.2% agarose gel and confirm RNA is of the highest quality, a prerequisite for successful library construction. Typically, small RNA libraries can be constructed with 300–500 μg of total RNA, for which about 1–2 g of rice tissue would typically be required.
13. Store the RNA at −80°C until use and avoid frequent freeze–thaw cycles.

3.2. Recovery of Small RNAs by Gel Electrophoresis

1. Pour a 15% PAGE gel by combining the components (15% polyacrylamide, 8 M urea, 0.5× TBE), after positioning a big-well comb, and let solidify at least 1 h.
2. Prepare radiolabeled size markers (20 nt and 24 nt) by labeling 20 and 24 nt RNA markers with polynucleotide kinase and ^{32}P-γ-ATP according to the manufacturer's instructions.
3. Add equal volume of 2× gel loading dye (Gel loading buffer II) to the RNA samples and the size markers and heat to 65°C for 5 min to denature the samples.
4. Immediately transfer and keep the samples and radiolabeled markers on ice until loading.
5. Wash the wells of PAGE gel thoroughly and electrophorese RNA samples along with radiolabeled RNA markers. Load the RNA samples and radiolabeled RNA size markers in the center wells. Leave an empty well between the sample RNAs and size markers to avoid contamination.

6. Run the gel at 250 V for about 2 h in 0.5× TBE buffer or until bromophenol blue dye (bottom dye) reaches 2/3 of the gel (see Note 4).
7. Stop the unit and dismantle the gel carefully.
8. Carefully place fluorescent markers on the gel randomly to help later aligning and wrap the gel with saran wrap.
9. Lay an X-ray film on top of the gel in darkroom and expose briefly for a minute or two to visualize sharp bands of 20 nt and 24 nt markers.
10. Position the gel on top of the X-ray film and align to the fluorescent labels to visualize 20 and 24 nt size markers. Cut the gel slices corresponding to these sizes (20–24 nt) with clean razors.
11. Transfer to a 1.5 mL tube (siliconized tubes), crush the gel into smaller pieces, and add three volumes of 0.3 M NaCl to the tube (e.g., for 100 mg gel piece add 300 μL of 0.3 M NaCl) (see Note 5).
12. Elute the RNA overnight at 4°C (in a cold room) by agitating the tube gently.
13. Transfer the supernatant to a new 1.5 mL tube and add 2.5 volumes 100% Ethanol.
14. Add 1 μL glycogen, mix well, and precipitate small RNAs overnight at −80°C.
15. Cut, elute, and precipitate the radiolabeled size markers as well in parallel.
16. Spin down the samples at 16,000 × *g* for 30 min at 4°C in a microfuge.
17. Remove the supernatant and air-dry the pellet briefly (5–10 min).
18. Resuspend RNA pellet in 10 μL of nuclease-free water and proceed to ligation with adaptors.

3.3. Ligation

3.3.1. Ligation with 5′ Adaptor

1. Set up the 5′ adaptor ligation as follows:

   In a 10 μL reaction volume combine the following:

Eluted small RNA	6.0 μL
(20 μM) 5′ RNA adaptor	1.0 μL
10× ligation buffer (NEB)	1.0 μL
T4 RNA ligation (NEB)	1.0 μL
RNase Out	1.0 μL

2. Incubate at 37°C for 1 h.

3.3.2. Purification of 5′ Adaptor-Ligated RNA

1. Prepare 15% polyacrylamide/urea gel and wash wells thoroughly using 0.5× TBE buffer.
2. Add an equal volume of 2× RNA gel loading dye to small RNAs ligated to the 5′ adaptor and to size markers. Denature the samples by heating at 65°C for 5 min and immediately keep on ice for 2 min.
3. Load the 5′adaptor-ligated RNA and 5′ adaptor-ligated marker in individual wells, leaving a blank well between the samples and markers.
4. Run the gel for 2 h at 250 V in 0.5× TBE buffer.
5. Stop the run and disassemble the gel as described before (as in step 8 of Subheading 3.2).
6. Prepare the gel for autoradiography and align the visible markers on the gel.
7. Cut the gel slice corresponding to appropriate size (approximately between 50 and 60 nt) from the RNA lanes (it is not necessary to cut the radiolabeled markers) and transfer to a 1.5 mL (low retention) tubes.
8. Crush the gel slice into smaller pieces and add 3 volumes of 0.3 M NaCl.
9. Elute and precipitate as described in steps 11–17 of the Subheading 3.2 and dissolve in 10 μL RNase-free water.

3.3.3. Ligation with 3′ Adaptor

1. Set up the 3′ adaptor ligation as follows:

   In a 10 μL reaction volume, combine the following:

5′ ligated small RNA	6.0 μL
(20 μM) 3′ RNA Adaptor (see Note 6)	1.0 μL
10× ligation buffer (NEB)	1.0 μL
T4 RNA Ligase (NEB)	1.0 μL
RNase Out	1.0 μL

2. Incubate at 37°C for 1 h and directly proceed to next section or freeze at −80°C until further use.

3.3.4. Reverse Transcription of Adaptor-Ligated Small RNAs

1. In a 1.5 μL Eppendorf (siliconized) tube, set up 10 μL RT reaction as follows:

5′ and 3′ adaptor-ligated small RNA	10.0 μL
20 μM Solexa RT primer	1.0 μL

2. Heat the mix at 65°C for 10 min.
3. Briefly spin to collect and cool on ice for 3–4 min.

4. Add sequentially

5× First-strand buffer	4 μL
0.1 M DTT	1 μL
2.5 mM dNTPs	3 μL
Superscript II	1 μL

5. Incubate in a water bath at 42°C for 1 h.
6. Heat inactivate by keeping at 70°C for 10 min and store at −20°C.

3.3.5. PCR Amplification

1. To amplify small RNAs ligated to adaptors, combine the following in 50 μL reaction for PCR amplification:

10× PCR buffer for KOD Polymerase	5.0 μL
2.5 mM dNTPs	1.0 μL
20 μM PCR primer 1	0.5 μL
20 μM PCR primer 2	0.5 μL
KOD polymerase enzyme (XL)	1.0 μL
RT reaction mixture	
(from step 6 of Subheading 3.3.4)	5.0 μL
Sterile distilled water to	50.0 μL

2. Incubate the PCR tubes in a thermal cycler at 94°C for 1 min, and then cycle 15–18 times at 94°C for 10 s, 60°C for 30 s, and 72°C for 15 s followed by a final extension step at 72°C for 10 min.

3.3.6. Gel Purification of PCR Products

1. Load the PCR product after mixing with 6× loading dye in a 3% Agarose gel in 1× TAE buffer.
2. Load the 25 bp size marker in the same gel.
3. Run the agarose gel at 100 V for 20–30 min.
4. In a transilluminator, visualize the gel and cut out the band corresponding to ~90 bp with a clean razor. Alternately, a 6% TBE PAGE gel can be used to elute the PCR product (see Note 7).
5. Elute the DNA from gel slice using Qiagen gel purification kit following the manufacturer's instructions.

3.3.7. Quality Control and Cloning of PCR Products

1. Load 1 mL of gel purified product onto a nanodrop and measure the concentration by reading the absorbance at 260 nm.
2. Measure the 260/280 ratio. The ratio should be approximately 1.8.

A sample of gel purified PCR product can be cloned in T-vector and sequenced to assess the quality of the library prior to sending samples for deep sequencing (see Note 8).

1. Briefly centrifuge the T-Easy Vector DNA (Promega) tube to collect contents at the bottom.
2. Set up ligation reaction as follows:

2× Rapid ligation buffer	5 μL
pGEM®-T Easy Vector (50 ng)	1 μL
T4 DNA Ligase (3 Weiss units/μL)	1 μL
PCR product	
(from step 5 of Subheading 3.3.6)	1–2 μL
Deionized water to	10 μL

3. Mix gently and incubate the reaction for 1 h at room temperature or at 4°C overnight.
4. Thaw JM109 High Efficiency Competent Cells in an ice bath and transfer 50 mL of cells to a 1.5 mL tube on ice.
5. Add 2–4 mL of ligation reaction to competent cells.
6. Gently flick the tubes and incubate on ice for 30 min.
7. Heat-shock the cells for 40 s in water bath at 42°C and return the tubes to ice.
8. Add 250 mL SOC medium to each tube and shake (~150–200 rpm) for an hour at 37°C.
9. Plate 100 mL of each transformation culture onto LB/ampicillin/IPTG/X-Gal plates.
10. Incubate plates overnight at 37°C.
11. Randomly select 20 white colonies, isolate plasmid DNA for sequencing, and verify that the clones contain small RNA inserts.

4. Notes

1. Washing the RNA pellet in 70% ethanol prior to dissolving in water can be omitted, as it may cause a small loss of small RNAs.
2. If there is less total RNA to begin with proceed directly to step 11, as small RNA enrichment would cause further loss of RNA.
3. The pellet obtained is high-molecular-weight RNA that includes most mRNAs and can be used for regular northern analyses if needed.

4. Do not run the gel longer than necessary as it will affect the recovery of RNA.
5. Low-retention siliconized tubes are recommended to minimize binding of small RNAs to tubes.
6. The 3'adapter oligonucleotide is blocked (inverted thymidine, idT) at its 3'-end to avoid self-ligation.
7. Isolating the right size band is critical for the success of this protocol and hence utmost care has to be taken at this step. If not careful, a band corresponding to the adaptor-ligated product without small RNA insert (around 70 bp) is more likely to contaminate.
8. Since T-Vector cloning requires a dA overhang in the PCR products, use a DNA polymerase with terminal transferase activity for PCR amplification. (KOD XL DNA Polymerase generates a mixture of PCR products with blunt end 3′-dA overhangs, suitable for cloning T-vector as well as blunt end vectors.)

Acknowledgments

This research was supported by the Oklahoma Agricultural Experiment Station and the USDA/NRI grant (2007-02019).

References

1. Jones-Rhoades MW, Bartel DP, Bartel B (2006) MicroRNAS and their regulatory roles in plants. Annu Rev Plant Biol 57:19–53
2. Mallory AC, Vaucheret H (2006) Functions of microRNAs and related small RNAs in plants. Nat Genet 38 Suppl: S31–S36
3. Sunkar R, Chinnusamy V, Zhu J, Zhu JK (2007) Small RNAs as big players in plant abiotic stress responses and nutrient deprivation. Trends Plant Sci 12:301–309
4. Voinnet O (2009) Origin, biogenesis, and activity of plant microRNAs. Cell 136:669–687
5. Katiyar-Agarwal S, Gao S, Vivian-Smith A, Jin H (2007) A novel class of bacteria-induced small RNAs in Arabidopsis. Genes Dev 21: 3123–3134
6. Lu S, Sun YH, Chiang VL (2008) Stress-responsive microRNAs3 in Populus. Plant J 55:131–151
7. Ruiz-Ferrer V, Voinnet O (2009) Roles of plant small RNAs in biotic stress responses. Annu Rev Plant Biol 60:485–510
8. Sunkar R, Girke T, Jain PK, Zhu JK (2005) Cloning and characterization of microRNAs from rice. Plant Cell 17:1397–1411
9. Tomari Y, Zamore PD (2005) Perspective: machines for RNAi. Genes Dev 19: 517–529

Chapter 10

Global Identification of Small RNA Targets in Plants by Sequencing Sliced Ends of Messenger RNAs

Yong-Fang Li and Ramanjulu Sunkar

Abstract

Small RNAs (microRNAs and other classes of endogenous small interfering RNAs) play important roles in a wide variety of biological processes. However, integration of small RNAs in diverse biological networks relies on the confirmation of their RNA targets. In plants, miRNAs negatively regulate mRNA targets by guiding a cleavage in the complementary site that leaves a 3′ polyadenylated RNA possessing monophosphate at its 5′ end. This chapter describes a detailed step-by-step protocol for cloning such sliced 3′ products in order to identify small RNA targets. Using this protocol, we have identified more than 150 small RNA targets in rice; some are conserved and others are non-conserved targets for rice small RNAs.

Key words: Small RNA, MicroRNAs, miRNA targets, Posttranscriptional gene regulation

1. Introduction

In plants, miRNAs and trans-acting siRNAs (a class of endogenous small interfering RNAs) are two major classes of small RNAs, which are known to regulate target genes at the posttranscriptional level. Both these classes of small RNAs mainly regulate their mRNA targets by AGO-mediated mRNA cleavage (1–3). Additionally, miRNAs can also inhibit translation (4–7). These small RNAs play important roles in regulating development, stress responses, and diverse other biological processes in plants (8–13). In plants, miRNA targets can be predicted by computational search against the mRNAs with perfect or near-perfect complementary to a mature miRNA (14). However, all such predictions require experimental validation. Plant miRNAs and tasiRNAs have been largely implicated in degradation of their RNA targets by slicing precisely between 10th and 11th nt from the 5′ end of the miRNA, leaving a monophosphate at 5′ end of the 3′ cleaved fragment.

Yinong Yang (ed.), *Rice Protocols*, Methods in Molecular Biology, vol. 956,
DOI 10.1007/978-1-62703-194-3_10, © Springer Science+Business Media, LLC 2013

This biochemical property has been exploited to validate mRNA transcripts targeted by miRNAs and tasiRNAs. Modified 5′ rapid amplification of cDNA ends (5′ RACE) is the most widely used method to detect *in vivo* 3′ sliced ends induced by miRNA or tasiRNA (3). A major limitation of modified 5′ RACE is that every single predicted target gene has to be verified independently, and thus is laborious, time consuming, and costly. High-throughput technologies were recently developed to identify and validate small RNA target transcripts at a global scale (15, 16). This technique takes the advantage of modified 5′-RACE, high-throughput sequencing, and bioinformatics, and is called as parallel analysis of RNA ends (PARE) or degradome analysis (15–18). Here we describe the protocol in detail on how to clone 3′ cleaved ends induced by small RNAs based on published procedures with slight modifications.

Briefly, Small RNA-induced slicing leaves the polyadenylated 3′ fragment possessing a 5′-monophosphate; this end is ligation competent. After RNA adapter possessing MmeI recognition site in its 3′ end is ligated with the sliced mRNAs, the transcripts are reverse-transcribed using a 3′ adapter oligo (dT) primer. The cDNA is amplified using primers corresponding to 5′ and 3′ adapter sequences and then digested using MmeI, which cleaves 20 nt downstream of its recognition site. The MmeI digestion products are gel-purified and are ligated to a 3′double-strand DNA adapter. The ligated products are amplified by PCR and the resulting library is gel-purified and submitted for SBS sequencing. These 20 nt sequences will serve as “Tags” to identify the annotated transcripts. Recently, 5′ RNA adaptor containing recognition site of EcoP15I instead of MmeI was used (19), and EcoP15I cleaves 27 nt downstream of its recognition site; accordingly, 27 nt long “Tags” are generated. However, this method is incapable of finding targets for miRNAs that tend to translationally repress targets without detectable slicing. Another drawback of this method is that any uncapped 3′ polyadenylated RNA with a 5′ phosphate could be potentially cloned; thus, comprehensive bioinformatics analysis is obligatory (15–18).

2. Materials

2.1. Reagents

1. Trizol® reagent (Invitrogen).
2. Chloroform.
3. Isopropanol.
4. Ethanol.
5. Nuclease-free water.
6. 3 M sodium acetate (pH 5.2).

7. Phenol/chloroform/iso-amyl alcohol (25:24:1).
8. T4 RNA ligase (New England biolabs).
9. 40 U/mL RNaseOUT (Invitrogen).
10. Glycogen (Invitrogen).
11. SuperScript III RT (Invitrogen).
12. dNTPs (Fisher).
13. KOD XL DNA polymerase (Novagen).
14. 2 U/mL MmeI (New England Biolabs).
15. 40% Acrylamide:Bis acrylamide (19:1) solution.
16. 10% Ammonium persulfate (APS).
17. Tetramethyl ethylene diamine (TEMED) (Bio-Rad).
18. 0.3 M NaCl.
19. S.O.C medium (BD Diagnostics): To make the medium, measure 20 g Bacto-tryptone, 5 g Bacto-yeast extract, 0.5 g NaCl, 2.5 mL 1 M KCl and bring to 1,000 mL with ddH_2O. Adjust pH to 7.0 with 10 N NaOH, autoclave to sterilize, and add 20 mL of sterile 1 M glucose (sterilized by passing through a 0.2 μm filter) immediately before use.
20. LB plates containing ampicillin (100 μg/mL), IPTG, and X-gal.
21. TOPO 10 competent cells.
22. Agarose.
23. 0.5× TBE buffer: Make by diluting 10X concentrated stock with water. The 10× TBE stock is composed of 108 g Tris base, 55 g Boric acid, 40 mL 0.5 M EDTA (pH 8.0). Bring to 1 L with dH_2O and autoclave for 30 min.
24. 1× TAE buffer: Make by diluting 50× concentrated stock with water. To make the 50× TAE stock, mix 242 g Trizma base, 100 mL of 0.5 M EDTA pH 8.0, 57.2 mL glacial acetic acid in 600 mL of ddH_2O. Bring to a final volume of 1 L with additional ddH_2O.
25. 6× DNA loading buffer: 25 mg bromophenol blue, 25 mg xylene cyanol, 4 g sucrose, adjust volume to 10 mL with H_2O.
26. 25 bp DNA ladder (Invitrogen).
27. Ethidium bromide (Bio-Rad), 10 mg/mL solution (see Note 1).
28. Rapid DNA ligation kit (Invitrogen).
29. MicroPoly(A) Purist™ kit (Ambion).
30. QIAquick® PCR pufification kit (QIAGEN).
31. pGEM T Easy vector system (Promega).

2.2. Oligos

1. 5′ RNA adapter: 5′ GUUCAGAGUUCUACAGUCCGAC 3′.
2. 3′ adapter oligo (dT) primer: 5′ CGAGCACAGAATTAAT ACGAC$(T)_{18}$V 3′.
3. 5′ adapter primer: 5′ GTTCAGAGTTCTACAGTCCGAC 3′.
4. 3′ adapter primer: 5′ CGAGCACAGAATTAATACGACT 3′.
5. Double-stranded DNA adapter: Top, 5′ p-TCGTATGCCTC TTCTGCTTG 3′; and bottom, 3′ NNAGCATACGGC AGAAGACGAAC 5′; p: phosphate group.
6. P5 primer: 5′ AATGATACGGCGACCACCGACAGGTTCAG AGTTCTACAGTCCGA 3′.
7. P7 primer: 5′ CAAGCAGAAGACGGCATACGA 3′.

3. Methods

3.1. Total RNA Isolation and polyA RNA Purification

Extract total RNA using Trizol® Reagent according to the manufacturer's instructions.

3.1.1. Total RNA Isolation

1. Grind ~5 g rice seedlings (or other interested tissues or plant material) to fine powder using liquid nitrogen. Divide the fine powder into two 50 mL tubes containing 25 mL of TRIzol® Reagent.
2. Vortex the tube vigorously for 15 s and incubate the homogenized samples for 5 min at room temperature.
3. Add 5 mL of chloroform to each tube, vortex the tubes vigorously for 15 s, and incubate tubes at room temperature for 2–3 min.
4. Centrifuge the tubes at 12,000 × *g* for 15 min at 4°C.
5. Divide the aqueous phase of each tube to two 50 mL tubes equally and add 6.25 mL of isopropanol to each tube. Vortex the tubes, incubate samples at room temperature for 10 min, and centrifuge at 12,000 × *g* for 15 min at 4°C.
6. Remove the supernatant and add 15 mL 70% ethanol to wash the RNA pellet.
7. Vortex the tubes and centrifuge at 12,000 × *g* for 5 min at 4°C.
8. Remove the supernatant and dry the RNA pellet for 5 min at room temperature.
9. Dissolve total RNA in nuclease-free water completely.
10. Check the RNA integrity by electrophoresis on a 1% agarose gel (see Note 2).

3.1.2. polyA RNA Purification

Purify polyA RNA with ~800 μg total RNA using MicroPoly(A) Purist™ kit (Ambion) according to the manufacturer's instructions. Use maximum 400 μg total RNA for each Oligo(dT) cellulose tube. All centrifugation steps are to be carried out at room temperature (see Note 3).

Binding ployA RNA to Oligo (dT) Cellulose

1. Bring 400 μg total RNA to 250 μL by adding nuclease-free water.
2. Add 250 μL 2× binding solution and mix thoroughly.
3. Transfer the 500 μL RNA sample to a Oligo(dT) cellulose tube, mix well, and heat the mixture at 65–75°C for 5 min.
4. Agitate the tube gently for 1 h at room temperature.
5. Centrifuge the tube at 4,000 × *g* for 3 min and remove the supernatant.
6. Add 500 μL wash solution 1 to the oligo(dT) cellulose pellet, vortex the tube well, transfer the oligo(dT) cellulose suspension to a spin column, centrifuge at 4,000 × *g* for 3 min, and discard the flow-through.

Wash the Oligo(dT) Cellulose

1. Put the spin column on a new tube; add 500 μL wash solution 1 to the column, vortex the tube, and centrifuge at 4,000 × *g* for 3 min.
2. Wash the Oligo(dT) cellulose twice with 500 μL wash solution 2 as last step.

Elution of polyA RNA

1. Put the spin column on a new tube; add 200 μL preheated (75°C) RNA storage solution to the column, vortex the tube, and centrifuge at 4,000 × *g* for 2 min.
2. Put the spin column on a new tube, and elute the polyA RNA with 200 μL preheated (75°C) RNA storage solution once more as last step (see Note 4).

Precipitation of polyA RNA

1. Add the following reagents in each tube containing 200 μL polyA RNA elution (four tubes in total):

   20 μL 3 M sodium acetate (pH 5.2)

   550 μL 100% ethanol

   1 μL glycogen
2. Mix well and leave the mixture at −20°C overnight.
3. Centrifuge at 12,000 × *g* for 30 min at 4°C.
4. Carefully remove the supernatant.
5. Add 1 mL 70% ethanol, and vortex well.
6. Centrifuge at 12,000 × *g* for 10 min at 4°C.
7. Carefully remove the supernatant, and dry the polyA RNA pellet for 5 min at room temperature.
8. Dissolve polyA RNA in 22 μL RNase-free water; use 1 μL to check polyA RNA quality by electrophoresis on 1% agarose gel and 1 μL to check RNA concentration.

3.2. RNA Adapter ligation and Purification of Ligated polyA RNA

3.2.1. RNA Adapter ligation

1. Assemble the following components in a nuclease-free tube on ice:

   5 μg polyA RNA (from step 4 of Subheading 3.1.2; see Note 5)

   6.8 μL 5′ RNA adapter (100 μM)

   Add nuclease-free water to 28 μL
2. Mix well and centrifuge briefly.
3. Heat the mixture for 5 min at 65°C and cool on ice for 2 min.
4. Add the following reagents to the tube:

   4 μL RNaseOut™ (40 U/μL)

   4 μL 10×T4 RNA ligase buffer

   4 μL T4 RNA ligase
5. Mix gently and spin briefly.
6. Incubate at 37°C for 1 h.

3.2.2. Purification of Ligated polyA RNA

1. Add 160 μL nuclease-free water to the ligation reaction and 200 μL phenol/chloroform/iso-amyl alcohol (25:24:1) and vortex well.
2. Centrifuge at 12,000×*g* for 5 min at room temperature.
3. Transfer the aqueous phase to a new 1.5 mL tube and add nuclease-free water to 250 μL.
4. Purify ligated polyA RNA using MicroPoly(A) Purist™ kit (Ambion) as described in Subheading 3.1.2 to remove free RNA adapters.
5. Dissolve ligated polyA RNA in 25 μL nuclease-free water.

3.3. Reverse Transcription

1. Assemble the following reagents on ice:

   25 μL purified ligated polyA RNA from steps in Subheading 3.2.2

   2.5 μL 3′ adapter (dT) primer (100 mM)

   2.5 μL dNTP mix (10 mM each)
2. Heat the mixture for 5 min at 65°C, and then cool on ice for 2 min.
3. Spin the tube briefly and add the following reagents to the tube:

   10 μL 5× first-strand buffer

   5 μL 0.1 M DTT

   2.5 μL RNaseOUT (40 U/μL)
4. Mix the tube gently and incubate at 42°C for 2 min.
5. Add 2.5 μL SuperScript III (200 U/ μL).
6. Incubate the tube at 42°C for 2 h.
7. Incubate the reaction at 70°C for 10 min to inactivate enzyme.

3.4. PCR Amplification and PCR Product Purification

3.4.1. PCR Amplification

1. Add the following reagents to the cDNA tubes on ice:

   15 μL 10 × buffer for KOD XL DNA polymerase

   3 μL dNTP mix (10 mM each)

   3 μL 5′ adapter primer (20 mM)

   3 μL 3′ adapter primer (20 mM)

   73 μL nuclease-free water

   3 μL KOD XL DNA polymerase (see Note 6)
2. Mix well and spin briefly.
3. Remove 10 μL to a PCR tube and aliquot the rest to three PCR tubes.
4. Conditions for PCR: 94°C for 4 min, followed by five cycles of 94°C for 50 s, 60°C for 50 s, and 72°C for 3 min, and then 72°C for 5 min (see Note 7). Run 15 cycles for the 10 μL PCR reaction.
5. Run the 10 μL reaction in a 1% agarose gel; a smear from 100 bp to 2 kb should be seen.

3.4.2. PCR Product Purification

Purify the PCR products using QIAquick® PCR purification kit (QIAGEN) according to the manufacturer's instructions.

1. Pool the three PCR reactions, add 700 μL buffer PB, and mix well.
2. Place a QIAquick spin column in a 2 mL collection tube; transfer the mixture to the QIAquick column and centrifuge at 12,000 × *g* for 30 s at room temperature.
3. Discard flow-through and put the column back into the same tube. Add 0.75 mL buffer PE to the column and centrifuge at 12,000 × *g* for 30 s at room temperature.
4. Discard flow-through and centrifuge the column for an additional 1 min at 12,000 × *g* and at room temperature.
5. Put the column in a clean 1.5 mL tube.
6. Add 50 μL nuclease-free water to the column and leave for 1 min and then centrifuge at 12,000 × *g* for 1 min at room temperature.
7. Repeat steps 5 and 6.
8. Condense the eluted DNA with speedVac to ~60 μL.
9. Check DNA concentration with nanodrop using 1 μL sample.

3.5. MmeI Digestion and Purification of MmeI Cleaved Fragments

3.5.1. MmeI Digestion

1. Add the following reagents to the cleaned PCR product (from Subheading 3.4.2):

   8 μL 10× NEB buffer 4.

   0.5 μL 8 mM S-adenosylmethionine (SAM) (final concentration is 50 μM).

Add nuclease-free water to make a final volume of 76 μL.

4 μL 2 U/μL MmeI (see Note 8).

2. Mix well and spin briefly.
3. Incubate at 37°C for 1 h.

3.5.2. Size Fractionation and Purification of the MmeI Cleaved Fragments

1. Prepare 8% polyacrylamide gel by combining the following reagents:

   10 mL 40% acryl:bis-acryl (19:1)

   2.5 mL 10 × TBE

   37.2 mL of nuclease-free water

   250 μL 10% APS

   50 μL TEMED
2. Add 6 × DNA loading buffer to the MmeI digested sample from Subheading 3.5.1.
3. Load the mixture to three wells, and load 1 μg 25 bp DNA ladder in one well (keep at least 1-well space from the cleaved samples).
4. Run gel at 200 V for 90 min.
5. Stain the gel with ethidium bromide (see Note 1).
6. Cut the MmeI cleaved fragments corresponding to 42 bp (the band is very weak, but visible), slice the gel into small pieces, and put into a 2 mL siliconized tube.
7. Add 500 μL 0.3 M NaCl, and elute the fragments overnight at 4°C with gentle agitation.
8. Centrifuge at 12,000 × *g* for 5 min at 4°C.
9. Transfer the supernatant to a new 2 mL siliconized tube and add 2 volumes of 100% ethanol and 1 μL glycogen.
10. Leave the tube at –20°C overnight.
11. Centrifuge at 12,000 × *g* for 30 min at 4°C.
12. Wash the pellet with 80% ethanol, and centrifuge at 12,000 × *g* for 5 min at 4°C.
13. Remove the supernatant carefully (see Note 9).
14. Dry the pellet for 5 min at room temperature.
15. Dissolve the pellet in 21 μL water.

3.6. Ligation of 3′ Double-Strand DNA Adapter

Add the following reagents to the gel-purified cleaved product from Subheading 3.5.2:

6 μL 5 × T4 DNA ligase buffer (Invitrogen)

1 μL 3′-double-stranded DNA adapter (125 μM)

2 μL T4 DNA ligase

Mix well and incubate overnight at 14°C.

3.7. Purification of Ligated Fragments

Run the ligation reaction on an 8% polyacrylamide gel, slice the ligation band (63 bp), and elute the ligation as described in Subheading 3.5.2.

3.8. Library Amplification and Purification

3.8.1. Test Amplification

1. Dissolve the cleaned ligation product from Subheading 3.7 in 50 μL water; use 2 μL to determine the optimal PCR cycles.
2. Mix the following reagents:

   2 μL ligation

   10 μL 10× KOD XL DNA polymerase buffer

   2 μL dNTP mix (10 mM each)

   2 μL P5 primer (20 mM)

   2 μL P7 primer (20 mM)

   80 μL nuclease-free water

   2 μL KOD XL DNA polymerase
3. Mix the tube well, spin down briefly, and aliquot to five PCR tubes.
4. Conditions for PCR: 94°C for 4 min, followed by 20 cycles of 94°C for 30 s, 60°C for 30 s, and 72°C for 30 s. Remove one tube at 8, 11, 14, 17, and 20 cycles.
5. Run the reactions on 8% polyacrylamide gel along with 1 μg 25 bp DNA ladder.
6. Stain the gel with EtBr, look for the expected band (~86 bp), and decide the optimal number of PCR cycles (see Note 10).

3.8.2. Library Amplification and Purification

1. Set up three 50 μL PCR reactions using all of the remaining ligation product and run the PCR using optimal cycle number.
2. Run the PCR products on 8% polyacrylamide gel; slice the corresponding bands (86 bp), elute, and precipitate library as described in Subheading 3.5.2.
3. Finally dissolve the library in 20 μL nuclease-free water.
4. Library can be sequenced using SBS sequencing technology.

3.9. Library Quality Check (Optional)

The library quality can be assessed by sequencing ~100 cloned PCR products.

3.9.1. Ligation

1. Set up the following reaction using pGEM®-T Easy ligation kit (Promega):

   Library product 5–10 ng

   2× ligation buffer 5 μL

   T-easy vector 1 μL

   Add water to 9 μL

   T4 DNA ligase 1 μL

2. Mix the reaction gently and incubate at room temperature for 30 min to 1 h.
3. Cool the reaction on ice for 5 min and then subject to transformation.

3.9.2. Transformation

1. Thaw competent cell (TOPO 10) on ice, and warm S.O.C. medium and LB-Amp plates.
2. Transfer 2 μL ligation reactions to a tube containing 50 μL competent cell and mix gently.
3. Incubate on ice for 30 min.
4. Heat-shock the cells at 42°C for 30 s.
5. Immediately cool the cells on ice for 2 min.
6. Add 500 μL room-temperature S.O.C. medium.
7. Inoculate the tube at 200 rpm for 1 h at 37°C.
8. Spread 50–100 μL cells on warmed LB-Amp + IPTG + X-gal plate.

3.9.3. Plasmid Isolation and Sequencing

Screen colonies by PCR using M13 forward primer and M13 reverse primer; choose colonies from which ~320 bp fragments can be amplified for sequencing and decide the library quality.

4. Notes

1. Ethidium bromide is carcinogenic. Always wear gloves and avoid contact with skin.
2. RNA integrity is critical for small RNA target library construction. Special care should be taken when working with RNA. Always check RNA integrity by running 1 μL total RNA in 1% agarose gel to visualize 28S and 18S ribosomal RNA.
3. If the total RNA contains a substantial amount of residual salt, precipitate RNA with ethanol and sodium acetate to remove residual salt before PolyA RNA purification.
4. We found that there is still an amount of polyA RNA left after the first elution; the second elution can increase polyA RNA yield.
5. polyA RNA amount is variable. Keep the ratio of RNA adaptor:polyA RNA at 136 pmol/μg.
6. KOD XL Taq polymerase can be replaced by other proofreading Taq polymerases with corresponding polymerase buffers.
7. The cycle number may vary with the initial polyA RNA amount, organism, or tissue.
8. Use 2 U of MmeI per μg DNA.

9. It is hard to see the pellet. Be cautious while removing the supernatant.
10. Reasonable number of PCR cycles to amplify the product is required, but saturation should be avoided.

Acknowledgments

This research was supported by the Oklahoma Agricultural Experiment Station and the USDA/NRI grant (2007-02019).

References

1. Carrington JC, Ambros V (2003) Role of microRNAs in plant and animal development. Science 301:336–338
2. Jones-Rhoades MW, Bartel DP (2004) Computational identification of plant microRNAs and their targets, including a stress-induced miRNA. Mol Cell 14:787–799
3. Llave C, Xie ZX, Kasschau KD, Carrington JC (2002) Cleavage of Scarecrow-like mRNA targets directed by a class of Arabidopsis miRNA. Science 297:2053–2056
4. Aukerman MJ, Sakai H (2003) Regulation of flowering time and floral organ identity by a MicroRNA and its APETALA2-like target genes. Plant Cell 15:2730–2741
5. Brodersen P, Sakvarelidze-Achard L, Bruun-Rasmussen M, Dunoyer P, Yamamoto YY, Sieburth L, Voinnet O (2008) Widespread translational inhibition by plant miRNAs and siRNAs. Science 320:1185–1190
6. Chen X (2004) A microRNA as a translational repressor of APETALA2 in Arabidopsis flower development. Science 303:2022–2025
7. Lanet E, Delannoy E, Sormani R, Floris M, Brodersen P, Crété P, Voinnet O, Robaglia C (2009) Biochemical evidence for translational repression by Arabidopsis microRNAs. Plant Cell 21:1762–1768
8. Axtell MJ, Bowman JL (2008) Evolution of plant microRNAs and their targets. Trends Plant Sci 13:343–349
9. Jones-Rhoades MW, Bartel DP, Bartel B (2006) MicroRNAs and their regulatory roles in plants. Annu Rev Plant Biol 57:19–53
10. Mallory AC, Vaucheret H (2006) Functions of microRNAs and related small RNAs in plants. Nat Genet 38:S31–S36
11. Shukla LI, Chinnusamy V, Sunkar R (2008) The role of microRNAs and other endogenous small RNAs in plant stress responses. Biochim Biophys Acta 1779:743–748
12. Sunkar R, Chinnusamy V, Zhu JH, Zhu JK (2007) Small RNAs as big players in plant abiotic stress responses and nutrient deprivation. Trends Plant Sci 12:301–309
13. Voinnet O (2009) Origin, biogenesis, and activity of plant MicroRNAs. Cell 136:669–687
14. Mallory AC, Bouche N (2008) MicroRNA-directed regulation: to cleave or not to cleave. Trends Plant Sci 13:359–367
15. Addo-Quaye C, Eshoo TW, Bartel DP, Axtell MJ (2008) Endogenous siRNA and miRNA targets identified by sequencing of the Arabidopsis degradome. Curr Biol 18:758–762
16. German MA, Pillay M, Jeong DH, Hetawal A, Luo S, Janardhanan P, Kannan V, Rymarquis LA, Nobuta K, German R et al (2008) Global identification of microRNA-target RNA pairs by parallel analysis of RNA ends. Nat Biotechnol 26:941–946
17. Addo-Quaye C, Miller W, Axtell MJ (2009) CleaveLand: a pipeline for using degradome data to find cleaved small RNA targets. Bioinformatics 25:130–131
18. German MA, Luo S, Schroth G, Meyers BC, Green PJ (2009) Construction of Parallel Analysis of RNA Ends (PARE) libraries for the study of cleaved miRNA targets and the RNA degradome. Nat Protoc 4:356–362
19. Addo-Quaye C, Snyder JA, Park YB, Li YF, Sunkar R, Axtell MJ (2009) Sliced microRNA targets and precise loop-first processing of MIR319 hairpins revealed by analysis of the Physcomitrella patens degradome. RNA 15:2112–2121

Chapter 11

Artificial MicroRNAs for Specific Gene Silencing in Rice

Norman Warthmann, Stephan Ossowski, Rebecca Schwab, and Detlef Weigel

Abstract

Artificial microRNAs (amiRNAs) have been shown to facilitate efficient gene silencing with high specificity to the intended target gene(s). For the plant breeder, gene silencing by artificial miRNAs will certainly accelerate gene discovery, because it allows targeting of all genes in a mapping interval, independent of the genetic background. In addition, beneficial knockout phenotypes can easily be transferred between varieties and across incompatibility barriers. This chapter describes the generation and application of amiRNAs as a gene silencing tool in rice.

Key words: Gene silencing, miRNA, Hairpin, Loss-of-function, Phenotypic complementation

1. Introduction

Detailed analysis of loss-of-function alleles has proven instrumental for the discovery of genes and characterization of gene function. In model plants such as *Arabidopsis* and rice, sequence-indexed insertion collections provide a large resource of potential null alleles that can often be accessed through convenient Web sites such as Salk Institute Genomic Analysis Laboratory (SIGnAL, http://signal.salk.edu), Rice Mutant Database (RMD, http://rmd.ncpgr.cn), and Taiwan Rice insertional mutants database (TRIM, http://trim.sinica.edu.tw/). For a comprehensive list of rice mutant resources *see* Krishnan et al. (1). While very useful, these collections have inherent limits. They are usually not available for non-model species or varieties, require stacking for knockout of redundant homologs, and do not easily allow for partial or regulated loss of gene function, which is particularly useful when null alleles are lethal. Transgenic approaches that employ directed gene silencing can substitute for null alleles without the limitation to a specific species or variety and in addition enable more refined studies of gene function, e.g., by tissue-specific and inducible gene silencing.

Yinong Yang (ed.), *Rice Protocols*, Methods in Molecular Biology, vol. 956,
DOI 10.1007/978-1-62703-194-3_11, © Springer Science+Business Media, LLC 2013

Mediators of transgene-induced gene silencing are single-stranded silencing RNAs (19–23 nucleotides) that bind to target transcripts through complementary base-pairing (2). MicroRNAs (miRNAs), one endogenous class of silencing RNAs, originate from characteristic hairpin-containing precursor transcripts and vectors that contain hairpin precursors are recognized as second-generation RNAi vectors (3). Their sequence can be modified such that the precursor releases miRNAs of desired sequence, so-called artficial miRNAs, in planta. Hence, these amiRNA vectors can serve as reverse genetics tools to direct gene silencing, also non-model systems. The unique applications of amiRNAs include the simultaneous silencing of several related genes including those closely linked in the genome (e.g., in tandem repeats), as well as allele- and splice form-specific, transient or tissue-specific gene silencing. In addition, they provide the possibility of phenotypic complementation with target transgenes that are no longer susceptible to amiRNA-mediated gene regulation after introduction of silent mutations into the target sites (4). The Web application Web MicroRNA Designer (WMD) facilitates the design of suitable amiRNA sequences for a variety of different plant species as well as the design of primer sequences needed to modify the amiRNA vectors. There are of course several ways of constructing amiRNA vectors. Here we describe the use of the tool (WMD) and the molecular steps necessary to engineer the gene silencing vectors as originally conceived by Schwab and colleagues (5) for *Arabidopsis* and adapted to rice by Warthmann and colleagues (6). We describe methods for the WMD-assisted design and construction of amiRNAs to silence one or several related genes of interest. Additionally, we detail how to generate and clone *aMIRNA* precursors by site-directed mutagenesis of a vector (shown here for pNW55) for the subsequent generation and analysis of transgenic plants.

2. Materials

2.1. Computational Prediction of amiRNA Sequences

Select an identifier(s), EST name(s), or sequence of the target gene(s) to be silenced.

2.2. Site-Directed Mutagenesis by PCR

1. Two general oligonucleotide primers A and B (see sequences in Table 1) that match the harboring plasmids (pBluescript derivatives) outside of the miRNA precursor are required. They have been placed such that the sizes of the resulting PCR products enable convenient purification and handling. All oligos are unmodified and can be ordered from any supplier.
2. Four oligonucleotides generated as the output of the amiRNA design program WMD (WMD-Oligo window) are needed to

Table 1
Oligonucleotide Sequences for NW55

Name	Position	Sequence 5′→3′
I	amiRNA forward	AG $(N)_{21}$C AGG AGA TTC AGT TTG A
II	amiRNA reverse	TG $(N)_{21}$C TGC TGC TGC TAC AGC C
III	amiRNA* forward	CT $(N)_{21}$T TCC TGC TGC TAG GCT G
IV	amiRNA* reverse	AA $(N)_{21}$ A GAG AGG CAA AAG TGA A
A (G-4368)	Outside forward	CTG CAA GGC GAT TAA GTT GGG TAA C
B (G-4369)	Outside reverse	GCG GAT AAC AAT TTC ACA CAG GAA ACA G

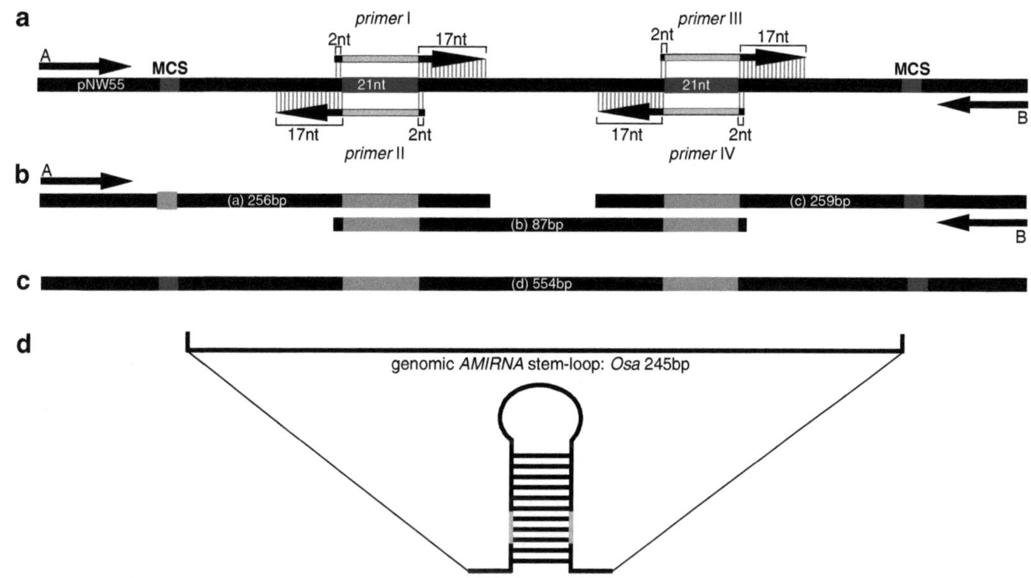

Fig. 1. Schematic representation of PCR reactions that generate *aMIRNA* precursors. (**a**) Illustration of the template plasmid (see Fig. 4) with oligonucleotide binding sites indicated. (**b**) PCR amplicons (a), (b), and (c). (**c**) (a), (b), and (c) are fused to (d) by PCR. (**d**) Only the central part encodes the *aMIRNA* precursor and is schematically shown at the *bottom*. *Symbols*: A, B, I, II, III, IV are oligonucleotide identifiers (*see* text); MCS indicates the multiple cloning sites; (a), (b), (c), and (d) are PCR fragments as indicated in the text.

produce an amiRNA transgene targeting a gene or a gene family of your interest. The directions for using WMD are given under Subheading 3.1. The four primer sequences are generated by and are given in 5′→3′ orientation. They are 40 nucleotides long and specific for the intended amiRNA and precursor. The 5′ most two and 3′ most 17 nucleotides match the template miRNA precursor, while the 21 nucleotides in between do not and will generate the amiRNA and amiRNA* in the amplicon (see Fig. 1). All oligos are unmodified and can be ordered from any supplier.

3. Template plasmid: For example, pNW55 (containing rice osa-MIR528).
4. Equipment and chemicals for PCR, agarose gel electrophoresis, molecular cloning and gel extraction.
5. A DNA polymerase with proofreading capability such as Pfu.

2.3. Cloning of aMIRNA Precursors

1. Commercial PCR product ligation kit (e.g., TOPO kits from Invitrogen) or any standard cloning plasmid (e.g., pGEM-7Z from Promega, linearized with *Sma*I, 10 mM ATP, restriction enzyme *Sma*I).
2. T4 DNA ligase.
3. Competent cells from a standard *E. coli* strain (e.g., DH5α, TOP10, MACH1).
4. LB plates containing appropriate antibiotics.
5. Plasmid extraction (Miniprep) solutions or kit (e.g., QIAprep from Qiagen).
6. Restriction enzymes to test for positive clones and to subclone aMIRNAs.
7. Binary plasmid containing a promoter of interest and a terminator.

2.4. Plant Transformation and Analysis of Transgenic Plants

1. A plant transformation system for the particular rice.
2. TRIzol® reagent (Invitrogen) or commercial RNA extraction kit (e.g., RNeasy from Qiagen).
3. Reverse transcriptase kit (e.g., SuperScript III from Invitrogen) and oligonucleotides for RT-PCR.
4. Optional: Kit for mRNA extraction (e.g., Oligotex from Qiagen) and 5′ RACE (e.g., GeneRacer from Invitrogen) for cleavage site mapping and/or oligonucleotides for site-directed mutagenesis to engineer silent mutations in transgenes containing the target gene(s) for phenotypic complementation.

3. Methods

3.1. Design of amiRNA Sequences

Sequences of amiRNAs need to be designed specifically for the gene(s) of interest and optimized for both effectiveness and specificity. The optimization for specificity (i.e., predicting and avoiding off-targets) depends on the availability of transcriptome sequence information of the particular species. We have developed the WMD, which is available at http://wmd3.weigelworld.org. It can predict suitable amiRNA sequences for gene silencing in a large number of plant species for which a whole-genome annotation or a significant EST/cDNA sequence information is available. Note 1 describes how to design amiRNA sequences for species that are not yet included in WMD. WMD can design amiRNAs to silence single genes as well as amiRNAs to simultaneously silence multiple genes. This is due to the fact that the pairing between an amiRNA and its target gene(s) does not have to be a perfect match. However, attempts to target multiple genes will only be successful, if these genes share at least some regions of high nucleotide sequence similarity. WMD creates amiRNA candidates that are complementary to the target gene(s) and ensures that no other annotated gene in the respective genome/transcriptome release (species) fulfils the criteria of miRNA target interaction (5, 7, 8). These criteria have been empirically determined and are expected to improve as the knowledge on miRNA biology and function grows. It is important to realize that it might not be possible to design an amiRNA against a gene (or several highly sequence-related genes), if all potential target sites to which the amiRNA would bind are also present in other (known) genes. Therefore, a search for highly similar genes is prudent before starting the design process. This can be done by a BLAST search implemented in the WMD Webpage. In cases where targeting of highly related genes does not interfere with the goal of the experiment, these genes can be specified as "accepted off-targets" to increase the chances of successful amiRNA design. WMD requires sequence information of the intended target gene(s) as input and selects 21mer amiRNA candidates from the reverse complement(s).

3.1.1. Using the amiRNA Design Tool WMD-Designer Window

To accommodate the different levels of sequence annotations and specific applications, there are several considerations to make (listed below) before starting the WMD-Designer.

1. Silencing single or multiple genes in *O. sativa* var. Nipponbare (TIGR5 annotation): For genomes that are well annotated (i.e., rice, *Arabidopsis*), it is sufficient to provide the gene identifier(s) including the splice variant (isoform) of interest (e.g., Os01g24680.1). Other annotated isoforms will not be considered off-targets, but the designer will not attempt to target all of them. Whether or not isoforms of a gene are targeted by a specific amiRNA can quickly be tested using the WMD "Target Search" tool (see Note 2).
2. Silencing single or multiple genes in other plant species: For plant species with little or no genome annotation, WMD has to operate on (non-redundant) EST collections. Here, locus identifier(s), e.g., EST name(s), need to be provided, whereby the sequence identifier has to be identical to the sequence identifiers used by WMD. The correct names are easily determined by a search with the "WMD-BLAST" tool. Often multiple EST sequences exist for a single locus. They all need to be identified with the WMD-BLAST tool first, then one of them

(preferably the longest) should be used as the target gene (identifier entered into the "Target genes" field), and the identifiers of all others that likely originate from the same locus are entered as "accepted off-targets" (separated by commas). When attempting to silence multiple genes from different loci simultaneously, all identifiers are entered in the "Target genes" field, and the others are added to the list of "accepted off-targets." When none of the redundant ESTs comprises the full-length target transcript, several can be concatenated and serve as target for the amiRNA design. It should be given a custom identifier that distinguishes it from all annotated ESTs, and all redundant ESTs need to be specified as "accepted off-targets."

3. Unannotated genes and sequence variants, e.g., in other rice strains: If the gene to be silenced is not contained in the WMD database for the respective species (e.g., *GUS*, *GFP*, viral genes, genes that are not yet annotated, etc.), or represents a sequence variant of an annotated gene (e.g., an allele from a different ecotype or cultivar), then the target gene sequence(s) should be entered into the "Target genes" window in FASTA format with a custom name that is different from any other annotated gene as header. Silencing of sequence variants requires the reference allele to be specified as "accepted off-target," unless allele-specific silencing is desired.
4. Splice-form or allele-specific silencing: If exclusively one splice form is to be targeted, then a DNA sequence (>21 nt, FASTA format), which is unique to the respective splice form, should be entered as the target into the "Target genes" window with a unique name as a header.

In all cases, the respective plant species/genome release is to be selected from the drop-down menu on the WMD-Designer page to ensure specificity within this set of sequences. Specification of the "minimal number of included targets" is necessary when more than two genes are to be silenced simultaneously. In addition to finding an amiRNA that silences all genes simultaneously, WMD will attempt to generate amiRNAs that target all possible subgroups of sizes greater or equal to the number given in "minimal number of included targets." The computation of amiRNAs will take between a few minutes and several hours. The results will be e-mailed to the provided e-mail address.

3.1.2. Processing amiRNA Design Results

The amiRNA result e-mail contains a hyperlink to a results Webpage on which the amiRNA candidate sequences are listed (Fig. 2). This list can be downloaded for future reference (Microsoft Excel format). See Note 3 in case the results page is empty. In principle, all amiRNA sequences returned by WMD fulfill the above-described criteria and are expected to silence the predicted target genes successfully. However, they comply differently to the parameters that we consider optimal in terms of base pair composition, hybridization

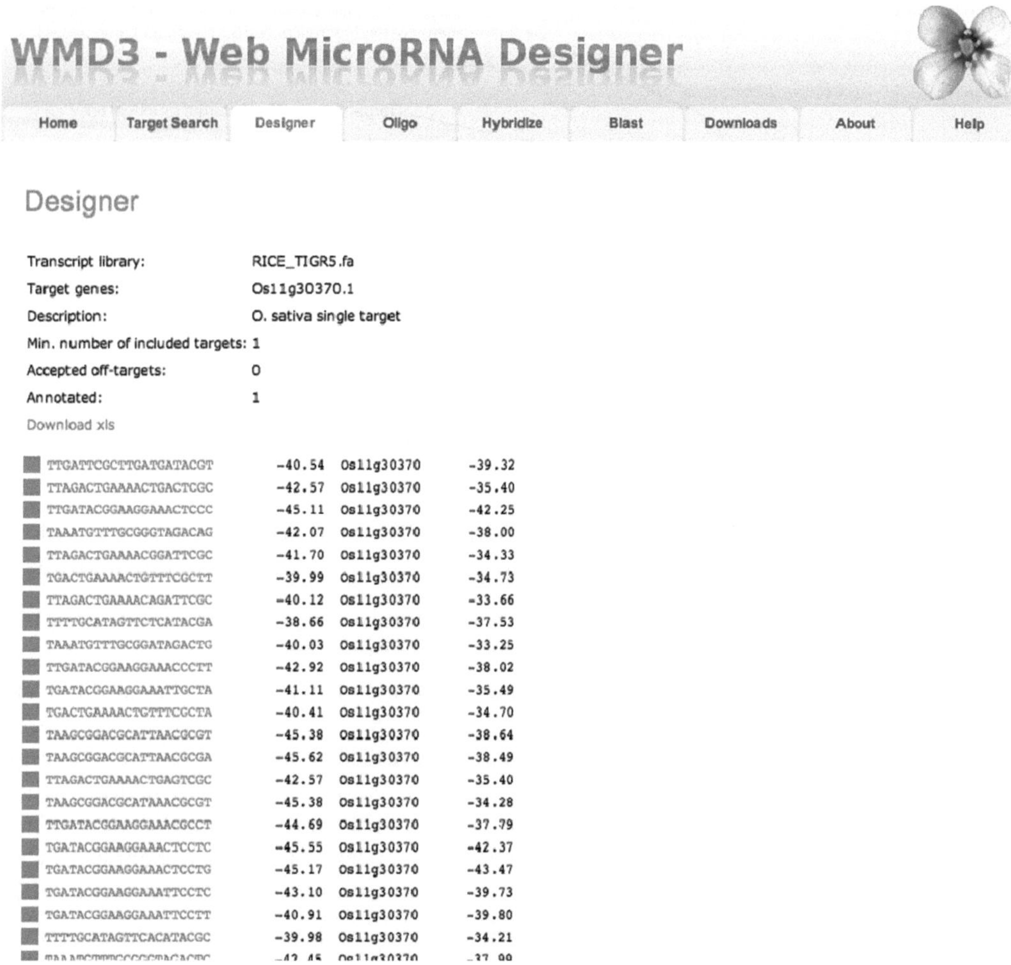

Fig. 2. Example of a WMD result page. Candidate amiRNA sequences are listed and ordered by efficiency and specificity criteria.

properties to the target gene(s), and specificity criteria (for details, see ref. 9). The possible amiRNAs are therefore ranked by a respective cumulative score. The highest ranking amiRNA candidates are presented on top of the list. Green color indicates a very favorable score, while orange and red often mark amiRNAs with potentially reduced efficiency or, more often, specificity. It is thus recommended to proceed from top to bottom of this list. The amiRNA sequences in the results page (Fig. 2) are hyperlinked to visualizations of alignments of the amiRNA to all potential target sequences in the WMD database, ordered by hybridization energy (for illustration see Fig. 3). The intended target gene ideally appears on top of the list. The WMD-Target Search tool can also show genes that align to the amiRNA with equal or fewer than five mismatches but do not fulfill other empirical rules, as indicated with respective

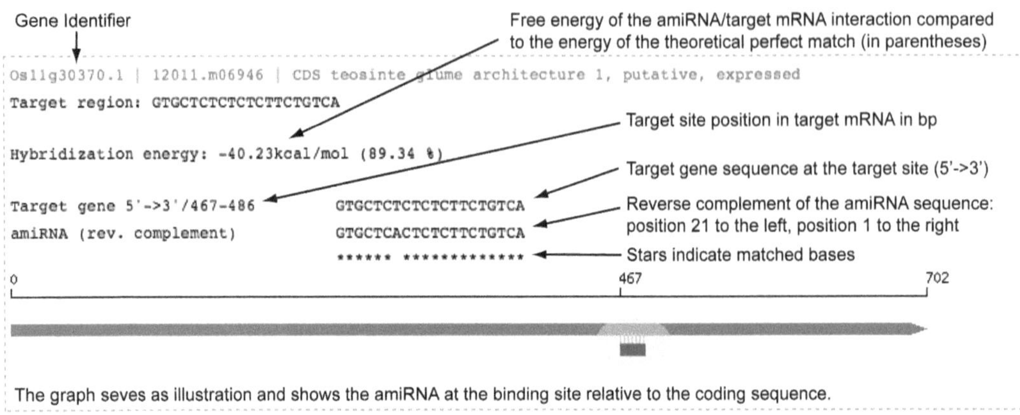

Fig. 3. Alignment of an amiRNA to its target gene. Explanation of the amiRNA-target alignment is presented by WMD.

notes in red bars. For the subsequent selection of one (or more) amiRNA for further experiments, we recommend the following:

1. It is preferable for all intended target genes not to have mismatches to the amiRNA at positions 2–12.
2. amiRNA candidates with one or two mismatches at the 3′ end of the amiRNA (positions 18–21) should be preferred, since it has been suggested that perfectly matching amiRNAs might trigger the so-called transitive siRNA formation, where amplification of sequences adjacent to the binding site is primed by the miRNA. These sequences could in turn themselves serve as silencing triggers and affect other, unintended genes (10).
3. The absolute hybridization energy of the binding between amiRNA and the target sequence should be less than −30 kcal/mole, and preferably be in the range between −35 and −40 kcal/mole. Additionally, good target site accessibility is preferable. This can be tested using RNAup instead of RNAcofold in the Target Search (see Note 6).
4. The amiRNA binding site should be located within the coding region of the target gene since UTRs are more likely to be misannotated.
5. At least two different amiRNAs per target gene or group of genes are selected for experimental work. If several are selected, the amiRNAs should bind the target mRNA at different locations, since secondary structure is suspected to influence miRNA efficacy.

3.2. Construction of aMIRNA Precursors by Site-Directed Mutagenesis

While exogenous small RNA duplices are often directly used to transfect animal cell cultures and induce gene silencing, their accumulation in plants requires the construction and subsequent expression of a precursor RNA in planta. To this end, amiRNAs are engineered into plasmid vectors that contain endogenous *MIRNA* precursors by site-directed mutagenesis. The resulting precursor RNAs are then processed in the plant cell by the endogenous miRNA machinery to release the amiRNAs. Several *Arabidopsis* precursor templates have been used successfully in *Arabidopsis* (5, 11–14) and other plants (11, 15), while NW55 containing part of rice *osa-MIR528* was specifically engineered for amiRNA production in rice (6). Please see Note 4 when working with a different plant species.

MIRNA precursors fold back on themselves to form a hairpin structure, and it is important to preserve this structure for successful processing. Therefore, engineering of amiRNAs into *MIRNA* precursor templates requires the exchange of not only the miRNA by the amiRNA sequence but also the pairing region in the hairpin, called the miRNA*, such that pairing positions as well as G:U pairs—if possible—are retained. The WMD software (WMD-Oligo window) thus generates four oligonucleotides per amiRNA sequence input to exchange two sequences by PCR on the precursor. Currently, the software supports precursors *osa-MIR528*, *ath-MIR319a*, and *cre-MIR1157* to be (selected from a drop-down menu); others will be included as they become available. Recently, a ligation-based construction of amiRNAs has been proposed (16), but this is not covered in this protocol.

Endogenous *MIRNA* precursors that have been cloned into plasmids serve as templates for PCR reactions to exchange miRNA and miRNA*. These precursors include the hairpin and short pieces of flanking sequence on either side, which are known to be part of the longer endogenous *miRNA* transcript. Plasmids that are currently available contain *ath-MIR319a* (plasmid pRS300) and *osa-MIR528* (plasmid pNW55). These plasmids should be obtained through www.addgene.org or, if unsuccessful, by sending a request to Detlef Weigel (weigel@weigelworld.org). A schematic representation of these *MIRNA*-containing plasmids is shown in Fig. 4; their complete sequences are available on http://wmd3.weigelworld.org. Using the four primers provided by the WMD program and the standard primers A and B, the *aMIRNA* precursor is amplified in three pieces (a–c) as shown in Fig. 1. The three pieces are subsequently fused to one amplicon (d) in another (single) PCR reaction by overlapping PCR.

1. Resuspend the template plasmid upon receipt, transform into competent *E. coli* cells (standard lab strain), spread on ampicillin-containing LB plates, inoculate an overnight culture from a single colony, and isolate the plasmid again using standard plasmid isolation procedures. Prepare a 1:100 dilution.

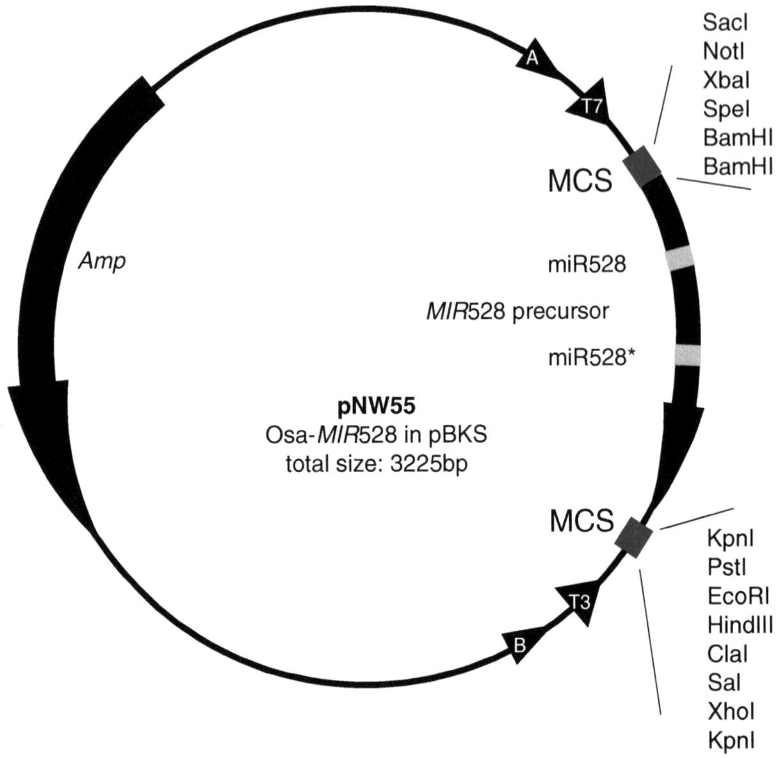

Fig. 4. Template plasmid pNW55 for construction of the amiRNA precursor, the *aMIRNA* foldback. Plasmid pNW55 containing the *osa-MIR528* precursor in pBluescript KS (cloned via *Sma*I site). The complete plasmid sequence is available at http://wmd3.weigelworld.org. *Abbreviations*: A, B oligonucleotide binding sites; T3, T7 RNA polymerase/oligonucleotide binding sites; *Amp* Ampicillin resistance gene; *MCS* multiple cloning site. The *aMIRNA* foldback and surrounding regions are detailed in Fig. 1.

Table 2
PCR reactions on template plasmid pNW55 (osa-MIR528)

Reaction	Forward oligo	Reverse oligo	Template	Length of PCR product
(a)	A	II	pNW55	256 bp
(b)	I	IV	pNW55	87 bp
(c)	III	B	pNW55	259 bp
(d)	A	B	(a) + (b) + (c)	555 bp

2. Setting up PCR reactions (a)–(c):
   All PCR reactions should preferentially be carried out with a proofreading polymerase (such as Pfu) to avoid PCR errors. Table 2 shows the oligonucleotide combinations for each PCR reaction together with the expected size of the product (Fig. 2).

Reactions (a)–(c):

2.0 μl	10× PCR buffer (with ~25 mM Mg^{2+})
2.0 μl	dNTPs (2 mM)
1.0 μl	Each oligonucleotide (10 μM; see PCR scheme)
1.0 μl	Template DNA (1:100 dilution of template plasmid)
0.2 μl	Polymerase
12.8 μl	Water
20 μl	Total

PCR conditions:

95°C	2 min	
95°C	30 s	
52°C	30 s	35 cycles
72°C	40 s	
72°C	7 min	

3. Isolate PCR fragments from a 2% agarose gel and purify with standard gel extraction procedures. PCR fragments from reactions (a) to (c) can be pooled already at this step. Elute in 20 ml of water.
4. Reaction (d): Fusion of fragments (a)–(c). Combine:

2.0 μl	10× PCR buffer (with Mg^{2+})
2.0 μl	dNTPs (2 mM)
1.0 μl	Oligonucleotides A and B (10 μM)
0.5 μl	Each purified gel fragment (a)–(c) or 1.5 μl of combined gel eluate
0.2 μl	Polymerase
12.3 μl	Water

Run in PCR machine using the following settings:

95°C	2 min	
95°C	30 s	
52°C	30 s	35 cycles
72°C	90 s	
72°C	7 min	

5. Isolate PCR fragment from a 1% agarose gel.

3.3. Cloning

To verify the sequence of the fusion-PCR product (d), it can be blunt end cloned into a standard cloning vector and DNA sequenced. It is important to keep in mind that this PCR fragment contains the T3 and T7 and the Multiple Cloning Sites of the template plasmid (see Fig. 3). Using T3 and/or T7 primers for sequencing may cause failed sequencing reactions if the vector of choice also contains T3 and/or T7 sites. PCR reactions with proofreading polymerases generate blunt-ended products. Some companies offer kits to directly clone blunt-ended DNA fragments (e.g., TOPO kits from Invitrogen); follow the manufacturer's recommendations (see Note 5 for the use of gateway-compatible plasmids).

Another simple and cheap protocol to clone blunt-ended PCR products is based on plasmids that are linearized with a restriction enzyme that produces blunt ends (e.g., *Sma*I). Since PCR products are not 5′ phosphorylated, the plasmid needs to retain its terminal phospho-groups after restriction and is directly used for ligations without prior dephosphorylation or purification. Re-ligation of the empty plasmid should be prevented by addition of *Sma*I to the ligation mix.

Ligation reaction:

1.0 μl	10× reaction buffer for *Sma*I (e.g., Fermentas)
0.5 μl	ATP (10 mM)
1.0 μl	Plasmid cut with *Sma*I, not dephosphorylated and not purified
1.0 μl	T4 DNA ligase (10 U/μl)
0.3 μl	*Sma*I
6.2 μl	Purified PCR fragment

The ligation mix is incubated at 16°C overnight, followed by ~2 h at 30°C (optimal temperature for *Sma*I restriction) prior to transformation into standard competent *E. coli* strains. Single colonies are cultured, and the recovered plasmid DNA can be test digested (e.g., with *Eco*RI and *Bam*HI to yield a 268 bp with pNW55) prior to sequence-verification with standard oligonucleotides (depending on the plasmid) or oligonucleotides A or B. Sequencing at this step is strongly recommended to ensure that the new plasmid was indeed transformed and not a potential carryover of the original plasmid. It may also be useful to know that the original miRNA (*osa-MIR528*) in the template plasmid NW55 harbors a uniquely occurring restriction site, which should (in most cases) be eliminated after successful PCR mutagenesis: *Sph*I.

aMIRNA precursors that are generated by site-directed mutagenesis do not contain a promoter or a terminator and therefore subsequent cloning steps are required. For functionality tests *in planta* and initial characterizations, strong ubiquitous promoters such as cauliflower mosaic virus (35S) or ubiquitin have proven very helpful. More detailed analyses can be carried out with tissue-specific promoters since amiRNAs function largely cell-autonomously (5). Because amiRNA-mediated gene silencing is quantitative (stronger promoters induce stronger effects), we do not recommend weak promoters when strongly expressed genes need to be silenced efficiently. They might, however, become useful when partial silencing is intended. Inducible and transient *aMIRNA* expression was successful with ethanol- and estrogen-inducible systems (5, 17). Promoters are often already contained in binary vectors, or can be inserted using standard cloning techniques. For rice we have so far only used the ubiquitin promoter. In *Arabidopsis*, we did not observe remarkable differences in phenotypic effects with different binary plasmids and therefore recommend using a plasmid system that is well established in the intended plant system.

All restriction sites of the pBluescript Multiple Cloning Sites flanking the *aMIRNA* precursor in the fusion PCR product (d) can be used to excise the amiRNA precursor (the *aMIRNA* transcript) from the sequencing plasmid. When sub-cloning, it is necessary to preserve the direction of the *aMIRNA* precursor since antisense transcripts are not expected to form the same secondary structure. Gateway-assisted cloning is also possible; the presence of AttB sites adjacent to the amiRNA precursor seems to not affect its processing (see Note 5).

3.4. Plant Transformation and Analysis of Transgenic Plants

3.4.1. Transformation of Agrobacterium and Plants

Most protocols for the generation of transgenic plants rely on an *Agrobacterium tumefaciens* strain delivering the above-described binary plasmid. Transformation of competent bacterial strains (e.g., LBA4404 or EHA105) is carried out with standard transformation protocols. Similarly, transfection of plants with the transgenic *Agrobacterium* strains should be carried out with established protocols and primary transformants require selection with appropriate selection markers. It is expected to observe phenotypic variation in primary transformants, which might in some cases resemble an allelic series of the respective mutant. Gene silencing with transgenes is sometimes not complete such that plants resembling null mutants of the respective target gene might not be recovered. See Note 6 in case you do not observe phenotypic changes in primary transformants.

3.4.2. Verifying Reduced Abundance of Target Transcripts

To confirm that phenotypic changes are indeed due to reduced abundance of the intended target gene product(s), their levels should be analyzed and compared to an untransformed or empty plasmid-transformed control. This can be performed in pools of primary transformants with similar phenotypes, or in individual plants. Plant amiRNAs, like many endogenous miRNAs, typically affect the accumulation of target mRNA, such that RT-PCR is often indicative of successful gene silencing. RNA is preferentially isolated from tissues with strong phenotypic effects, either with commercial kits or with TRIzol® reagent (Invitrogen). Commercial reverse transcription kits can be used for cDNA synthesis. RT-PCR products preferentially span the amiRNA-guided cleavage site. See Note 7 when you observe phenotypic abnormalities, but no change in target mRNA levels.

To estimate the specificity of gene silencing, it is recommended to also test for the accumulation of closely related transcripts which contain regions of partial sequence complementarity to the amiRNA (five or fewer mismatches, determined with the WMD-Target search tool; see Note 2). Reduced levels can be the result of direct amiRNA targeting, but also of feedback regulation when the two genes participate in the same genetic pathway. To discriminate between the two possibilities, it is necessary to specifically test for the accumulation of cleaved target(s) by 5′ RACE-PCR, since amiRNAs should trigger the cleavage of the target transcripts—always opposite of positions 10 and 11 of the amiRNA.

3.4.3. Cleavage Site Mapping by RACE-PCR

Cleavage products are presumably short lived; however, they can often be detected and by a slightly modified RACE-PCR procedure. Even the exact point of cleavage can be determined. RACE-PCR typically uses mRNA as starting material, which can be isolated from total RNA with commercial kits. Whereas standard protocols for 5′RACE typically start by de-capping full-length mRNAs, this step is omitted for cleavage product detection. The isolated mRNA is directly ligated to the RNA linker oligonucleotide followed by reverse transcription, typically carried out with an oligo-dT primer. This then serves as a template for PCR amplification using a forward oligonucleotide that binds the introduced linker sequence and gene-specific reverse oligonucleotides complementary to a region ~200–300 nucleotides downstream of the putative amiRNA binding site in the gene of interest. The abundance of cleavage product can be very low and sometimes a second, nested PCR may be necessary. Amplified products should be ligated into standard cloning vectors and sequenced to determine where the linker had been ligated and hence where the target transcript had been cleaved. Cleavage is predicted to occur at the amiRNA binding site between the two base pairs opposing positions 10 and 11 of the amiRNA.

3.4.4. Genetic Complementation

Since target sites of amiRNAs are small and distinct, it is possible to engineer silent mutations in this region of the target gene such that the transcript is no longer susceptible to amiRNA-mediated regulation, e.g., by increasing the number of mismatches. Introducing this transgene under its endogenous or a stronger promoter should suppress the amiRNA-induced phenotypes. This approach has successfully been applied to bypass regulation by endogenous miRNAs (4), and it can provide powerful evidence that the observed phenotypes are caused only by downregulation of the intended target and not of additional other genes. Silent mutations are typically introduced in as many positions as possible within the amiRNA binding site by PCR-based site-directed mutagenesis, in a similar way as *aMIRNAs* are produced (see Subheading 3.2).

4. Notes

1. When the plant species of interest is not yet included in WMD, but significant sequence information is publicly available, you may contact wmd@tuebingen.mpg.de to have the species added to the online tool. Alternatively, you can download the AmiRNA/WMD3 application from the wmd3 Webpage, install it locally, and include any species dataset you are interested in. Obviously, the specificity calculations can only take the available set of sequences into account, so there is always the possibility that amiRNAs affect additional genes that are not included or only partially annotated in the current sequence release of the respective species.
2. The WMD-Target Search application rapidly identifies target genes for miRNAs and other small RNAs in a given transcript collection/genome annotation. It uses sequence matching algorithms based on enhanced suffix arrays (vmatch, http://vmatch.de (18)) and/or based on hash-indexing (genomemapper (19)), which enable the identification of all genes in the collection with a defined number of mismatches to the search sequence. In addition, the WMD-Target Search applies the empirically determined parameters of miRNA target selection (4, 8) to filter for putative target genes. The output includes an alignment of the small RNA (reverse complement) to putative targets as illustrated in Fig. 3. With default settings, WMD-Target Search output lists only one hit in one isoform per gene; optionally you can have WMD display one hit per every isoform to examine whether all splice forms are targeted, or "all hits" in order to see whether there are multiple target sites within a gene (Advanced Search Options).
3. Failure of WMD to produce suitable aMIRNAs can have several reasons:
   (a) The input sequence might have been too short to contain suitable target sites.
   (b) The WMD-Designer is not able to compute a specific amiRNA against a target gene of interest, if its nucleotide sequence is very similar to that of one or several other genes at all potential target sites. They can be easily

identified using WMD-BLAST and one or several will have to be silenced together by adding them as additional targets or as "accepted off-targets." It might still be possible to conduct conclusive experiments by choosing off-target(s) that do not interfere with the experimental design or by evaluating the effects of several amiRNA constructs in planta with different off-targets.

(c) Some transcript collections contain redundant ESTs and multiple ESTs might span the locus of interest. Here, all genes/ESTs that are highly related to the gene of interest should be identified with WMD-BLAST and included in the WMD-Designer input as "accepted off-targets."

(d) WMD can only compute a multi-gene amiRNA that targets several genes, if they share regions of high nucleotide sequence similarity. Simultaneous silencing of multiple related genes might fail if the genes are not similar enough, or one or more is/are not different enough from other genes (see Note 2). Try to reduce the minimal number of included target genes or silence them individually.

4. *A. thaliana MIRNA* precursors have successfully been used for amiRNA production in other plants (e.g., tomato and tobacco; (11)), but precursor functionality across species has not yet been systematically investigated. Therefore, adapting the cloning protocol to *MIRNA* precursors endogenous to the respective plant species of interest might be the optimal approach (6, 20).

   MIRNA precursors have been identified and characterized from several different plants (*see* miRBase, http://www.microrna.sanger.ac.uk; (21)), often by homology to known miRNAs. As backbones for amiRNA production, we recommend to either use a precursor that has been shown to be expressed and processed, i.e., by northern blot, or—when this information is not available—to use a highly conserved precursor, e.g., *MIR164* or *MIR319*. Oligonucleotides I through IV output by WMD will then obviously need to be adapted to the precursor used and also to reconstruct the proper hairpin structure such that bulges remain at their respective positions.

5. Cloning with the Gateway® technology seems not to interfere with amiRNA production. In the following we list possibilities to enter the Gateway® system:

   (a) The *MIRNA* precursor fragment can be excised from the sequencing plasmid (see Subheading 3.3) with restriction enzymes and then ligated into a Gateway® entry plasmid.

   (b) The fusion product (d) of Subheading 3.2. can be ligated into a Gateway® entry plasmid as is.

(c) Alternatively, the fusion PCR of the fragments (a), (b), and (c) to (d) of Subheading 3.2 can be carried out with oligonucleotides that already contain AttB sites at their 5′ ends. These primers do not necessarily need to bind to the primer binding sites A and B. Primers that bind to sequences in the Multiple Cloning Sites have successfully been used to obtain shorter inserts and to eliminate undesired restriction sites. The resulting PCR product with AttB sites at both ends can then be ligated into a vector of choice (e.g., pGEM T easy), which then serves as entry plasmid for a subsequent recombination reaction.

6. Missing phenotypic changes in transgenic plants that (over) express amiRNAs can have several reasons:
   (a) The phenotypes might not be detectable in the growth conditions tested.
   (b) The loss-of-function phenotype of the gene of interest might be masked by redundancy. A (literature) search for related genes with similar expression patterns might help to identify potentially redundant genes to be used as additional targets.
   (c) The target gene might not be sufficiently downregulated to detect phenotypic changes. It is critical to achieve high amiRNA expression in the tissue(s) of target gene expressions, but even promoters such as the one from the CaMV35 gene are not entirely ubiquitous.

   Experience shows that sometimes amiRNAs do not silence the intended target gene(s) and the reasons need yet to be determined. It is possible that their target sites are not accessible to the miRNA due to extensive local secondary structures in the mRNA of the target gene, similar to what has been observed for siRNAs in animal systems (22). Hence we integrated "RNAup" (23) into WMD. When calculating the RNA–RNA binding, "RNAup" also considers the folding of the respective RNA molecules to themselves and can therefore be used to predict the accessibility of the target sites in the target mRNA. At present we recommend to construct at least two amiRNAs per target gene or group of target genes, with target sites located in different regions of the target transcript(s).
   (d) If even very potent amiRNAs cause only small effects on the transcript levels, the target genes might be under negative feedback regulation. Those genes may be silenced effectively on a transcriptional level (e.g., by promoter methylation (24)), however not by posttranscriptional gene silencing.

7. Typical (a)miRNA-mediated gene silencing includes the cleavage of target transcripts followed by degradation of the cleavage

products, leading to a reduction in transcript abundance, which can be measured by RT-PCR. However, for some endogenous miRNAs, e.g., *ath*-miR172, translational inhibition is at least as important as miRNA-guided cleavage (25, 26). Thus, phenotypic changes can be present even though mRNA levels might not have appreciably changed. When available, translational effects can be monitored on the protein level, by western blotting with target-specific antibodies. In published cases, transcripts that were regulated on the translational level were still cleaved by the miRNAs (8, 25–27), and cleavage products were detected by RACE-PCR (see step 3 of Subheading 3.4).

Acknowledgements

We thank Hao Chen and Philippe Hervé, who generated and evaluated the first transgenic rice plants (Nipponbare and IR64) carrying *aMIRNA* transgenes at the International Rice Research Institute in the Philippines (IRRI). Markus Riester contributed to earlier versions of WMD and Joffrey Fitz codeveloped AmiRNA/WMD3; we are further thankful to everybody who contributed by sharing technical expertise and discussion, namely, Alexis Maizel, Javier Palatnik, Heike Wollmann, and Wolfgang Busch. Work on small RNAs in the Weigel laboratory is supported by European Community FP6 IP SIROCCO (contract LSHG-CT-2006-037900) and by the Max Planck Society.

References

1. Krishnan A, Guiderdoni E, An G et al (2009) Mutant resources in rice for functional genomics of the grasses. Plant Physiol 149:165–170
2. Chapman EJ, Carrington JC (2007) Specialization and evolution of endogenous small RNA pathways. Nat Rev Genet 8: 884–896
3. Tang G, Galili G, Zhuang X (2007) RNAi and microRNA: breakthrough technologies for the improvement of plant nutritional value and metabolic engineering. Metabolomics 3:357–369
4. Palatnik JF, Allen E, Wu X et al (2003) Control of leaf morphogenesis by microRNAs. Nature 425:257–263
5. Schwab R, Ossowski S, Riester M, Warthmann N, Weigel D (2006) Highly specific gene silencing by artificial microRNAs in *Arabidopsis*. Plant Cell 18:1121–1133
6. Warthmann N, Chen H, Ossowski S, Weigel D, Herve P (2008) Highly specific gene silencing by artificial miRNAs in rice. PLoS One 3:e1829
7. Allen E, Xie Z, Gustafson AM, Carrington JC (2005) microRNA-directed phasing during trans-acting siRNA biogenesis in plants. Cell 121:207–221
8. Schwab R, Palatnik JF, Riester M, Schommer C, Schmid M, Weigel D (2005) Specific effects of microRNAs on the plant transcriptome. Dev Cell 8:517–527
9. Ossowski O, Schwab R, Weigel D (2008) Gene silencing in plants using artificial microRNAs and other small RNAs. Plant J 53: 674–690
10. Vaucheret H (2005) MicroRNA-dependent trans-acting siRNA production. Sci STKE 2005:pe43
11. Alvarez JP, Pekker I, Goldshmidt A, Blum E, Amsellem Z, Eshed Y (2006) Endogenous and synthetic microRNAs stimulate simultaneous, efficient, and localized regulation of multiple targets in diverse species. Plant Cell 18: 1134–1151

12. Niu QW, Lin SS, Reyes JL et al (2006) Expression of artificial microRNAs in transgenic *Arabidopsis thaliana* confers virus resistance. Nat Biotechnol 24:1420–1428
13. Parizotto EA, Dunoyer P, Rahm N, Himber C, Voinnet O (2004) In vivo investigation of the transcription, processing, endonucleolytic activity, and functional relevance of the spatial distribution of a plant miRNA. Genes Dev 18:2237–2242
14. Qu J, Ye J, Fang R (2007) Artificial miRNA-mediated virus resistance in plants. J Virol 81:6690–6699
15. Khraiwesh B, Ossowski S, Weigel D, Reski R, Frank W (2008) Specific gene silencing by artificial MicroRNAs in Physcomitrella patens: an alternative to targeted gene knockouts. Plant Physiol 148:684–693
16. Chen S, Songkumarn P, Liu J, Wang G (2009) A versatile zero background T-vector system for gene cloning and functional genomics. Plant Physiol 150:1111–1121
17. Michniewicz M, Zago MK, Abas L et al (2007) Antagonistic regulation of PIN phosphorylation by PP2A and PINOID directs auxin flux. Cell 130:1044–1056
18. Abouelhoda MI, Kurtz S, Ohlebusch E (2004) Replacing suffix trees with enhanced suffix arrays. J Discrete Algorithm 2:53–86
19. Schneeberger K, Hagmann J, Ossowski S et al (2009) Simultaneous alignment of short reads against multiple genomes. Genome Biol 10:R98
20. Molnar A, Bassett A, Thuenemann E et al (2009) Highly specific gene silencing by artificial microRNAs in the unicellular alga Chlamydomonas reinhardtii. Plant J 58:165–174
21. Griffiths-Jones S (2004) The microRNA registry. Nucleic Acids Res 32:D109–D111
22. Ameres SL, Martinez J, Schroeder R (2007) Molecular basis for target RNA recognition and cleavage by human RISC. Cell 130:101–112
23. Muckstein U, Tafer H, Hackermuller J, Bernhart SH, Stadler PF, Hofacker IL (2006) Thermodynamics of RNA-RNA binding. Bioinformatics 22:1177–1182
24. Matzke M, Kanno T, Huettel B, Daxinger L, Matzke AJ (2006) RNA-directed DNA methylation and pol IVb in *Arabidopsis.* Cold Spring Harb Symp Quant Biol 71:449–459
25. Aukerman MJ, Sakai H (2003) Regulation of flowering time and floral organ identity by a microRNA and its *APETALA2*-like target genes. Plant Cell 15:2730–2741
26. Chen X (2004) A microRNA as a translational repressor of *APETALA2* in *Arabidopsis* flower development. Science 303:2022–2025
27. Gandikota M, Birkenbihl RP, Hohmann S, Cardon GH, Saedler H, Huijser P (2007) The miRNA156/157 recognition element in the 3′ UTR of the Arabidopsis SBP box gene *SPL3* prevents early flowering by translational inhibition in seedlings. Plant J 49:683–693

Chapter 12

Rice Proteomic Analysis: Sample Preparation for Protein Identification

Ganesh Kumar Agrawal, Nam-Soo Jwa, Young-Ho Jung, Sun Tae Kim, Dea-Wook Kim, Kyoungwon Cho, Junko Shibato, and Randeep Rakwal

Abstract

Rice is one of the most important food and cereal crop plants in the world. Rice proteomics began in the 1990s. Since then, considerable progress has been made in establishing protocols from isolation of rice proteins from different tissues, organs, and organelles, to separation of complex proteins and to their identification by mass spectrometry. Since the year 2000, global proteomics studies have been performed during growth and development under numerous biotic and abiotic environmental conditions. Two-dimensional (2-D) gel-based proteomics platform coupled with mass spectrometry has been retained as the workhorse for proteomics of a variety of rice samples. In this chapter, we describe in detail the different protocols used for isolation of rice proteins, their separation, detection, and identification using gel-based proteomics and mass spectrometry approaches.

Key words: Rice, Gel-based proteomics, One/two-dimensional polyacrylamide gel electrophoresis, Mass spectrometry

1. Introduction

There is no questioning as to the scientific and social/economic popularity and interest of rice (*Oryza sativa* L.) as a model crop plant. Although the rice plant has been the subject of numerous physiological, biochemical, and molecular biology studies, it has gained in popularity as a genome model since the acceptance of the Japonica strain as a standard for genome research during the First International Symposium of Rice Molecular Biology held in Kurashiki, Okayama, Japan, in 1986. This was followed by the sequencing of its (Japonica) (and Indica) genomes in 2002 (1, 2), setting the pace for advancements in rice functional genomics (3). Although proteomics had an earlier beginning (initial studies involved one-dimensional gel electrophoresis (1-DGE) and two-DGE

Yinong Yang (ed.), *Rice Protocols*, Methods in Molecular Biology, vol. 956,
DOI 10.1007/978-1-62703-194-3_12, © Springer Science+Business Media, LLC 2013

(2-DGE) coupled with electroblotting and *N*-terminal amino acid sequencing), the availability of the rice genomes along with various genetic resources of rice accelerated proteomics usage by the community of rice (and plant) proteomics. Since then (2002 onwards), there has been no looking back for the rice proteomers (reviewed in refs. 4–8). Proteomer means the one who studies proteome and this term was recently introduced by our group (see ref. 6). The number of publications on rice proteomics has increased considerably, with many new groups generating new results and new ideas, ultimately setting the standard for proteomics research and rejuvenating rice proteomics as a whole (reviewed in refs. 4–9). This progress has been mainly due to developments and/or improvements in proteomics techniques (from protein extraction to separation by 2-DGE and mass spectrometry (MS)) along with serious efforts by rice proteomers to dig deeper into the proteome of interest. For a glimpse of the recent progress in rice proteomics in the context of plant proteomics as a whole, readers are referred to a recent comprehensive book entitled "Plant Proteomics: Technologies, Strategies, and Applications" (http://as.wiley.com/WileyCDA/WileyTitle/productCd-0470369833.html) edited by two coauthors of this chapter Ganesh and Randeep (10).

As mentioned above, 2-DGE coupled with MS (11)—generally known as a 2-D gel-based proteomics platform—has been and is the method of choice for rice proteomers. Moreover, 2-D gel-based proteomics (12) is considered a mature and well-established technique and remains the most widely and routinely used technique in the plant proteomics field mainly due to continuous improvements in separation and detection technologies, such as narrow-range isoelectric point gel (IPG) strips and fluorescence-based protein labeling and stains (13–15).

However, it should also be noted that it is impossible to cover complete information on the proteomic nature of plant samples due to its variability by genetic and environmental factors. This point of view becomes more critical in certain fields of study such as rice breeding. To study genetics of the proteome variation among a large number of breeding lines, it is essential to prepare a protein simplification method that can reveal significant genetic variation of each sample.

In this chapter, we have attempted to provide step-by-step the protocol used for rice protein analysis using the gel-based proteomics approaches starting from sample preparation techniques to separation and identification of peptides/proteins by MS analysis. We focus on total protein isolation from a variety of rice plant tissues and organs. It is almost impossible to provide protocols for isolation of pure plant organelles. We recommend a comprehensive protocol book on plants entitled "Plant Proteomics: Methods and Protocols" (16) for isolation of organelles and for downstream analysis using gel-based proteomics. Nevertheless, once pure organelles are isolated, they are re-suspended into the recommended solution suitable for gel-based proteomics separation and detection analyses. We would also like to mention that even the gel-based proteomics approaches contain a number of techniques, such as image analysis of 2-D gels using ImageMaster 2D Platinum software, which are difficult to discuss in detail as they can form a separate chapter. In such cases, readers are encouraged to utilize the user manual provided by supplier. We should emphasize that the detection of all distinct spots and manual editing of those spots are important steps in the 2-D gel image analysis. The same goes for MS and database analysis of trypsin- or other enzymatic-digested peptides. For simplicity and clarity for readers, students, and young researchers, we have provided the techniques in flowcharts wherever required.

2. Materials

2.1. Seed and Seed Treatment, Plant Material, and Storage

1. Dry mature seeds of rice (*Oryza sativa* L.) cv. Nipponbare (or any other cultivar) (see Note 1).
2. Sterilizing solution: Fourfold (v/v) diluted sodium hypochlorite (Wako Pure Chemicals, Tokyo, Japan) solution in Milli Q (MQ) water or distilled water.
3. Sterile glass beaker (300 mL).
4. Darkroom (25°C).
5. Soil (see Note 2).
6. Seedling pot (14 mm × 5 mm × 10 mm::L × W × H).
7. Growth chamber (Biotron LPH-220S; NK Systems, Osaka, Japan) equipped with white fluorescent light (wavelength 390–500 nm, 150 μmol/m^2 s), temperature (25°C), and relative humidity (70%).
8. Leaves (3rd and/or 4th) of 2-week-old rice seedlings (see Note 3).
9. Liquid nitrogen (N_2).
10. Sterile/clean mortar and pestle.
11. Aluminum foil, 1.5 and 2 mL microfuge tubes, and 14 or 50 mL Falcon tubes.
12. Collect the materials, freeze immediately with liquid N_2, and store at –80°C.

2.2. General

1. Fresh *N*,*N*,*N′*,*N′*-Tetramethyl-ethylenediamine (TEMED; Wako) is important for polymerization of gel (see Note 4).
2. 10% (w/v) ammonium persulfate (APS) in deionized water. Use freshly prepared solution (see Note 5).
3. Sodium dodecyl sulfate (SDS).

4. Bromophenol blue (BPB).
5. DualColor PrecisionPlus protein standard protein markers (Bio-Rad).
6. Unless otherwise stated, all solutions were prepared in deionized water (18.2 MΩ conductivity).
7. Filter sterilize the solutions using either 25 mm syringe filter (0.2 μm, nylon; Product no. 09-719C, Fisher Scientific, Houston, TX) or Steritop (0.22 μm; Product no. SCGPS02RE, Millipore Corporation, Billerica, MA) filter depending on solution volume.
8. Kimwipes.
9. Recipro shaker.
10. Centrifuge.
11. Centrifugal evaporator.
12. Staining trays and boxes. Keep clean and do not use the same trays and boxes for Western blots.
13. Large gel-staining tray, size 26 cm × 20 cm × 10 cm.
14. Pipettes and sterile tips.
15. Water bath with temperature control.
16. Calibrated glass beakers (300 mL).
17. Vortex.

2.3. Total Protein Extraction: Direct Extraction in Lysis Buffer Method

1. Microfuge tubes (2 mL).
2. Lysis Buffer (LB): 8 M urea, 4% (w/v) 3-((3-Cholamidopropyl) dimethylammonio)-1-propanesulfonate (CHAPS), 2% (v/v) pharmalyte pH 3–10, and 0.3% (w/v) dithiothreitol (DTT).

2.4. Total Protein Extraction: Direct Extraction in LB Containing Thiourea Method

LB containing Thiourea (LB-T): 7 M urea, 2 M thiourea, 4% (w/v) CHAPS, 2% (v/v) pharmalyte pH 3–10, and 0.3% (w/v) DTT.

2.5. Total Protein Extraction: Direct Extraction in LB-T Containing Tris Method

LB-T containing Tris (LB-TT): 7 M urea, 2 M thiourea, 4% (w/v) CHAPS, 18 mM Tris–HCl pH 8–10, 14 mM Trizma base, 0.2% (v/v) Triton X-100, ethylenediaminetetraacetic acid (EDTA)-free proteinase inhibitor tablets (Roche Diagnostics GmbH), 1% (v/v) pharmalyte pH 3–10, and 50 mM DTT (see Note 6 and Fig. 1).

2.6. Total Protein Extraction: TCA–Acetone Extraction Method

1. Trichloroacetic acid (TCA)–Acetone Extraction (TCAAEB): 10% (w/v) TCA/acetone with 0.07% (v/v) 2-mercaptoethanol (2-ME). To prepare TCAAEB solution, dissolve 10 g TCA in 100 mL acetone containing 70 μL 2-ME (see Note 7).

1. **Add 4 g CHAPS in a 300 (or 500) mL very clean/sterile Glass Beaker (calibrated if possible).**
2. Dissolve in **50 mL MQ** water by slowly shaking using wrist movement; or use a stirrer to mix but not too strongly.
3. **42 g** of **Urea** was measured and added in parts, and mixed by gently turning the beaker.
4. Occasionally dip the beaker end in a 40-50°C water bath; for 15 sec or so, not too long at one time.
5. Add **15.2 g Thiourea** and mix gently again/stirrer.
6. Into the dissolved mixture solution add **1.8 mL of 1 MTris-HCl** and mix/stirrer.
7. Add **169.5 mg Trizma base** and mix nicely/stirrer.
8. Take **2 tablets of EDTA-free proteinase inhibitor** and dissolve (need to use stirrer for around 10 min).
9. To this mixture add slowly **0.2 mL of Triton X-100** and mix avoiding air bubbles.
10. To the almost completely dissolved buffer solution add **771.5 mg of DTT** and mix/stir.
11. Finally, add **1 mL of Ampholyte** and mix up to **100 mL with MQ water** (be careful to rinse any powders in the beaker walls by MQ).
12. Vacuum filter the solution with a **0.45 or 0.22 micron SteriCup filter** unit or Syringe filter unit.
13. Store **1 mL filtered LB-TT** in Eppendorf tubes at **-80°C**.
14. Thaw at room temperature (RT) before use.

Fig. 1. Preparation of LB-TT solution and storage.

2. Wash buffer (WB): 0.07% (v/v) 2-ME, 2 mM EDTA, EDTA-free proteinase inhibitor tablet in acetone. To prepare WB solution, mix 70 μL 2-ME, 400 μL of 500 mM EDTA (pH 8.0) with two tablets of EDTA-free proteinase inhibitor in 99.5 mL acetone (see Note 8).

2.7. Total Protein Extraction: Phenol–Ammonium Acetate/Methanol Method

1. Extraction buffer: 100 mM Tris–HCl (333 μL of 1.5 M Tris–HCl stock, pH 8.8), 10 mM EDTA (10 mL of 0.5 M EDTA stock, pH 8.0), 900 mM sucrose (15.40 gw/v), and 0.4% 2-ME (v/v; 200 μL). Store at 4°C (see Note 9).
2. Phenol buffered with Tris–HCl: Prepare in the laboratory (Fig. 2) or use a commercial product, pH 8.8. Store at 4°C (see Note 10).
3. Extraction buffer plus phosphatase and protease inhibitors: Add 5 mM sodium vanadate (or sodium metavanadate), 5 mM sodium fluoride , 25 mM glycerophosphate disodium salt pentahydrate, and protease inhibitor cocktail complete mini EDTA-free tablet (1 tablet/10 mL solution) to the extraction buffer and mix until dissolved (see Note 11).
4. Phenol extraction buffer: Add equal volume of extraction buffer and buffered phenol in a 15 mL Falcon tube and mix before use.
5. Ammonium acetate/methanol solution: 100 mM (0.385 gw/v) ammonium acetate in 50 mL of 100% methanol. Store at 4°C.
6. 80% (v/v) acetone in deionized water. Store at 4°C.
7. 70% (v/v) ethanol in deionized water. Store at 4°C.
8. LB-TT (see Subheading 2.5 for recipe).

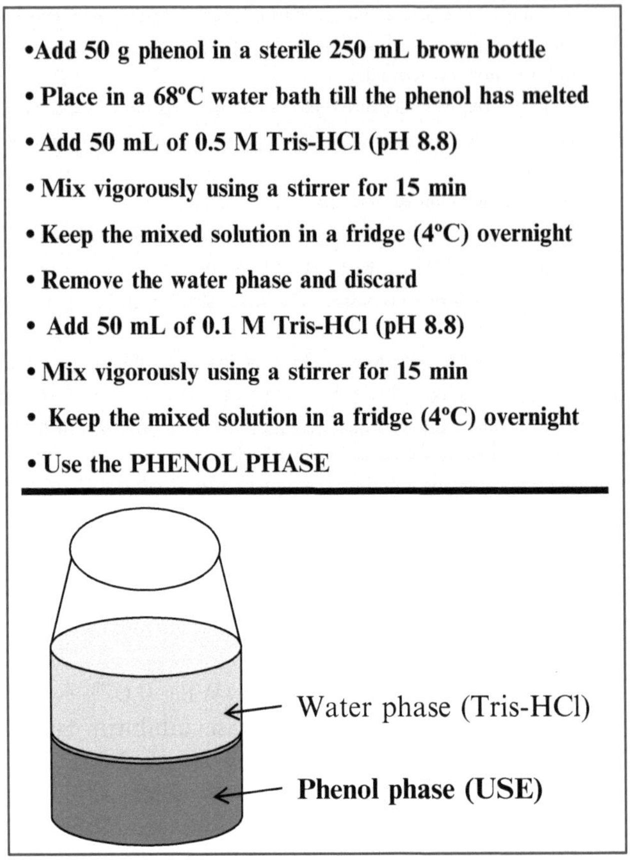

Fig. 2. Preparation of Tris-buffered phenol solution and storage.

2.8. Total Protein Extraction: Tris-Buffered Saline Method

1. Tris-buffered saline (TBS): 10 mM Tris–HCl, 150 mM NaCl pH 7.4, and 0.1% (v/v) Tween 20 (TBS-T20) (see Note 12).
2. SDS-sample buffer (SDS-SB): 62 mM Tris–HCl pH 6.8, 10% (v/v) glycerol, 2.5% (w/v) SDS, and 5% (v/v) 2-ME (pH 6.8 with HCl).
3. ProteoExtract protein precipitation kit (Catalog no. 539180, Calbiochem, Merck KGaA, Darmstadt, Germany).
4. LB-TT (see Subheading 2.5 for recipe).

2.9. Polyethylene Glycol Fractionation

1. Rice seedlings (*O. sativa* L. cv. Jinheung) (4th and 5th leaf stages).
2. MgNP-40 extraction buffer: 0.5 M Tris–HCl, pH 8.3, 2% (v/v) NP-40, 20 mM $MgCl_2$, 1 mM phenylmethylsulfonyl fluoride (PMSF), 2% (v/v) 2-ME, and 1% (w/v) polyvinyl polypyrrolidone (PVPP).
3. 50% Polyethylene Glycol (PEG) 4000 solution.
4. Methanol containing 0.1 M ammonium acetate.
5. Water-saturated phenol, pH 8.8.

2.10. RuBisCO Depletion Using Seppro® IgY-Rubisco Spin Column (Genway)

1. A Seppro™ RuBisCO IgY spin column kit (GenWay Biotech).
2. Dilution buffer (TBS): 10 mM Tris–HCl, 150 mM NaCl pH 7.4, and 0.1% Tween 20 (TBS-T20).
3. Stripping buffer: 250 mM glycine, pH 2.5.
4. Neutralization buffer: 250 mM Tris–HCl, pH 8.0.

2.11. Protein Quantification Kit

1. Coomassie Plus protein assay kit (Thermo Scientific). Use exactly as directed by the manufacturer.
2. Bovine serum albumin (BSA). Store in single-use aliquots of 0.1 mL at −80°C.
3. Disposable cuvettes.
4. UV–VIS Spectrophotometer.

2.12. Protein Quantification: A Modified Bradford Method

1. 10× SDS-PAGE running buffer: 0.25 M Tris-base, 1.92 M glycine, 2% (w/v) SDS. Make in deionized water and keep the solution in a brown bottle. Store at room temperature (RT). Use 1× diluted running buffer for electrophoresis (see Note 13).
2. 0.5× SDS running buffer. Store at RT.
3. 1 mg/mL BSA in 0.5× SDS running buffer. Store 100 μL aliquots at −20°C.
4. Bradford dye: Dilute the Bradford protein assay dye (Bio-Rad) fivefold in deionized water before use.
5. Microtiter plate spectrophotometer.

2.13. One-Dimensional Gel Electrophoresis (1-DGE/SDS-PAGE)

1. Electrophoresis equipment.
2. 30% (w/v) acrylamide solution: Dissolve 150 g acryalmide and 4.0 g bis-acrylamide (also called *N*,*N*′-methylenebisacrylamide) in deionized water and bring the final volume to 500 mL. Filter sterilize using a 0.22 μm cup filter (1 L capacity) and store in a brown (dark) bottle at 4°C for approximately 1 month.
3. Separation buffer (pH 8.8): For 500 mL solution, add 90.85 g Trizma base (1.5 M) in deionized water and adjust pH with HCl followed by filter sterilization (0.2 μm). Store at 4°C.
4. Stacking buffer (pH 6.8): For 500 mL solution, add 30.28 g Trizma base in deionized water, adjust pH with HCl, and sterilize using a 0.2 μm filter. Store at 4°C.
5. 10% (w/v) SDS in deionized water. Filter sterilize and store at RT.
6. 10% (w/v) APS in deionized water. Use freshly prepared solution.
7. Pre-cast gels (Catalog no. 2331820, e-PAGEL E-T12.5L, 14 wells, ATTO).

8. 1× SDS-PAGE running buffer (dilute tenfold using 10× stock solution).
9. SDS-PAGE: 4%T, 2.6%C stacking gels, pH 6.8 and 12.5%T, 2.6%C separating gels, pH 8.8. The percent T is the total monomer concentration expressed in grams per 100 mL and the percent C is the percentage cross-linker.

2.14. Isoelectric Focusing of Total Proteins: First Dimension in 2-DGE

1. IPGphor unit and strip holders (GE Healthcare Bio-Sciences AB, Uppsala, Sweden) and/or Protean IEF Cell, Protean IEF system 24 cm disposable trays (Catalog no. 1654043), and IPG focusing tray (Bio-Rad).
2. 24 cm Linear IPG strips pH 4–7 (GE Healthcare).
3. 24 cm IPG buffer pH 4–7 (GE Healthcare).
4. DeStreak reagent (GE Healthcare).
5. Cover fluid (Catalog no. 17-1335-01, GE Healthcare).
6. Electrode wicks (Protean IEF system; Catalog no. 1654071, Bio-Rad).

2.15. SDS-PAGE: Second Dimension in 2-DGE

1. Horizontal electrophoresis unit (Nihon Eido) and Ettan DALTsix/twelve electrophoresis unit (GE Healthcare).
2. Equilibration buffer (EB): 6 M urea, 4%w/v SDS, 0.002% BPB (w/v; 4 mg), 30%v/v glycerol, and 50 mM Tris–HCl pH 8.8. Sterilize with 0.2 μm filters and store at −20°C in 50 mL Falcon tubes. Before use, thaw at room temperature and add 2% DTT. Mix nicely by constant shaking at RT. This is the reduction buffer.
3. Alkylation buffer: To the above EB (after thawing at RT) add 2.5% iodoacetamide to the thawed EB, just before use. Mix nicely by constant shaking at RT.
4. Agarose sealing solution or overlay agarose: Add 1.0 g medium EEO agarose type II to 100 mL of 1× SDS-SB (pH 6.8, minus 2-ME) and then add a few grains of BPB.
5. 12.5% homogenous or 12–14% gradient pre-cast polyacrylamide gels (PAGs; GE Healthcare).
6. Standard protein markers (Bio-Rad or any other standard protein marker).

2.16. Total Protein Detection with CBB

1. Coomassie Brilliant Blue (CBB) solution: To prepare 3 L add 1,500 mL of methanol (50% final volume), 3.0 g of brilliant blue R-250 (0.1% final volume), 300 mL of acetic acid (10% final volume). Bring to 3 L volume by adding Milli Q or deionized water. Store in dark brown bottles at RT.
2. Destaining solution: To prepare 3 L, add 1,100 mL methanol (36.66% final volume) and 300 mL acetic acid (10% final volume) and bring to volume with deionized water.

2.17. Total Protein Detection with Colloidal CBB

1. Colloidal CBB solution: To prepare 4 L, add 800 mL of ethanol (20% final volume), 3.2 g of brilliant blue G-250 (0.08% final volume), 64 mL of phosphoric acid (1.6% final volume), 320 g of ammonium sulfate (8% final volume) and bring to volume with deionized water. Store in brown bottle at RT.
2. Gel storage solution: To prepare 250 mL, use 25 mL of CBB solution, 2.5 mL of freshly prepared 100× sodium azide (final concentration, 0.02%), and 222.5 mL of deionized water. Use 250 mL per 24 cm large-format gel, shake for 5 min, and store at 4°C for up to 3–4 months.

2.18. Total Protein Detection with Silver Nitrate

PlusOne Silver Staining Kit, Protein (Catalog no. 17-1150-01, GE Healthcare; use as per the direction of the manufacturer) (see Note 14).

2.19. Phosphoprotein Detection

1. PeppermintStick phosphoprotein molecular weight standards (Product no. P33350, Invitrogen, Carlsbad, CA). Store in aliquots at –20°C (see Note 15).
2. PhosphoQUANTI SolidBlue Gel Dye Kit (TOYO INK, Tokyo, Japan).
3. Pro-Q Diamond Phosphoprotein Gel Stain (Product no. P33301, Invitrogen, Carlsbad, CA). Store at 4°C (see Note 16).
4. Fixation solution: 50% (v/v) methanol, 10% (v/v) acetic acid in deionized water.
5. Washing solution: Deionized water.
6. Staining solution: Threefold diluted Pro-Q DPS (v/v) in deionized water.
7. Destaining solution: 50 mM sodium acetate-acetic acid, pH 4.0, 20% (v/v) acetonitrile (CH_3CN) in deionized water. To prepare 1 L of destaining solution, combine 50 mL of 1,000 mM sodium acetate, pH 4.0, 750 mL of deionized water, and 200 mL of CH_3CN (see Note 17).

2.20. Image Acquisition for Proteins

1. ProteomeScan2000 digital scanner (System Biotics KK, Yokohama, Japan) (see Note 18).
2. Fluorescence image capture device such as the FLA 5000 laser scanner (Fuji Medical Systems, Stamford, CT).
3. Protein image capture device such as the ScanMaker 9800XL (Microtek, Carson, CA).
4. FluoroPhore Star 3000 (ANATECH).

2.21. Image Acquisition for Phosphoproteins

1. Fluorescence image capture device (FLA 5000 laser scanner, Fuji Medical Systems).
2. FluoroPhore Star 3000 (ANATECH).

2.22. Image Analysis for Proteins or Phosphoproteins

1. ImageMaster 2D Platinum software version 5 or version 6 (GE Healthcare) (see Note 19).
2. Progenesis PG200 gel image analysis software (Nonlinear Dynamics).
3. LUDESI image analysis service (http://www.ludesi.com).

2.23. 1-D Bands and 2-D Spots Excision

1. Spot picker (1.5/3.0 mm; Catalog no. P2D1.5/P2D3.0, The Gel Company, San Francisco, CA) (see Note 20).
2. Razor blade (sterile) (see Note 21).

2.24. In-Gel Digestion for CBB or Silver-Stained Proteins

1. Destaining solutions for silver nitrate-stained gels: To destain each band use 100 μL of 30 mM potassium ferricyanide and 100 mM sodium thiosulfate. To destain each spot use distaining solution of 15 mM potassium ferricyanide and 50 mM sodium thiosulfate.
2. 0.2 M ammonium bicarbonate (pH 7.8; hereafter called AMBIC).
3. 0.1 M AMBIC (pH 7.8).
4. Trypsin solution: 2 μg/μL trypsin (Product no. V5111, sequencing grade; Promega, Madison, WI) in 158 μL of Solution A (100 mM AMBIC and 0.2 M $CaCl_2$).
5. Centrifugal concentrator (CC-105; TOMY, Tokyo, Japan).

2.25. Mass Spectrometry and Data Analysis for Protein Identification of CBB-Stained 1-DGE Bands by MALDI-TOF-MS and Q-TOF-MS/MS

1. MALDI-TOF-MS: Model MALDI-R (Micromass).
2. Q-TOF/TOF MS (Micromass).
3. Matrix solution: 70% (v/v) acetonitrile (CH_3CN), 0.1% (v/v) trifluoroacetic acid, and 10 mg/mL alpha-cyano-4-hydroxycinamic acid.
4. Stainless steel target plate (Micromass, Manchester, UK).
5. C18 nanoscale (porous) column (IN2GEN, Seoul, South Korea).
6. C18 ZipTip (Millipore, Bedford, MA).
7. Data analysis: MassLynx (Ver. 3.5).

2.26. Mass Spectrometry and Data Analysis for Protein Identification from Silver-Stained 2-D Gel Spots on an Agilent nESI-LC-MS/MS System

1. Mass spectrometry: Agilent 1100 Nano LC-1100 system combined with a microwell-plate sampler and thermostatted column compartment for preconcentration (LC Packings, Agilent Technologies, Santa Clara, CA).
2. Analysis column (Zorbax 300SB-C18, 150 mm × 75 μm, 3.5 μm).
3. Preconcentration column (Zorbax 300SB-C18, 5 mm × 300 μm, 5 μm).
4. Mobile phase:
   (a) 0.1% HCOOH in water.
   (b) 0.1% HCOOH in CH_3CN.

5. Data analysis: Agilent Spectrum Mill MS Proteomics Workbench (Spectrum Mill version; Agilent).

2.27. Mass Spectrometry and Data Analysis for Protein Identification from Silver-Stained 2-D Gel Spots Using Hitachi NanoFrontier eLD Mass Spectrometry

1. Mass spectrometry: NanoFrontier *eLD*-IT-TOF-MS/MS (Hitachi High-Technologies Corporation, Hitachi, Japan).
2. Separation column (NTCC-360, 75 μm × 100 mm; Nikkyo Technos, Japan).
3. Trap column (MONOLITH TRAP, 0.05 × 150 mm; Kyoto Monotech Co., Japan).
4. Mobile phase A: 0.3% HCOOH in H_2O:CH_3CN = 98:2.
5. Mobile phase B: 0.3% HCOOH in H_2O:CH_3CN = 2:98.
6. Data analysis: Nano Frontier data processing software (P/N:3807051-01, Hitachi High-Technologies Corporation).

2.28. In-Gel Digestion, Mass Spectrometry, and Data Analysis for Phosphoproteins

1. 100 mM AMBIC, pH 8.0 in deionized water. Store at RT.
2. In-gel wash solution: 50% CH_3CN and 50 mM AMBIC, pH 8.0. To prepare 100 mL of in-gel wash solution mix 50 mL of 100 mM AMBIC, pH 8.0 with 50 mL of CH_3CN. Use freshly prepared solution.
3. In-gel trypsin solution: Use sequencing grade trypsin (Promega, Madison, USA; product number V5111). One vial contains 20 μg. Dissolve 20 μg in 5.1 mL of 50 mM AMBIC, pH 8.0. Use freshly prepared solution. Keep on ice until use.
4. In-gel extraction solution: 60% (v/v) CH_3CN and 1% (v/v) formic acid in deionized water.
5. CentriVap Console (Catalog no. 7812010, Labconco, Kansas City, MO, USA).
6. 0.1% (v/v) formic acid in deionized water.

3. Methods

Preparation of a high-quality protein sample free from interfering compounds, such as carbohydrates and nucleic acids, is essential to properly separate, detect, quantify, and identify proteins or phosphoproteins using gel-based proteomics approaches. Moreover, rice leaves and tissues contain potentially problematic phenolic compounds and pigments. We have experienced more phenolic compounds in stressed tissue or organs. On the other hand, rice panicles and seeds are rich in carbohydrates (starch), some sugars, and pigments, making protein extraction even more complicated. Furthermore, each tissue requires an optimized protocol. The protein extraction protocols described here are suitable for a wide range of tissues/organs of the rice plant. Among the total protein extraction methods, the TCA/acetone extraction and phenol–ammonium acetate/methanol methods have been widely accepted for different plant species and have been successfully used to extract total protein from leaves and other green rice tissues, including developing seeds. These isolated proteins were suitable for 2-DGE-based proteomics analysis and establishing the high-resolution 2-D gel reference maps by various groups around the world (reviewed in refs. 4–9). Techniques like PEG fractionation have been developed and successfully used to remove RuBisCO from rice leaf-extracted proteins in order to enrich low-abundance proteins (6, 17). A number of affinity columns are now commercially available to deplete/enrich proteins including affinity column for depleting RuBisCO which has been described here step-by-step.

Extracted proteins are then separated by 1-DGE or 2-DGE. Use of the precast IPG strip format is now essential for good and reproducible separation of proteins in the first dimension. Researchers can now also avail of precast SDS-PAGs for the second dimension, although expense is a factor in most labs sticking to the hand-cast gels. It must be noted here that both biological and experimental replications are essential for downstream 2-D gel image analysis and quantification. Following image analysis of 2-D gels and expression profiling, excised 2-D spots are in-gel digested usually with the enzyme trypsin. Extracted peptides are then subjected to MS and database analysis for identification of proteins and phosphoproteins. The 1-DGE has almost same steps.

3.1. Seed Treatment, Germination and Growth Condition, and Storage

1. Pour 30 mL sterilization solution into a 50-mL falcon tube containing rice seeds.
2. Sterilize the seeds for 30 min with constant shaking on a rotary shaker.
3. Wash five times with 30 mL deionized water for 5 min each.
4. Pour sterilized and washed seeds along with deionized water into a glass beaker.
5. Wrap the glass beaker with aluminum foil and place it in a darkroom at 25°C.
6. Replace deionized water of the glass beaker everyday for 3–4 days.
7. Prepare seedling pot with soil and place it in a tray containing water.
8. Sow each germinated seed in a neat row (4 rows of 10 seeds each) in soil, and cover them again with soil.
9. Place the tray (containing seedling pots) inside a growth chamber.
10. Incubate for 2 weeks with watering (distilled water) everyday from top of seedlings and in the tray below. Two-week-old rice seedling model system is figuratively presented in Fig. 3 (see Note 22).

Fig. 3. Steps involved in preparing the 2-week-old rice seedling model system. (**a**) Sterilizing rice seeds; (**b**) germination; (**c**) seeding in soil; and (**d**) culture in seedling pots in a growth incubator to 2-week-old stage.

11. Harvest 3rd and/or 4th leaves, stems (leaf sheath), seed, and root.
12. Keep plant material as such in aluminum foil or microfuge tubes/cryotubes and freeze immediately in liquid N_2. Store at –80°C.

3.2. Preparation of Tissue Powders for Protein Extraction

Grind 100 mg of plant materials to a fine powder with liquid N_2 in a mortar and pestle (see Note 23).

3.3. Total Protein Extraction: Direct Extraction in LB, LB-T, and LB-TT Method

1. Add 1.5 mL LB, LB-T, or LB-TT to a 2.0 mL microfuge tube containing 100 mg finely powdered tissue sample.
2. Vortex sample solution immediately (see Note 24).
3. Centrifuge for 15 min at 15,000 × g (4°C) and remove supernatant to a new 1.5 mL microfuge tube. Store in single-use aliquots at –80°C for further analysis (see Note 25).
4. Precipitate proteins using ProteoExtract protein precipitation kit as per supplier's recommendation.
5. Resuspend the pellet in LB-TT and proceed for 2-DGE analysis (see step 2 of Subheading 3.12 and Note 26).

Finely powdered tissue sample

↓

Transfer 50-100 mg of powdered tissue in a 2 mL microfuge tube

↓

Add **1600** μL of **TCA/Acetone Extraction Buffer (TCAEEB)** stored at 4°C

↓

Vortex vigorousy (30 sec) and incubate at -20°C for 1 h

↓

Centrifuge at 15,000 rpm at 4°C for 15 min and discard the supernatant

↓

Add **1,600** μL of **Wash Buffer** (cold) and completely resuspend the pellet with vortexing and/or sonicating

↓

Centrifuge at 15,000 rpm at 4°C for 15 min and discard the supernatant

↓

(Repeat this step three times)

↓

Dry the pellet at RT for 30 min / Lyophilize

↓

Place at -80°C for over night

↓

Resuspend the pellet in **200-500** μL **LB-TT** by pipetting, vortexing and sonicating, and incubate at RT for 30 min – 1h

↓

Centrifuge at 15,000 rpm at 20°C for 15 min (SUP is total protein)

↓

Store at -80°C / Protein determination

Fig. 4. The TCA/acetone extraction protocol.

3.4. Total Protein Extraction: TCA–Acetone Method

Detail of this method is depicted as a flowchart in Fig. 4. Briefly,

1. Add cold acetone containing TCA (TCAAEB) to 100 mg of finely powdered tissue sample in a 2.0 mL microfuge tube to precipitate proteins in the cold.
2. Wash precipitated proteins with wash buffer, air-dried (or lyophilized), and resuspend the protein pellet into LB-TT.
3. Proceed for 2-DGE analysis (see step 2 of Subheading 3.12 and Note 26).

3.5. Total Protein Extraction: Phenol–Ammonium Acetate/Methanol Method

Detail of this method is depicted as a flowchart in Fig. 5. Briefly,

1. Add phenol extraction solution to a 2.0 mL microfuge tube containing 200 mg of finely powdered tissue sample. Mix gently at RT.
2. Centrifuge at 15,000 rpm for 15 min at 4°C. Transfer clear supernatant to a new 2.0 mL microfuge tube.
3. Add ammonium acetate–methanol solution and incubate at –20°C overnight to precipitate proteins.
4. Centrifuge and discard the solution.

Finely powdered tissue sample

↓

Transfer 100 - 200 mg of powdered tissue in a 2 mL microfuge tube

↓

Add **500** μL of **Extraction Media** and **500** μL of **Tris (pH 8.8) Buffered Phenol** stored at 4°C

↓

Mix gently using an inverting shaker at RT for 30 min

↓

Centrifuge at 15,000 rpm at 4°C for 20 min

↓

Transfer **top phenol phase** and re-extract the remaining portion with **500** μL of **Extraction Media** and **500** μL of **Tris-Buffered Phenol**

↓

Invert mix × 15 times

↓

Centrifuge at 15,000 rpm at 4°C for 20 min

↓

Remove **top phenol phase** and combine with the first extraction
Divide into **300** μL each phenol phase into two new 2 mL tubes

↓

Add **1,500** μL of **0.1 M Ammonium Acetate in Methanol** (cold/-20°C)

↓

Vortex and incubate at -20°C overnight

↓

Centrifuge at 15,000 rpm at 4°C for 20 min and discard the supernatant

↓

Add **1,000** μL of **0.1 M Ammonium Acetate in Methanol** and completely resuspend the pellet with vortexing and/or sonicating

↓

Place at -20°C for 30 min
Centrifuge at 15,000 rpm at 4°C for 10 min and discard the supernatant **(x 2 times)**

↓

Add **1,000** μL of **80% Acetone** (cold/-20°C) and completely resuspend the pellet with vortexing and/or sonicating

↓

Place at-20°C for 30 min.
Centrifuge at 15,000 rpm at 4°C for 10 min **(x 2 times)**

↓

Add **1,000** μL of **70% Ethanol** (cold/-20°C) and completely resuspend the pellet with vortexing and/or sonicating

↓

Place at -20°C for 30 min
Centrifuge at 15,000 rpm at 4°C for 20 min

↓

Dry the pellet at RT for 10-15 min or at 37°C for 5 min

↓

Add 200-500 μL of LB-TT and incubate at RT (vortex/sonicate x 5 times) for 30 min to 1 h

↓

Centrifuge at 15,000 rpm at 20°C for 15 min (SUP is total protein)

↓

Store SUP in aliquots at -80°C / Protein determination

Fig. 5. The phenol–ammonium acetate/methanol extraction protocol.

5. Wash the protein pellet with the same ammonium acetate–methanol solution. Repeat this step three times.
6. Semidry the tubes in a 37°C oven to remove excess liquid.
7. Resuspend protein pellet in LB-TT and proceed for 2-DGE analysis (see step 2 of Subheading 3.12 and Note 26).

3.6. Total Protein Extraction: Tris-Buffered Saline Method

1. Add cold TBS-T20 to a 2.0 mL microfuge tube containing 100 mg of finely powdered tissue sample.
2. Vortex and sonicate to disperse the powder sample.
3. Add additional equal volume of SDS-SB to increase efficiency of protein extraction. Vortex and sonicate repeatedly and then give heat treatment at 95°C for 1 min in a heat block. After heat treatment, cool sample to RT.
4. Centrifuge at 15,000 rpm for 15 min at 4°C.
5. Transfer supernatant to a new 2.0 mL microfuge tube.
6. Precipitate proteins using ProteoExtract protein precipitation kit.
7. Resuspend the pellet in LB-TT and proceed for 2-DGE analysis (see step 2 of Subheading 3.12 and Note 26).

3.7. PEG Fractionation

1. Incubate finely powdered tissue sample with Mg/NP-40 extraction buffer (10 mL/g tissue).
2. Vortex for 1 min at RT and centrifuge at 12,000 rpm for 15 min.
3. Collect clear supernatant and subject to PEG fractionation.
4. Add PEG to a final concentration of 15% (w/v) using a 50% (w/v) stock solution and incubate on ice for 30 min.
5. Centrifuge at 13,000 rpm for 15 min at 4°C.
6. Save the protein pellet. Transfer supernatant to a new tube and add 4 volumes of 100% (v/v) acetone.
7. Precipitate proteins for 2–3 h at –20°C and then centrifuge at 4,000 rpm for 15 min.
8. Save protein pellet and transfer supernatant to a new tube.
9. Resuspend protein pellets in 10 mL of Mg/NP-40 extraction buffer and agitate at RT.
10. Add an equal volume of water-saturated phenol and vortex to mix.
11. Add 0.7 M sucrose, vortex, and centrifuge at 4,000 rpm for 15 min.
12. Collect the phenolic phase and precipitate proteins by adding 4 volumes of methanol containing 0.1 M ammonium acetate at –20°C for 3 h followed by centrifugation at 4,000 rpm for 10 min.
13. Decant solution carefully and wash protein pellet three times with methanol containing 0.1 M ammonium acetate. Store in 80% acetone at –20°C until use.
14. RuBisCO depletion in the 15% PEG-fractionated supernatant is shown in Fig. 6.

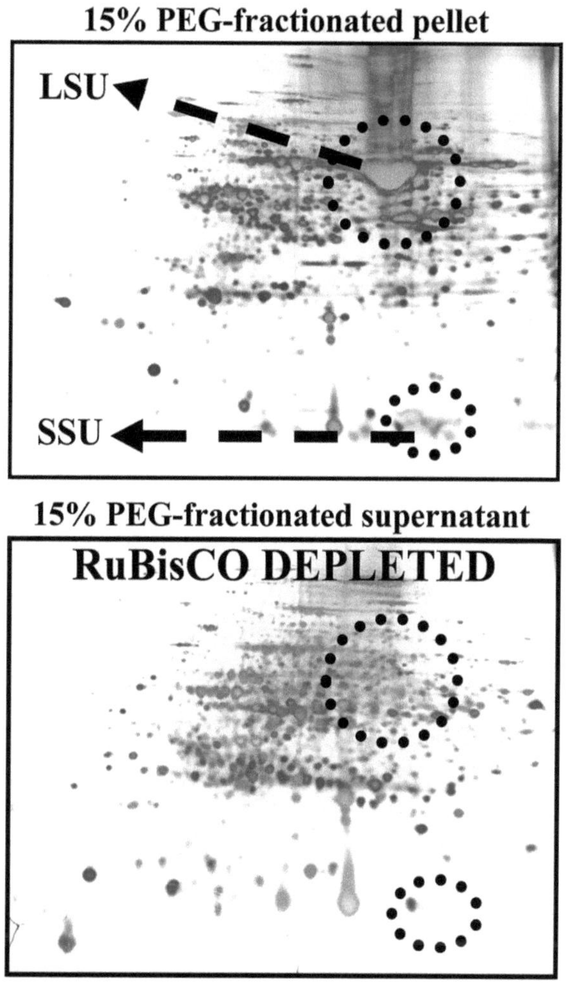

Fig. 6. 2-DGE analysis of rice leaf proteins fractionated by PEG showing successful depletion of RuBisCO subunits (*LSU* large subunit, *SSU* small subunit).

3.8. RuBisCO Depletion Using Seppro® IgY-Rubisco Spin Column (Genway)

1. Add 1 mL of TBS buffer containing protease inhibitor cocktail (Catalog no. P9599, Sigma-Aldrich) to 20 mg of powdered leaf tissue and vortex.
2. Pass supernatant through a 0.45 μm spin filter.
3. Collect supernatant and determine protein concentration.
4. Dilute proteins (leaf extract, 200 mg) with the same TBS buffer to a final volume of 500 μL.
5. Add 800 μL of GenWay bead suspension to a spin column provided by supplier.
6. Centrifuge the column at 2,000 rpm for 30 s to remove suspension buffer.
7. Resuspend the beads in 500 μL of protein sample (from step 4) and incubate at RT for 15 min with agitation.

8. Centrifuge at 2,000 rpm for 30 s to collect unbound protein solution.
9. Wash beads four times with 500 μL of dilution buffer, agitate for 1 min each time, and centrifuge at 2,000 rpm for 30 s to collect unbound or nonspecifically bound beads (see Note 27).
10. Pool the unbound protein fractions and neutralize with 500 μL of neutralization buffer (250 mM Tris–HCl, pH 8.0).
11. Add TCA/acetone to a final concentration of 80% (v/v) to precipitate proteins that are nearly free from RuBisCO.
12. Wash pellet thrice with wash solution (80% ice-cold acetone, 0.07% 2-ME). Store in acetone at −20°C until use.

3.9. Protein Quantification: Kit

Use the Coomassie Plus protein assay kit exactly as directed by the manufacturer.

3.10. Protein Quantification: A Modified Bradford Method

1. Dilute protein sample with 0.5× SDS running buffer in a 1.5 mL microfuge tube, and mix (see Note 28).
2. Use 0.5× SDS running buffer as blank and add 6 μL to the first column of the microtiter plate.
3. Add 1, 2, 4, and 6 μL of 1 mg/mL BSA standard in triplicate to wells and bring the final volume to 6 μL with 0.5× SDS running buffer.
4. Add 1 μL of tenfold diluted protein sample in triplicate to wells plus 5 μL of 0.5× SDS running buffer to make a total volume of 6 μL (see Note 29).
5. Add 200 μL of diluted Bradford dye to wells with protein.
6. Mix and incubate at RT for 5 min.
7. Vortex the microtiter plate on the spectrophotometer and measure absorbance at 595 nm (see Note 30).
8. Calculate the protein concentration.

3.11. 1-DGE (SDS-PAGE): Casting and Assembly of the Vertical Electrophoresis Gel Unit (Hand-Cast)

1. Incubate 10 mg protein for 3 min at 95°C in SDS-SB and cool to RT. After incubation on the bench (ambient RT of 25°C) for 10 min, centrifuge the mixture. The sample is ready to load onto the well of a PAG.
2. To cast a PAG (standard size), follow the description on the illustration provided in Fig. 7a, b.
3. Add all solutions starting from acrylamide till 10% SDS in series in a 50 mL glass beaker. Mix nicely by turning the beaker counterclockwise. Avoid bubbles.
4. Add 10% APS while mixing the solution.
5. Immediately add TEMED, and continue to mix for 30 s. Move quickly to cast the gel by pouring solution till the mark on the

a

12.5% Separation gel				3.9% Stacking gel		
1 gel	**2 gels**	**4 gels**		**1 gel**	**2 gels**	**4 gels**
11.26 mL	22.52 mL	40.56 mL	**30% Gel acrylamide**	532 µL	1.06 mL	1.92 mL
6.77 mL	13.54 mL	24.36 mL	**(pH 8.8) Buffer (pH 6.8)**	1.0 mL	2.0 mL	3.6 mL
8.55 mL	17.1 mL	30.81 mL	**MQ**	2.50 mL	5.0 mL	9.0 mL
270.4 µL	540.8 µL	973.1 µL	**10% SDS**	40.0 µL	80.0 µL	144.0 µL
135.6 µL	271.2 µL	488.2 µL	**10% APS**	30.0 µL	60.0 µL	108.0 µL
27.2 µL	54.4 µL	98.0 µL	**TEMED**	2.0 µL	4.0 µL	7.2 µL

b

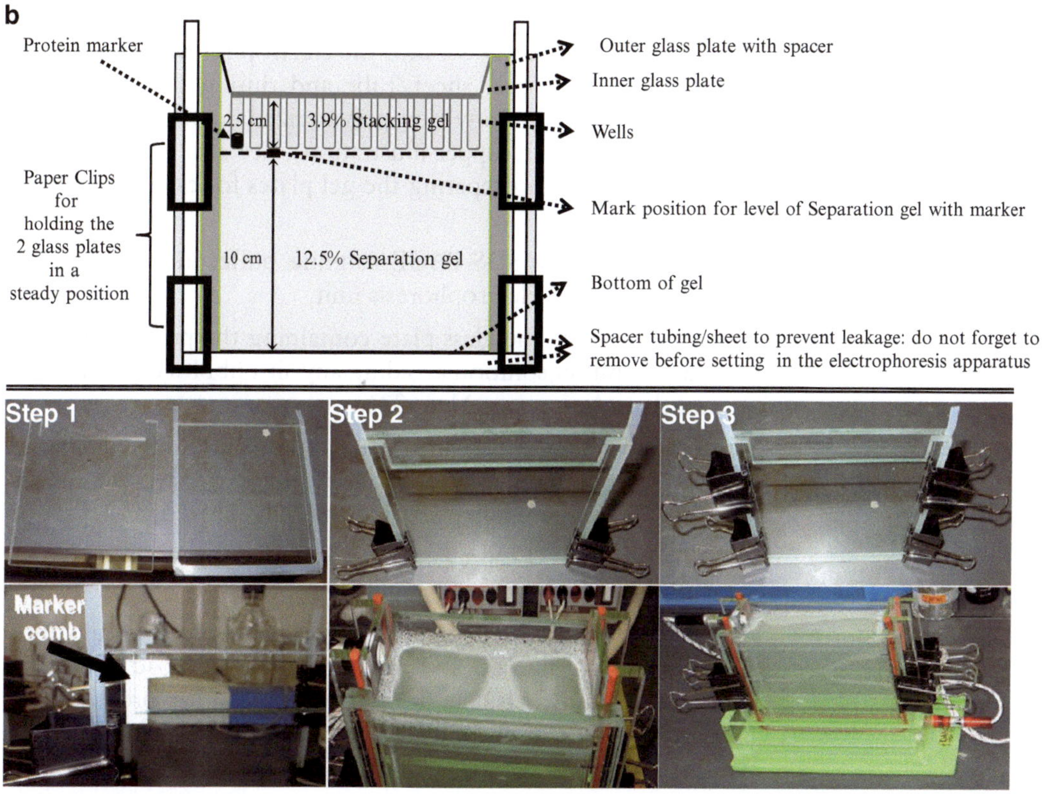

Fig. 7. 1-DGE (SDS-PAGE) gel electrophoresis. Preparation of separating and stacking gel solutions for casting the gel (**a**) solutions; (**b**) 1-DGE/2-DGE.

glass plate indicating the boundary between separation and stacking gels.

6. Very slowly overlay each gel with 1.0 mL deionized water.
7. Stand the gel steady on the bench for 1–1.5 h at RT to polymerize the separation gel.
8. Bring the unit near a sink, carefully pour off the water, and rinse the space with deionized water. Keep it back on the bench.

9. Quickly prepare the stacking gel solution as indicated in Fig. 7a and following the sequence of steps as above. Do not make a mistake: The stacking gel buffer (pH 6.8) needs to be used for this step.
10. Pour the stacking gel solution into the remaining gel space till just below the top of the inner glass plate.
11. Carefully insert the comb into the stacking gel solution. Overlay the wells with 1 mL isobutanol (Wako) (see Note 31).
12. Stand the gel steady on the bench to polymerize for 30 min to 1 h.
13. Once polymerized, pour off isobutanol and wash several times with deionized water to remove any traces of isobutanol.
14. The gel is ready to be set into the electrophoresis apparatus by removing the spacer sheet/tube and rinsing the gel (within glass plate) free of any non-polymerized acrylamide or solutions (see Fig. 7b, upper illustration). See also Fig. 7b, lower panel images, for preparing the gel plates for casting and running the gel.
15. First add the 1× SDS-PAGE running buffer into the bottom chamber of the electrophoresis unit.
16. Carefully place the glass plate containing the polymerized gel into the buffer solution avoiding any air bubbles below the gel and the solution (see Note 32).
17. Clamp tightly the gel with electrophoresis unit upper chamber using the same paper clips used for holding the two glass plates bound together while preparing the polymerized gel.
18. Fill the enclosed top chamber space also with 1× SDS-PAGE running buffer.
19. Load the processed protein samples (see step 1) carefully into wells using a fine-tip and calibrated pipette. Load marker at desired volume (usually 5.0 μL of ready-to-use protein standards from Bio-Rad for a standard gel shown in Fig. 7b, upper illustration).
20. Connect the apparatus to a power supply.
21. Run the electrophoresis unit at constant current of 40 mA for around 3 h or till the BPB line reaches the bottom of the gel.

3.12. 2-DGE: IEF in the 1st Dimension

1. Mix 100 μg total soluble protein with LB-TT containing 0.5% (v/v) IPG buffer and 1% DeStreak™ Reagent (GE Healthcare) and bring to a final volume of 450 μL.
2. Carefully place (using pipette) the prepared protein solution into the IPG strip holder.
3. Place IPG strips (pH 4–7; 24 cm, GE Healthcare) carefully (using forceps) onto the protein samples avoiding air bubbles between the sample and the gel.

4. Allow the IPG strips to passively rehydrate with the protein samples for 1 h and 30 min, followed by overlaying the IPG strips with 1.0 mL cover fluid (mineral oil).
5. Follow the five-step active rehydration and focusing protocol as described in Fig. 8. The whole procedure is controlled at 20°C.

3.13. 2-DGE: Reduction/Alkylation of Proteins in the IPG Strips

1. Remove the IPG strips from the IPG focusing tray by holding one end of the strip with forceps and blotting off excess cover fluid using a Kimwipe. Do not touch the gel end of the IPG strip with Kimwipe.
2. Use Protean IEF system 24 cm disposable tray to place the IPG strip in a well carrying 2.5 mL reduction solution, facing the gel side up (see Note 33).
3. Incubate on rocking platform at medium speed for 15 min at RT.
4. Transfer the IPG strips to a new disposable tray with 2.5 mL alkylation solution in each well.
5. Incubate again on rocking platform at medium speed for 15 min at RT.
6. Transfer the IPG strips again to a new disposable tray carrying 2.5 mL 1× SDS-PAGE running buffer in each well for a brief wash. The IPG strips are now ready for second-dimension SDS-PAGE.

3.14. 2-DGE: SDS-PAGE in the 2nd Dimension (Hand-Cast)

1. Prepare SDS-PAG (see Subheading 3.11). Do not use the well comb.
2. Add a marker comb for loading protein markers (Fig. 7b, lower panel).
3. Follow the steps as indicated in Fig. 8.
4. Carry out SDS-PAGE at constant current of 40 mA/per 18 cm gels for around 4.5 h (see Subheading 3.11).
5. If using precast gels (homogenous or gradient), follow the manufacturer's instructions (i.e., Ettan DALTsix/twelve, GE Healthcare) (see Subheadings 3.15 and 3.16).

3.15. 2-DGE: Assembling the Ettan DALTtwelve Gel Caster Unit (Hand-Cast)

1. Tilt the DALTtwelve unit back so that it rests on its support legs (see Note 34).
2. Place a thicker separator sheet against the back wall to easily remove the last cassette from the gel caster unit after polymerization.
3. Clean each glass plate carefully with deionized water and ethanol using Kimwipe.
4. Fill the caster by alternating cassettes with separator sheets. End with a separator sheet, and then use the thicker separator

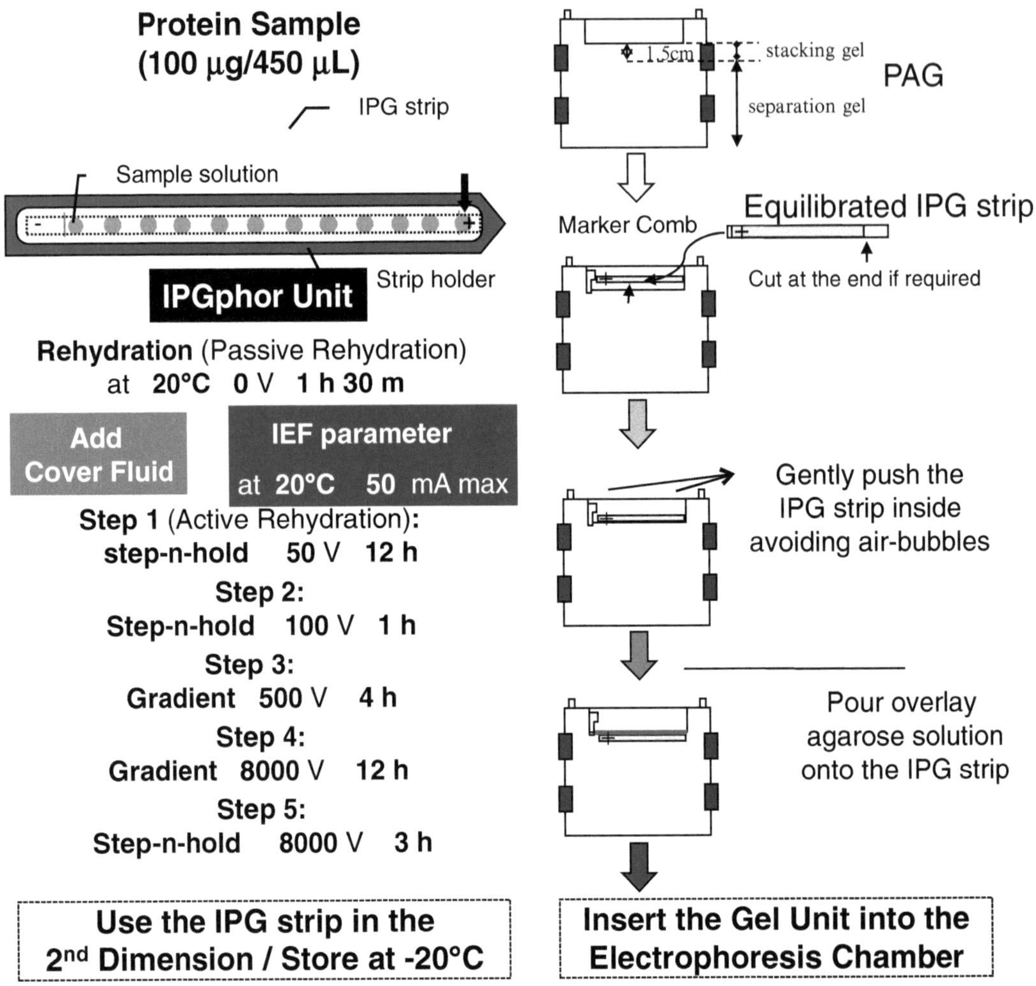

Fig. 8. 2-DGE (IEF and SDS-PAGE) proteomics approach. A graphical depiction of 2-D gel electrophoresis starting from IEF to SDS-PAGE (illustrated).

sheets to bring the level of the stack even with the edge of the caster.

5. Lubricate the foam gasket with a small amount of GelSeal and place it in the groove on the faceplate.
6. Turn four black-knobbed screws into the four threaded holes across the bottom until they are well engaged. Usually two to three full turns are enough.
7. Place the faceplate carefully onto the caster with the bottom slots resting on their respective screws. Screw the four remaining black-knobbed screws into the holes at the sides of the faceplate and tighten all eight evenly. Assembled unit is now ready to use to cast gels (see Note 35).

3.16. 2-DGE: Casting 12% SDS-PAG into the Ettan DALTtwelve Gel Caster (Hand-Cast)

1. To cast 12 gels, add 300 mL of acrylamide stock, 188 mL of 1.5 M Tris–HCl pH 8.8, 7.5 mL of 10% SDS, and 252 mL of deionized water in 1 L sidearm flask. This is the separating gel solution.
2. Place the flask on stir plate, add a medium size stirring bar, and stir the solution.
3. Connect sidearm to vacuum, cover top opening with solid rubber stopper, and apply vacuum for 30 min.
4. Turn off vacuum, remove rubber stopper, and disconnect hose.
5. Add 3.6 mL of 10% APS while stirring the solution.
6. Add 120 μL TEMED and continue to stir for 30 s. Move quickly to cast the gels.
7. Slowly pour the gel solution into the caster through the hydrostatic balance chamber until it is about 2 cm below the desired gel height.
8. Pour the displacement solution into the chamber until it is 0.25 cm below the surface of glass plates and then immediately place the feed tube into the grommet to stop the flow.
9. Very slowly overlay each gel with 1.5 mL deionized water.
10. Cover the upper portion of the unit with plastic wrap to prevent dehydration.
11. Allow polymerization for 16 h.
12. Bring the unit near a sink, carefully disassemble, and scrape off access acrylamide into special waste container.
13. Remove the gel cassette from the caster by pulling forward on the separator sheets and rinse the outer surface of each gel cassette with deionized water to remove any polyacrylamide particles. Place gel cassettes in the cassette rack.
14. Prepare stacking gel solution by combining 10.64 mL of acrylamide stock, 20 mL of 0.5 M Tris–HCl pH 6.8, 0.8 mL of 10% SDS, and 48.8 mL of deionized water in a 200 mL beaker.
15. Stir solution thoroughly on a magnetic stir plate.
16. Add 0.6 mL of 10% APS while stirring the solution.
17. Add 40 μL TEMED and continue to stir for 30 s.
18. Place enough stacking gel solution on the top of the separating gel using a plastic transfer pipette to give 2 cm height after polymerization.
19. Overlay each gel with 1 mL isobutanol (see Note 31).

20. Allow to polymerize approximately for 1 h.
21. Once polymerized, pour off isobutanol and wash several times with deionized water to remove any traces of isobutanol.
22. Add 1× SDS running buffer on the top of the stacking gel to prevent dehydration (see Note 36).

3.17. 2-DGE: SDS-PAGE on Ettan DALTsix/twelve Unit

1. Remove the IPG strip from the tray by holding one end of the strip with forceps and placing the IPG strip carefully on the surface of stacking gel (see Note 37).
2. Place a small square of electrode wick (0.5 cm × 0.5 cm) containing 1–2 μL of protein standard (for example, use PeppermintStick phosphoprotein standards in case of phosphoprotein analysis) next to the acidic end of the strip (see Note 38).
3. Quickly overlay the IPG strip and the electrode wick with 2–3 mL agarose overlay solution (see Note 39).
4. Pour about 7.5 L 1× SDS running buffer to the lower chamber of the separation unit (see Note 40).
5. Once agarose is solidified, insert the gel cassette into the separation unit through the buffer seal slots flanked by rubber gaskets (see Note 41).
6. Pour about 2.0 L 2× SDS running buffer until solution reaches the marked upper level on the separation unit.
7. Run the electrophoresis unit at 2 W/gel until dye migrates off the gel (see Note 42).

3.18. Protein Detection with CBB R-250, Colloidal CBB, or Silver Nitrate

1. Remove the PAG from the unit and carefully peel out from the glass plate and place in an appropriate tray containing deionized water.
2. Wash the gel with deionized water and discard the water. Add staining solution to detect proteins.
3. For staining with CBB R-250, shake the gel for 30 min on a shaker. Destain with destaining solution four times for 30 min each or till clear background is observed (see Note 43).
4. For staining with colloidal CBB, shake the gel for 16 h in 250 mL solution to detect total proteins (see Note 44).
5. For silver nitrate staining, follow the protocol as provided by the manufacturer (GE Healthcare).

3.19. Image Acquisition

1. Scan the gel at 300 dpi resolution and 16-bit grayscale pixel depth using ProteomeScan2000 digital scanner or ScanMaker 9800XL.
2. Save data in tagged image file formats (TIFF).

3.20. Image Analysis for Total Proteins

Use ImageMaster 2D Platinum software version 5.0 or 6.0 (GE Healthcare) for downstream analysis of acquired gel images as per supplier's instructions. Confirm each high-quality protein spot manually and if required edit manually. Progenesis PG200 software or LUDESI also works well for image analysis.

3.21. Phosphoprotein Detection with Pro-Q DPS and Image Acquisition

1. Transfer the gel to a gel-staining tray.
2. Wash the gel twice with deionized water for 10 min each. Incubate at RT with constant shaking on an orbital shaker at a speed of 35 rpm in all steps of this modified Pro-Q DPS procedure, unless stated otherwise (see Note 45).
3. Decant deionized water and immerse the gel in 200 mL fixation solution. Decant fix solution and repeat (see Note 46).
4. Immerse the gel in 250 mL washing solution for 15 min. Decant wash solution and repeat.
5. Incubate the gel in 150 mL staining solution in the dark for 2 h and decant solution.
6. Immerse the gel in 250 mL destaining solution and incubate in the dark for 30 min. Decant destaining solution and repeat this step three more times. The total required destaining time is 2 h.
7. Wash the gel with 250 mL deionized water in the dark for 5 min. Decant deionized water and repeat.
8. Scan the gel using a laser imager with 532-nm excitation and 580-nm bandpass emission filter (Fujifilm FLA 5000). Collect and analyze data at 100-μm resolution, 16-bit TIFF files (Image Gauge Analysis software, Fuji) (see Notes 47 and 48).
9. Use Image Gauge Analysis software (Fuji) and ImageMaster 2D Platinum software version 5 or 6 (GE Healthcare) to display and analyze data.

3.22. Excision of 1-D Band and 2-D Spot

1. Stained protein bands/spots are excised from the CBB colloidal or silver Nitrate-stained 1-D and 2-D gels and transferred to sterilized Eppendorf tubes (1.5 mL).
2. The bands and spots are kept at –30°C till processed for further analysis.

3.23. In-Gel Digestion of 1-D Bands and MS Analysis by MALDI-TOF-MS and Q-TOF-MS/MS

1. Use peptide-mass fingerprinting (PMF) methods (18) and tandem mass spectrometry (Q-TOF-MS/MS) to identify protein(s) in 1-D band.
2. Cut the excised gel bands with razor blade into two smaller pieces and destain with 100 μL of destain solution with shaking for 5 min.
3. Discard the solution, add 200 mM AMBIC solution to cover gel pieces, and incubate at RT for 20 min.

4. Discard solution. Dry gel pieces in a speed vacuum concentrator for 5 min and then rehydrate them with 20 μL of 50 mM AMBIC containing 0.2 μg modified trypsin for 45 min on ice.
5. Add 30 μL of 50 mM AMBIC and continue the digestion reaction overnight at 37°C.
6. Desalt and concentrate peptides using C18 nanoscale (porous C18) column.
7. For PMF analysis, elute peptides with 0.8 μL of matrix solution.
8. Spot the eluted peptides onto a stainless steel target plate and perform mass analyses of peptides using MALDI-TOF-MS (Model MALDI-R). For calibration, use internal mass of trypsin autodigestion product (*m/z* 2211.105).
9. For analyses by MS/MS, dilute 15 μL of the peptide solutions from the digestion supernatant with 30 μL of 5% formic acid. Load onto the column and wash with 30 μL of 5% formic acid.
10. Elute peptides with 2.0 μL methanol/H_2O/formic acid (50/49/1, v/v/v) directly into a precoated borosilicate nano-electrospray needles.
11. Load eluted peptides for MS/MS analysis using nano-ESI on a Q-TOF/TOF MS (Micromass).
12. Process acquired MS/MS data using a MassLynx (Ver. 3.5).
13. Use the processed data against the suitable nonredundant rice database using the MASCOT search engine (Matrix Science, Inc., London, UK; http://www.matrixscience.com).
14. Assign search parameters for the database and then protein assignment criteria to identify proteins or phosphoproteins.

3.24. In-Gel Digestion of 2-D Gel Spots and MS Analysis on Agilent nESI LC-MS/MS

1. Incubate 2-D gel spot or spots in destaining solution to remove stain from the gel. Repeat the step if required.
2. Replace destaining solution with sterilized water and wash the gel pieces twice.
3. Discard solution. Add 0.2 M AMBIC, pH 7.8 solution and incubate at RT for 20 min.
4. Discard solution. Wash with the same volume of CH_3CN and 0.1 M NH_4HCO_3, pH 7.8 solution by slow vortexing at RT for 20 min.
5. Remove solution and add the same volume of CH_3CN to dehydrate gel pieces by slow vortexing at RT for 20 min. Add 0.1 M AMBIC solution pH 7.8 to rehydrate them.
6. Repeat dehydration process twice. Dry the gel pieces in a vacuum centrifuge.
7. Add digestion buffer containing 10 mg/mL of trypsin (Promega, sequencing grade) and incubate in ice for 45 min to swell the gel pieces.

8. Add 20 μL of 50 mM NH_4HCO_3 (pH 7.8) and incubate further at 37°C for 12 h.
9. Collect supernatant and elute digested peptides with CH_3CN solution. Pool the fractions together and desalt through a C18 ZipTip (Millipore, Bedford, MA) according to the manufacturer.
10. Analyze prepared sample using an Agilent 1100 Nano LC-1100 system combined with a microwell-plate sampler and thermostatted column compartment for preconcentration (LC Packings, Agilent).
11. Load sample (2–5 μL) on the column (Zorbax 300SB-C18, 150 mm × 75 μm, 3.5 μm) using a preconcentration step in a microprecolumn cartridge (Zorbax 300SB-C18, 5 mm × 300 μm, 5 μm).
12. Use the solution 0.1% HCOOH in water (A) and 0.1% HCOOH in acetonitrile (B) to generate a linear gradient from 15 to 45% B for 25 min.
13. Search the acquired tandem MS spectra against the rice protein database using the Agilent Spectrum Mill MS Proteomics Workbench (Spectrum Mill version; Agilent Technologies, Santa Clara, CA).

3.25. In-Gel Digestion of 2-D Gel Spots and MS Analysis on a Hitachi NanoFrontier eLD-IT-TOF-MS/MS

1. Dice each gel slice into small pieces (1 mm × 1 mm).
2. Add 100 μL of 100% CH_3CN and vortex for 5 min.
3. Extract the supernatant and discard.
4. Dry the gel particles for 3 min in a vacuum centrifuge.
5. Add enough 10 mM DTT in 100 mM AMBIC to cover the gel pieces and reduction was carried out for 1 h at 65°C.
6. Replace the DTT solution by roughly the same volume 50 mM iodoacetamide in 100 mM AMBIC and incubate for 45 min at RT in the dark.
7. Remove the supernatant.
8. Wash the gel pieces with 500 μL of 100 mM AMBIC for 10 min while vortexing, and dehydrate with 500 μL of 100% CH_3CN.
9. Repeat the wash/dehydration step one time.
10. Remove the supernatant and dry the gel particles for 3 min in a vacuum centrifuge.
11. Add trypsin solution to barely cover the gel pieces.
12. Incubate for 12 h at 37°C.
13. Add 5 μL 5% formic acid.
14. Transfer the digest solution into a clean 0.5 mL microfuge tubes.
15. To the gel pieces, add 40 μL of 100 mM AMBIC for 15 min while vortexing and transfer the supernatant into the tube.

Add 20 μL of 50% CH_3CN in 5% formic acid for 15 min while vortexing and transfer the supernatant into a tube.

16. Repeat one time.
17. Concentrate the recovered peptides by reducing the final volume of the extracts to 3 μL in a vacuum centrifuge.
18. Add 20 μL of 0.3% formic acid.
19. Load the prepared sample to eLD-IT-TOF-MS/MS.
20. Use Nano Frontier Data Processing software P/N:3807051-01 (Hitachi High-Technologies Co.) for generating peaklist.
21. Search the acquired data against the nonredundant rice database using the MASCOT (version 2.1, Matrix Science, Inc.) as search engine (see Note 49).

3.26. Excision of Phosphoprotein, In-Gel Digestion, and MS Analysis

1. Overlay the images of phosphoproteins and proteins as false colors using Adobe photoshop by aligning the phosphoprotein and protein markers.
2. Manually excise the desired phosphoprotein spots using protein spots as landmarks with the help of 0.15 mm spot picker and transfer to a 96-well MultiScreen plate.
3. Use a vacuum manifold system to process in-gel digestion in a medium-throughput manner.
4. Destain gel plugs with 200 μL in-gel wash solution for 15 min at RT with gentle agitation on a microplate shaker at a speed of 50 rpm. Evacuate the solution from the bottom of the filter plate using a vacuum manifold system. Repeat this step two more times or until all stain is removed.
5. Dehydrate the gel plugs with 100 μL CH_3CN for 5 min and remove CH_3CN by vacuum evacuation.
6. Remove residual CH_3CN by blotting the plate gently with Kimwipes.
7. Place a 96-well V-bottom sample collection plate underneath the MultiScreen plate.
8. Add 50 μL of in-gel trypsin solution to wells to rehydrate the gel plugs, cover the plate with adhesive film, and place the cassette inside a plastic sealable bag.
9. Incubate at 37°C for 16 h.
10. Add 100 μL of in-gel extraction solution to wells and incubate for 10 min at RT with gentle agitation.
11. Centrifuge at 2,000 × g for 2 min to collect trypsin-digested peptides into V-bottom polypropylene collection plate.
12. Repeat steps 10 and 11 once more.
13. Dry the pooled extracted peptides using CentriVap Console and store at −80°C.

14. Resuspend the dried pellet in 50 μL of 0.1% formic acid.
15. Load sample for mass spectral analysis on a high-mass accuracy LC-MS/MS instrument following standard procedures (see Note 50).
16. Search MS/MS data against the suitable database using either the SEQUEST or MASCOT algorithm.
17. Assign search parameters for the database and then protein assignment criteria to identify phosphoproteins.

4. Notes

1. Dry mature seeds are stored at 10°C in case of long-term storage.
2. We used Nursery soil for rice seedling growth and transplantation (JA Zen-Noh, Japan; http://www.zennoh.or.jp/) at one of our laboratories. However, any other soil can be used to achieve the plant growth and development desired in your experiments.
3. Use gloved hands and/or clean/sterile equipment to collect samples. Use gloves at all times for the experiments that precede mass spectrometry.
4. The quality of TEMED may decline with time after opening, and therefore gels take a longer time to polymerize. It is recommended to buy a small bottle of TEMED and to store at 4°C.
5. The quality of APS may decline with time after preparation, so store the prepared solution (2 mL each time) at 4°C.
6. Store the prepared and sterilized extraction buffer at −80°C in 1 mL aliquots in microfuge tubes.
7. TCAAEB solution can be stored at 4°C for up to 1 month.
8. Wash buffer solution can be stored at 4°C for up to 1 month.
9. 2-ME should be added just before preparation of "phenol extraction buffer."
10. Prepared buffered phenol solution should be kept at −20°C in 40 mL aliquots in 50 mL polypropylene tubes.
11. It takes approximately 30 min to dissolve the chemicals.
12. Prepare a 10× stock solution of TBS (stored at 4°C) and freshly add Tween 20 in 1× working solution.
13. Dilute the 10× running buffer with deionized water before use.
14. Stick to the incubation times as mentioned clearly in the protocol supplied along with the kit. Do not forget to add glutaraldehyde or formaldehyde to buffers wherever needed.
15. PeppermintStick phosphoprotein molecular weight standards carry two phosphorylated proteins (ovalbumin and

bovine β-casein of 45.0 and 23.6 kDa, respectively) and four non-phosphorylated proteins (β-galactosidase, BSA, avidin, and lysozyme of 116.25, 66.2, 18.0, and 14.4 kDa, respectively). Such protein markers are essential to include in any Pro-Q DPS-based experiments to exclude false positive identification by normalizing the detected phosphoprotein bands or spots against positive and negative phosphoprotein markers. Authors' experience suggests that any protein standard with ovalbumin and/or casein can be used for this purpose.

16. In our experience, storage of Pro-Q DPS at 4°C prolongs its stability by at least twofold.
17. It is recommended to prepare a stock solution of 50 mM sodium acetate pH 4.0. Use acetic acid to adjust the pH to 4.0. Store at RT.
18. There are a number of different scanners being used by different laboratories. Our groups have been using the different scanners mentioned in this subsection. All these scanners work well. Here, the fluorescence image capture device is used when protein is labeled with fluorescence dye such as Cy3, Cy5, SYPRO-RYBY, or Pro-Q DPS.
19. Like scanners, a number of image analysis software is available (three are mentioned here). In addition to these, phoretics has been also widely used. Most recently, Delta 2D is getting popular. Each software has advantages and disadvantages.
20. Mostly used for picking 2-D protein spots in 2-DGE proteomics approach.
21. Mostly used for gel slicing in 1-DGE proteomics approach.
22. Plant materials, such as leaf (in vitro model system) or whole plants (in vivo model system), can be treated with various chemicals or stress factors.
23. Prechill the mortar and pestle, microfuge tubes, and spatulas for transferring the sample tissue powders in liquid nitrogen for at least 15 min before starting the grinding of the sample tissues.
24. It is important to completely disperse the sample powder in solution. If powder is not dispersed, use additional vortexing or sonication.
25. The protein solution can be cleaned up or further purified or concentrated prior to 1-DGE and 2-DGE by using the protein precipitation kit (Calbiochem; as per the instructions of the manufacturer) or the 2-D Clean Up Kit (GE Healthcare).
26. Proteins re-solubilized in LB-TT can be directly separated with SDS-PAGE.
27. The bound RuBisCO can be eluted off the column with five 500 μL volumes of stripping buffer (250 mM glycine pH 2.5).

Each time, agitate the column for 3 min and then centrifuge at 2,000 rpm for 30 s.

28. We usually prepare 10- and 20-fold dilutions.
29. It is advisable to check whether 1 μL produces a color level that comes within the range of colors produced by BSA standards. If not, a 20-fold dilution is needed.
30. Based on loading of blanks and protein standards, a program can be created in the instrument with parameters that vortex the plate before measuring the absorbance.
31. Isobutanol is added to obtain uniform and straight wells. It is also added to obtain uniform and straight layer on the gel which is necessary in order to place the IPG strip.
32. Insertion of the gel in slating position into the lower chamber helps in removing any air bubbles that may appear below the gel.
33. Alternatively, Petri dishes can be used for incubation. Care should be taken to place the IPG strips so that the gel-side is facing inwards and the backing faces the wall of the dish.
34. We recommend reading the user manual for Ettan DALTsix/twelve electrophoresis system.
35. Make sure that the sealing gasket is compressed evenly by the faceplate and forms a tight seal with the caster. Do not overtighten the screws.
36. It is important to keep the ready gels moist to prevent any dehydration. Gels can also be stored for a week at 4–8°C if wrapped with plastic wrap and moist paper towels to prevent dehydration.
37. Add 1× SDS running buffer on top of the stacking gels. It helps in sliding the strip between the plates and in positioning the strip on the gel surface. Again, avoid trapping air bubbles between strip and gel. The acidic end of the strip should be on the left. The gel face of the strip should not touch the opposite glass plate.
38. We recommend dipping the electrode wick containing PeppermintStick standards into the agarose sealing solution to prevent diffusion of standards into the liquid. Usually we place the electrode wick near the acidic end of the IPG strips. It is important to remove excess liquid before placing the electrode wick.
39. Do not use agarose overlay solution that is too warm. Again, there should not be any air bubbles.
40. The Ettan DALTtwelve electrophoresis unit requires a total of about 9.5 L of SDS running buffer: about 7.5 L 1× SDS running buffer for the lower chamber and about 2.0 L of 2× SDS running buffer for the upper chamber.
41. Unoccupied slots should be filled with the blank cassettes. Use 1× SDS running buffer from a squirt bottle to wet the surface

Table 1
A ProQ DPS protocol for phosphoprotein detection in large-format 2-D polyacrylamide (PAG) gels. Same steps are involved in 1-D gel staining

Step	Solution	Amount	Time (min)	Dark incubation	Reuse of solution
1	Fixation	250 mL	2 × 30	Not required	Yes
2	Washing	250 mL	2 × 15	Not required	No
3	Staining	150 mL	120	Required	Yes
4	Destaining	250 mL	4 × 30	Required	Yes
5	Washing	250 mL	2 × 5	Required	No

of gel or blank cassettes before inserting the cassettes, as it helps to slide easily into the unit.

42. It usually takes overnight (about 15–16 h) to finish the run.
43. The staining/destaining solution should cover the gel. As a rule, destaining solution (for CBB R-250) should be twice the volume of the staining solution. For CBB R-250 stain (not for colloidal CBB), a tightly folded Kimwipe (white) can be added to the destaining solution to accelerate the destaining process by absorbing excess dye in the solution.
44. Decant colloidal CBB solution and add 250 mL of gel storage solution to store the gel at 4°C for up to a few months.
45. To decant the solution, it is important to wear powder-free nitrile gloves. Hold the tray with one hand and use the other to hold the gel in the tray. Tilt the tray to decant most of the solution. Be careful not to press the gel hard; otherwise it may break or have a finger impression.
46. Table 1 summarizes all steps including the required volumes of solutions for large-size gels, total incubation times and conditions, and whether solutions can be reused (for details, see ref. 19).
47. Gels should be imaged immediately after Pro-Q DPS staining.
48. Any flatbed fluorescent imager can be used for image capture. Pro-Q DPS has ~555/580 nm excitation/emission maxima. Therefore, stained gels can be best imaged using excitation lasers or LEDs with a range of 532–560 nm coupled with ~580 nm long-pass or ~600 nm band-pass emission filters.
49. Additionally, the nonredundant protein databases—National Center for Biotechnology Information (NCBI) and SwissProt for green plants—can also be searched to identify proteins.
50. Any LC-MS/MS mass spectrometry system and associated software can be used for protein identification.

Acknowledgments

GKA and RR appreciate the kind support of Dr. Akihiro Kubo (NIES, Japan) for providing the space to grow and conduct rice stress experiments. GKA appreciates the Japan Society for the Promotion of Science (JSPS; ID Number S-10182) for his stay and research at Plant Genome Research Unit (NIAS, Tsukuba, Japan). This work was supported by the National Research Foundation of Korea (NRF) grant funded by the Korean Government (MEST) (No. 2011-0000139, No. R0602536). This work was also supported by a grant from the BioGreen 21 Program (No. PJ007993), Rural Development Administration, Republic of Korea. K.C. was supported at NIES as an Eco-Frontier Fellow (09-Ba086-02). RR acknowledges the great support of Professors Yoshihiro Shiraiwa (Chairperson, Faculty of Life and Environmental Sciences, University of Tsukuba) and Seiji Shioda and Dr. Tetsuo Ogawa (Department of Anatomy I, Showa University School of Medicine) in promoting interdisciplinary research and unselfish encouragement. For all correspondence, contact Ganesh Kumar Agrawal or Randeep Rakwal at plantproteomics@gmail.com.

References

1. Goff SA, Ricke D, Lan TH et al (2002) A draft sequence of the rice genome (*Oryza sativa* L. ssp. japonica). Science 296:92–100
2. Yu J, Hu S, Wang J et al (2002) A draft sequence of the rice genome (*Oryza sativa* L. ssp. *indica*). Science 296:79–92
3. International rice genome sequencing project (2005) The map-based sequence of the rice genome. Nature 436:793–800
4. Rakwal R, Agrawal GK (2003) Rice proteomics: current status and future perspectives. Electrophoresis 24:3378–3389
5. Agrawal GK, Rakwal R (2006) Rice proteomics: a cornerstone for cereal food crop proteomes. Mass Spectrom Rev 25:1–53
6. Agrawal GK, Jwa NS, Iwahashi Y, Yonekura M, Iwahashi H, Rakwal R (2006) Rejuvenating rice proteomics: facts, challenges, and visions. Proteomics 6:5549–5576
7. Agrawal GK, Rakwal R (2008) Rice proteome at a glance. In: Agrawal GK, Rakwal R (eds) Plant proteomics: technologies, strategies, and applications. Wiley, Hoboken, NJ, pp 165–178
8. Agrawal GK, Jwa NS, Rakwal R (2009) Rice proteomics: ending phase I and the beginning of phase II. Proteomics 9:935–963
9. Agrawal GK, Rakwal R (2011) Rice proteomics: a move toward expanded proteome coverage to comparative and functional proteomics uncovers the mysteries of rice and plant biology. Proteomics 11:1630–1649
10. Agrawal GK, Rakwal R (eds) (2008) Plant proteomics: technologies, strategies, and applications. Wiley, Hoboken, NJ
11. Han X, Aslanian A, Yates JR 3rd (2008) Mass spectrometry for proteomics. Curr Opin Chem Biol 12:483–490
12. Righetti PG, Antonioli P, Simo C, Citterio A (2008) Gel-based proteomics. In: Agrawal GK, Rakwal R (eds) Plant Proteomics: Technologies, Strategies, and Applications. Wiley, Hoboken, NJ, pp 11–30
13. Gorg A, Weiss W, Dunn MJ (2004) Current two-dimensional electrophoresis technology for proteomics. Proteomics 4:3665–3685
14. Miller I, Crawford J, Gianazza E (2006) Protein stains for proteomic applications: which, when, why? Proteomics 6: 5385–5408
15. Görg A, Drews O, Luck C, Weiland F, Weiss W (2009) 2-DE with IPGs. Electrophoresis 30:S122–S132

16. Thiellement H, Zivy M, Damerval C, Mechin V (eds) (2007) Plant Proteomics: Methods and Protocols. Humana, Totowa, NJ
17. Kim ST, Cho KS, Jang YS, Kang KY (2001) Two-dimensional electrophoretic analysis of rice proteins by polyethylene glycol fractionation for protein arrays. Electrophoresis 22:2103–2109
18. Jensen ON, Mortensen P, Vorm O, Mann M (1997) Automation of matrix-assisted laser desorption/ionization mass spectrometry using fuzzy logic feedback control. Anal Chem 69:1706–1714
19. Agrawal GK, Thelen JJ (2005) Development of a simplified, economical polyacrylamide gel staining protocol for phosphoproteins. Proteomics 5:4684–4688

Chapter 13

Quantification of Jasmonic and Salicylic Acids in Rice Seedling Leaves

Kyoungwon Cho, Oksoo Han, Shigeru Tamogami, Junko Shibato, Akihiro Kubo, Ganesh Kumar Agrawal, and Randeep Rakwal

Abstract

Jasmonic acid (JA) and salicylic acid (SA) are critical signaling components involved in various aspects of plant growth, development, and defense. Their constitutive levels vary from plant to plant and also from tissue to tissue within the same plant. Moreover, their quantitative levels change when plant is exposed to biotic and abiotic stresses. To better understand the JA- and SA-mediated signaling and metabolic pathways, it is important to precisely quantify their levels in plants/tissues/organs. However, their extraction and quantification are not trivial and still technically challenging. An effort has been made in various laboratories to develop a simple and standard procedure that can be utilized for quantification of JA and SA. Here, we present the experimental procedure and our decade of experience on extracting and quantifying them in an absolute manner in leaves of rice seedlings. We must mention that this method has been applied to both monocotyledonous and dicotyledonous plants for absolute quantification of JA and SA. As collaboration is the key towards rapid progress in science and technology, we are always open to sharing our experience in this field with any active research group with an aim to improve the procedure further and eventually to connect the importance of their (JA and SA) quantitative levels with networks of signaling and metabolic pathways in plants.

Key words: *Oryza sativa*, Jasmonic acid, Salicylic acid, Extraction, Quantification

1. Introduction

Plants have evolved activated defense mechanisms to protect themselves from various biotic and abiotic stresses. Jasmonic acid (JA) and salicylic acid (SA) are the known key mediators of signaling pathways, leading to activation of plant defense systems and responses (1–5). Jasmonic acid is biosynthesized from linolenic acid by subsequent enzyme reaction of 13-lipoxygenase, allene oxide synthase, allene oxide cyclase, 12-oxo-phytodienoic acid reductase, and three β-oxidation steps (6, 7). The SA is known to be synthesized

Yinong Yang (ed.), *Rice Protocols*, Methods in Molecular Biology, vol. 956,
DOI 10.1007/978-1-62703-194-3_13, © Springer Science+Business Media, LLC 2013

via benzoic acid from phenylalanine and via isochorismate from chorismate (8, 9). The production of JA and SA has been reported to be antagonistically inhibited by SA and JA, respectively (1, 10–13). SA or aspirin blocks JA biosynthesis in tomato leaves (14). On the other hand, the SA production is inhibited by JA in the wounded transgenic tobacco plants (15). Moreover, a kinetic analysis of endogenous JA and SA in the wounded rice leaves provided an evidence of a negative cross talk in the production of JA and SA during the early wound response in rice (13), an important monocotyledonous plant model system (16). Therefore, JA and SA quantities are tightly regulated in plants.

The temporal changes in JA and SA levels largely determine the resistance level of plants against biotic and abiotic stresses. For example, ozone (O_3)-induced changes in JA and SA levels were different between O_3-sensitive and -tolerant genotypes in *Arabidopsis*, tobacco, and birch (17). JA and SA also have a protective and promotive function on O_3-induced lesion formation (17). Moreover, endogenous SA in rice was found to be involved in protection from oxidative damage caused by biotic and abiotic stress (18). Most recently, a tight correlation was reported between endogenous JA production due to airborne methyl jasmonate and emission of volatile organic compounds in *Achyranthes*, a wild-plant species (19). Taken together, these and other related studies indicate that quantification of JA and SA in an absolute manner in time and space will further our understanding on the complex signaling and metabolic networks mediated by JA and SA in plants during both plant growth and development and biotic and abiotic stresses.

In this chapter, we describe the experimental protocol, routinely used in our laboratories, for absolute quantification of JA and SA in rice seedling leaves along with our long-term experience in this field. Moreover, we must emphasize that the same procedure has been successfully applied to different tissues, organs, and plants to accurately measure these two signaling components.

2. Materials

2.1. Seeds

1. Dry mature seeds of rice (*Oryza sativa* L.) cv. Nipponbare (see Note 1).

2.2. Seed Treatment

1. Sterilizing solution: Fourfold diluted sodium hypochlorite solution in Milli Q water.
2. Distilled or Milli Q water.

2.3. Seed Germination and Growth Condition

1. Sterile glass beaker (300 mL).
2. Darkroom (25°C).

3. Soil.
4. Seedling pot (14 mm × 5 mm × 10 mm::L × W × H).
5. Growth chamber equipped with white fluorescent light (wavelength 390–500 nm, 150 μmol/m^2 s), temperature (25°C), and relative humidity (70%).

2.4. Sample Preparation and Storage

1. 3rd and/or 4th leaves of 2-week-old rice seedlings.
2. Liquid nitrogen (N_2).
3. Sterile/clean mortar and pestle.

2.5. Preparation of (9,10-$^2H_2$) Dideuterated-Dihydro-JA, Internal Standard of JA

1. JA (Sigma, USA).
2. Methylene chloride.
3. Deuterium (D_2) gas.
4. 5% charcoal of palladium (Pd/C).
5. Three 50 mL wide-neck round-bottomed glass flasks.
6. Rubber balloon.
7. 50 mL Syringe.
8. Glass filter.
9. Rotary evaporator.
10. 100 mL Pear-shaped flask.
11. Septum.
12. Magnetic stirrer and stirrer bar.

2.6. JA Extraction: Classical Method

1. 100 mL Conical flask.
2. Acetone–citric acid solution: Mix 70 mL of 50 mM acetone with 30 mL of citric acid.
3. 250 ng of (9,10-$^2H_2$) Dideuterated-Dihydro-JA (D_2DH-JA) as internal standard (see Note 2).
4. Diethyl ether.
5. Glass separation funnel.
6. Solid-phase extraction cartridge (Waters, MA, USA) containing 360 mg of aminopropyl sorbent.
7. Washing solution: Trichloromethane:2-propanol::2:1; v/v.
8. Elution solution: Diethyl ether:acetic acid::98:2; v/v.
9. Derivative: Sigma-Sil-A trimethylchlorosilane:hexamethyldisilazane:pyridine (1:3:9) reagent (Product No. 1391, Sigma, USA).

2.7. SA Extraction: Classical Method

1. 15 mL Test tube.
2. 70 mM potassium phosphate (pH 7.4).
3. Internal standard: Propyl paraben.
4. Deionized water.

5. 3N hydrochloric acid.
6. Chloroform ($CHCl_3$).
7. Vortexer.
8. Anhydrous sodium sulfate.
9. Derivative: 10 μL Sigma-Sil-A.

2.8. GC-MS Analysis

1. GC (Varian STAR 3400cx, Varian, Walnut Creek, CA, USA).
2. Column (Rtx-1; 30 m × 0.32 mm ID, 0.25 μm film thickness; Product No.10124, Restek, USA).
3. Mass detector (Varian SATURN 4D, Varian, Walnut Creek, CA, USA): Electron Ionization (EI) mode or Selected Ion Monitoring (SIM) mode.

2.9. LC-MS/MS Analysis

LC-MS/MS system: HPLC, Agilent 1100 (Hewlett-Packard, Waldbronn, Germany) with *SCIEX* API-2000 LC-MS/MS (Concord, Ontario, Canada). The column used was Atranthis dC_{18}, 3 μm, 2.1 mm × 15 cm (Waters, MA, USA).

3. Methods

3.1. Seed Treatment

1. Pour 30 mL sterilization solution into a 50-mL falcon tube containing rice seeds.
2. Sterilize the seeds for 30 min with constant shaking on a rotary shaker.
3. Wash the seeds five times with Milli Q water (30 mL) for 5 min each.

3.2. Seed Germination and Growth Condition

1. Pour the sterilized and washed seeds along with Milli Q water into a glass beaker.
2. Wrap the glass beaker with aluminum foil and place it in a darkroom at 25°C.
3. Replace the Milli Q water of the glass beaker everyday for 3 days (see Note 3).
4. Prepare seedling pot with soil, and place it in a tray containing water.
5. Sow each germinated seed in a neat row (4 rows of 10 seeds each) in the soil, and cover them with soil.
6. Place the tray (containing seedling pots) inside a growth chamber.
7. Incubate for 2 weeks, with watering everyday from top of the seedlings and in the tray below.

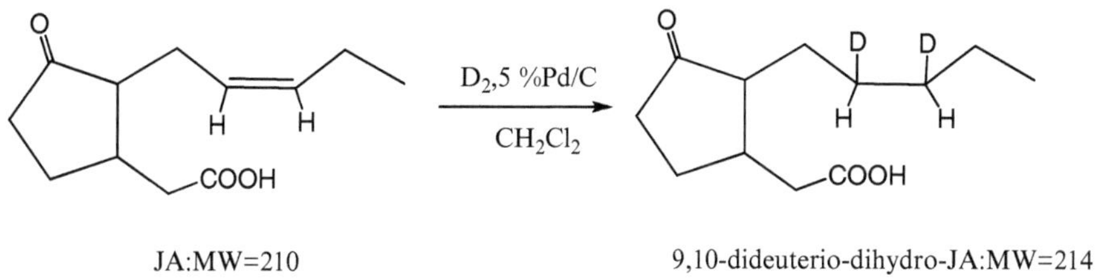

Fig. 1. Structure of JA and (9,10-$^2H_2$) Dideuterated-Dihydro-JA (D_2DH-JA).

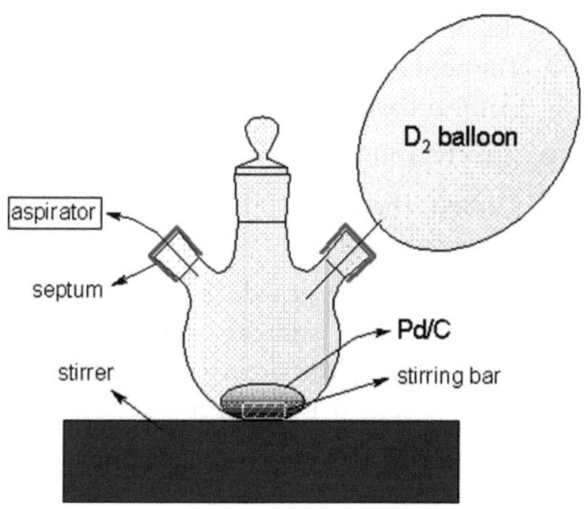

Fig. 2. Preparation of D_2DH-JA, the internal standard for JA.

3.3. Sample Preparation and Storage

Harvest 3rd and/or 4th leaves, freeze immediately in liquid N_2, and store at –80°C (see Note 4).

3.4. Preparation of (9,10-$^2H_2$) Dideuterated-Dihydro-JA (Internal Standard of JA) by Catalytic Deuteration of JA

Here, the D_2DH-JA internal standard is generated by hydrogenation of JA under a deuterium atmosphere using 5% Pd/C as follows (Figs. 1 and 2) (see Note 5).

1. Add 10 mg of 5% Pd/C into the reaction flask (Fig. 2).
2. Evacuate air in the reaction flask using an aspirator (Fig. 2).
3. Flush D_2 gas into the reaction flask slowly and carefully (Fig. 2).
4. Repeat steps 2 and 3 once more.
5. Add 200 mg of JA dissolved in 30 mL CH_2Cl_2 into the reaction flask using a syringe.
6. Stir the reaction mixture gently overnight, followed by filtration with glass filter and celite.
7. Evaporate the filtered solution (D_2DH-JA/CH_2Cl_2) using a rotary evaporator to remove CH_2Cl_2 (see Note 6).
8. Weigh D_2DH-JA and dissolve in carbon tetrachloride to a final concentration of 20 μg/mL.

3.5. JA Extraction: Classical Method

1. Bring out the leaf material from −80°C into a precooled (in liquid N_2) mortar.
2. Grind the leaves to a very fine powder by repeated addition of liquid N_2.
3. Transfer 1 g of powdered leaf sample into a precooled (in liquid N_2) sterile 2 mL microfuge tube (MCT-200; Axygen, CA, USA) and store at −80°C (see Note 7).
4. Take out one tube with 1 g of prepared leaf powder from −80°C and transfer powder to a 100 mL conical flask containing 210 ng D_2DH-JA as internal standard.
5. Immediately add 10 mL of acetone–citric acid solution to the conical flask to resuspend the powder.
6. Stir the solution gently for 12 h at 4°C using a magnetic stirrer.
7. Collect the clear supernatant carefully into a glass beaker and store at 4°C.
8. Resuspend the powder with 10 mL of the same acetone–citric acid solution, stir the solution gently for another 6 h, and transfer the clear supernatant into the same glass beaker.
9. Repeat step 8 but stir the solution only for 3 h followed by transfer of the clear supernatant to the same glass beaker (see Note 8).
10. Evaporate the organic solvent-(~30 mL acetone) of the collected supernatant by leaving the beaker at room temperature in a fume hood or by using a rotary evaporator (see Note 9).
11. Mix the remaining aqueous phase solution with 7 mL of diethyl ether to extract organic phase using a separating funnel (see Note 10).
12. Equilibrate the solid-phase extraction cartridge containing 360 mg of aminopropyl sorben with 10 mL of diethyl ether using a 10 mL glass syringe.
13. Load the equilibrated cartridge with extracted organic phase using a glass syringe.
14. Wash the cartridge with 10 mL of washing solution.
15. Elute bound JA and internal standard with 10 mL of elution solution (see Note 11).
16. Evaporate the eluted solution using a rotary evaporator.
17. Silylate the residue with 10 μL Sigma-Sil-A reagent at 80°C for 5 min.
18. Inject the silylated product (1 μL) into a GC-MS (see Note 12).

3.6. JA Extraction: Improved Method

1. Weigh 1g of plant leaves and extract with 60 mL of acetone (see Note 13).
2. Concentrate the extracted solution to approximately 2 mL using a rotary evaporator.

3. Add known amount of internal standard and 50 mL of Milli Q water or HPLC-grade water (see Note 14).
4. Add 20 mL of $CHCl_3$ to extract the organic phase (see Note 15).
5. Collect the aqueous phase in a rotary flask and repeat step 4 once more (see Note 16).
6. Concentrate the collected organic phase using a rotary evaporator.
7. Dissolve the concentrated residue with 2 mL of 80% methanol.
8. Load the prepared solution onto the SepPak C18 light cartridge (Waters, MA, USA) preequilibrated with 80% MeOH and collect the flow through.
9. Concentrate the flow through solution to 200 μL by passing N_2 gas above the solution (see Note 17).
10. Inject (2 μL out of 200 μL) into the LC-MS/MS system to quantitate JA.

3.7. SA Extraction: Classical Method

1. Bring out the leaf material from −80°C into a precooled mortar with liquid N_2.
2. Grind the leaves to a very fine powder by repeated addition of liquid N_2.
3. Weigh 0.3 g powdered leaf sample in precooled (in liquid N_2) sterile/new 2 mL microtube and store at −80°C (see Note 7).
4. Bring out one tube from −80°C and suspend the fine leaf powder (0.3 g) and internal standard propyl paraben (500 ng) in a 15 mL glass test tube containing 4 mL of 70 mM potassium phosphate (pH 7.4) solution.
5. Add subsequently 1 mL of deionized water, 1 mL of 3 N hydrochloric acid, and 2 mL $CHCl_3$ to the test tube and mix the sample nicely each time.
6. Centrifuge at 2,500 × *g* for 15 min at 4°C.
7. Transfer the chloroform phase to a rotary flask (see Note 18).
8. Evaporate the solvent using a rotary evaporator.
9. Silylate the residue with 10 μL Sigma-Sil-A reagent for 5 min at 80°C.
10. Inject 1 μL of the silylated product into a GC-MS for SA quantification (see Note 19).

3.8. GC-MS Analysis of Extracted JA and SA by Classical Method

1. Inject 1 μL TMS-derivatized standard (JA, D_2DH-JA, SA, or PP) into GC-MS.
2. Set the splitless mode on GC and equip with an Rtx-1 column.
3. Set injector and detector temperatures to 200 and 280°C, respectively.

4. Program column temperature to a step gradient as follows: 80°C for 10 min, 80–150°C at 2°C/min, 150°C for 1 min, 150–270°C at 25°C/min, and 270°C for 9 min.
5. Collect spectra of TMS-derived standard in EI mode (Fig. 3) (see Note 20).
6. Repeat steps 1–4 once more, and collect spectra in the SIM mode based on its fragmentation pattern in EI mode (Fig. 4) (see Note 21).
7. Inject 1 μL TMS-derivatized sample into GC-MS for quantification of JA and SA in the leaf sample.
8. Collect the spectra in the SIM mode to enhance sensitivity and selectivity of the analysis (see Note 21).
9. Calculate the amounts of endogenous JA and SA in the leaf sample by comparing the peak areas of JA and SA with the corresponding internal standards.

3.9. LC-MS/MS Analysis of the Extracted JA Sample Using Improved Method

1. Inject 2 μL of each internal standard (0.1 ppm in MeOH) into the LC-MS/MS system to ensure its sensitivity (see Note 22).
2. Collect the spectra in multiple ion monitoring method (MRM) that is used for the analysis and monitoring ion pairs determined by synthetic standards as follows: JA = m/z 209/59; 9,10-Dehydro-JA (as an internal standard) = m/z 207/59; and dihydro-JA and D_2-JA = m/z 211/59 (Fig. 5).
3. Inject 2 μL of the sample into the LC-MS/MS system (see Note 23).
4. Perform data analysis using PE SCIEX MassChrome Version 1.1.1.

4. Notes

1. Dry mature seeds are stored at 10°C in case of long-term storage.
2. Internal standard is prepared by catalytic deuteration of JA and dissolved in carbon tetrachloride.
3. If seeds do not germinate after 3 days, one should extend the incubation for another 1 or 2 days or till seeds germinate.
4. The leaves can also be used for any treatment experiment before freezing them in liquid N_2.
5. An appropriate internal JA or SA standard of known quantity is always necessary to mix with the sample material in the beginning of extraction procedure for their absolute quantification.

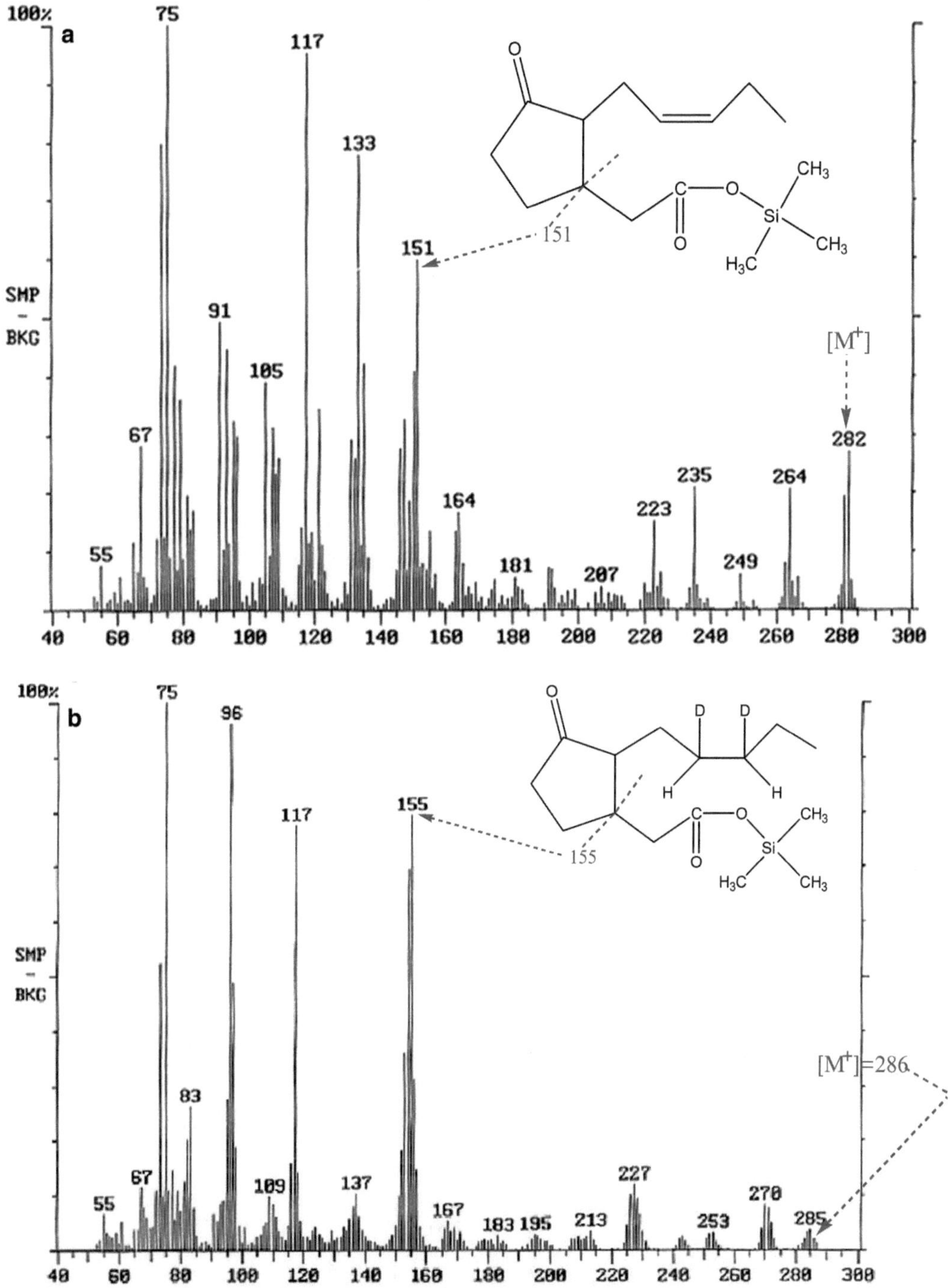

Fig. 3. Mass spectra of (**a**) TMS-JA, (**b**) TMS-D_2DH-JA, (**c**) TMS-SA, and (**d**) TMS-PP in EI mode.

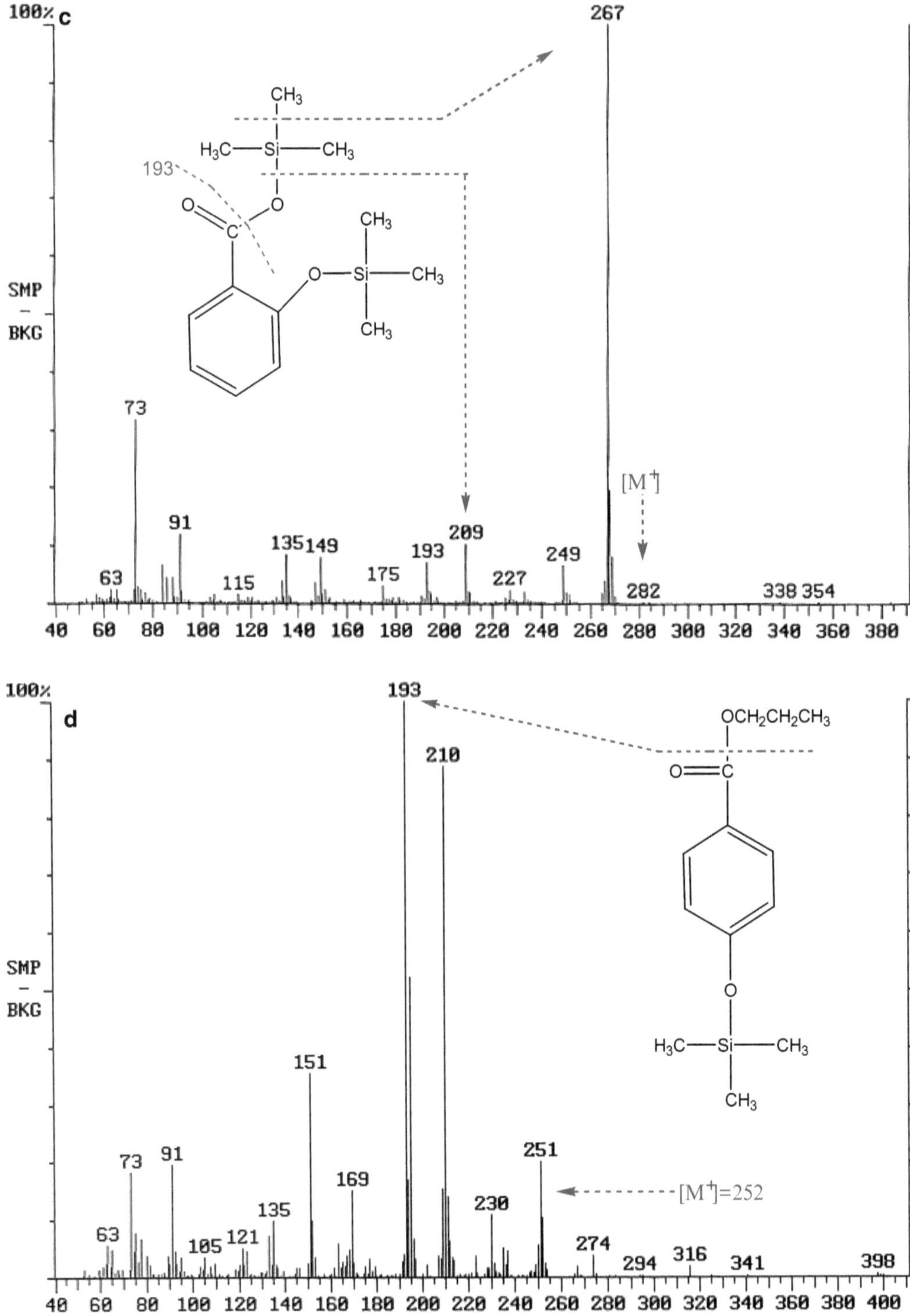

Fig. 3. (continued)

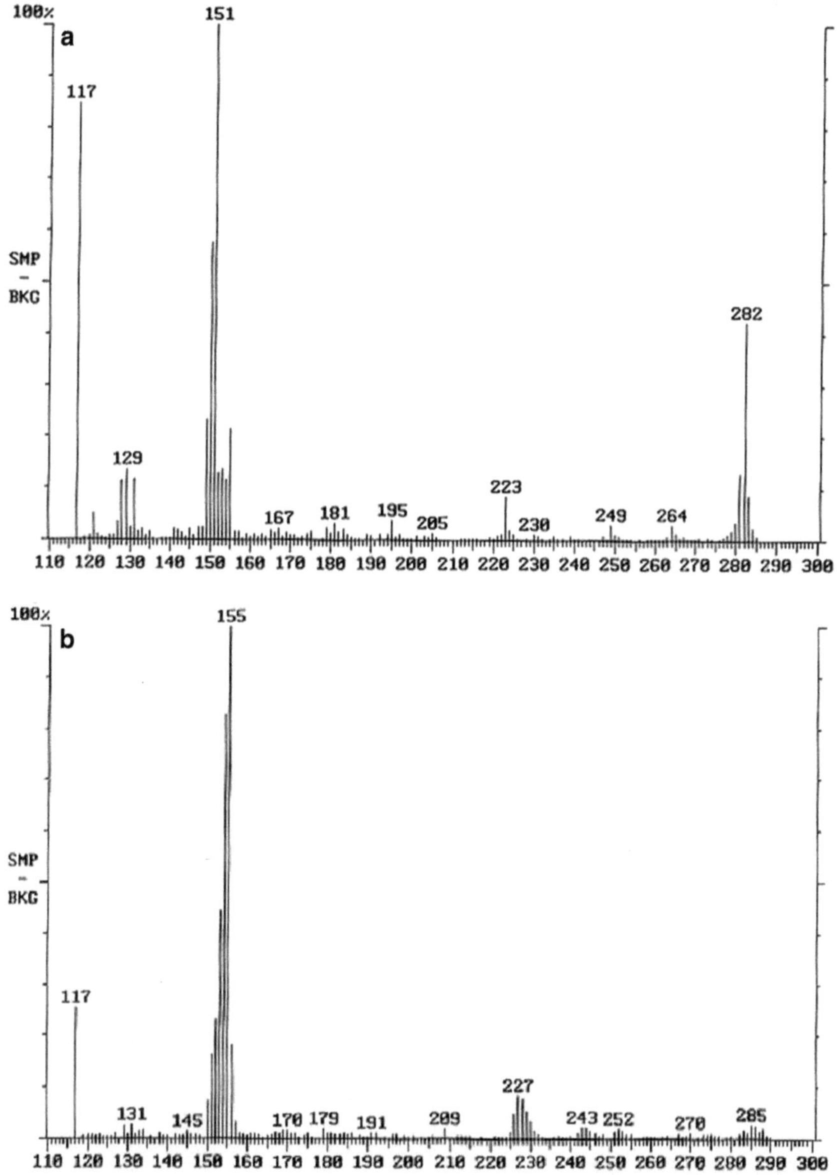

Fig. 4. Mass spectra of (**a**) TMS-JA, (**b**) TMS-D_2DH-JA, (**c**) TMS-SA, and (**d**) TMS-PP in SIM mode. Based on their fragmentation pattern in EI mode, ion values of 117, 151, 282 for JA, 117, 155, 286 for D_2DH-JA, 193, 267, 282 for SA, and 193, 252 for PP are selected.

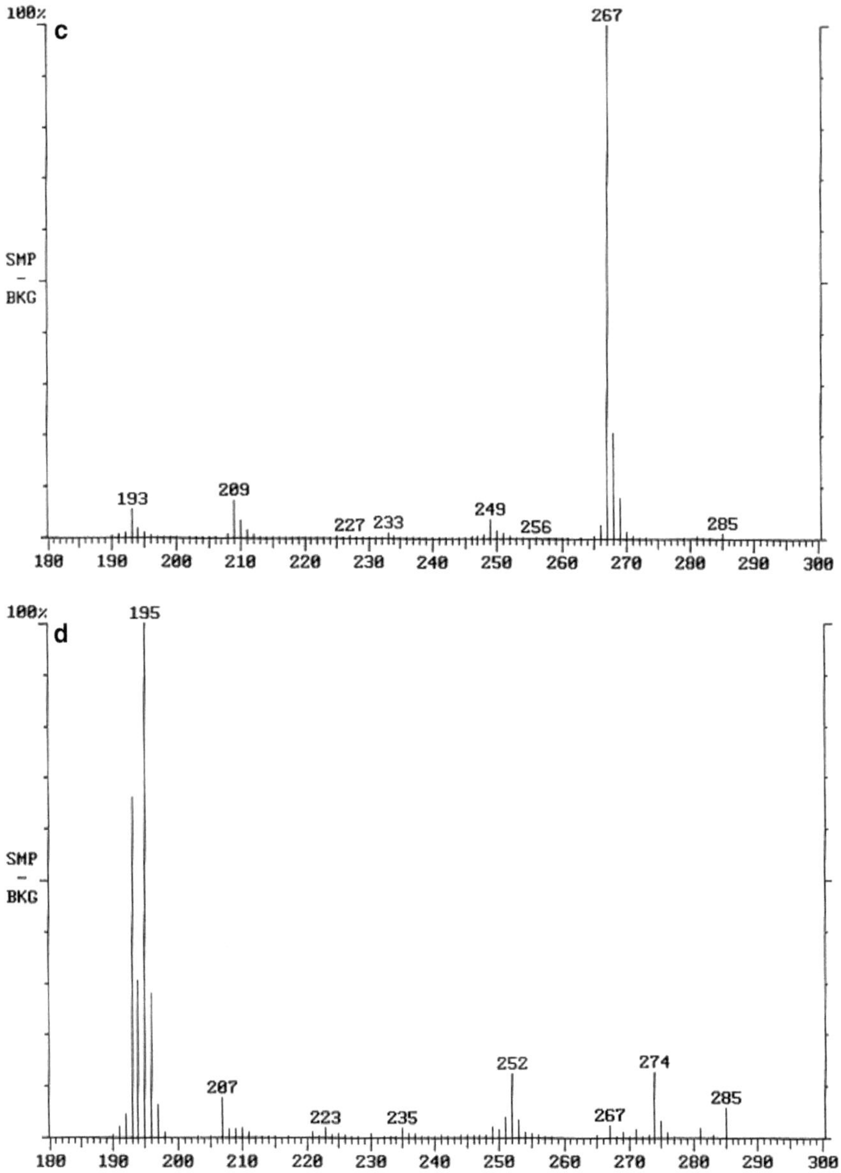

Fig. 4. (continued)

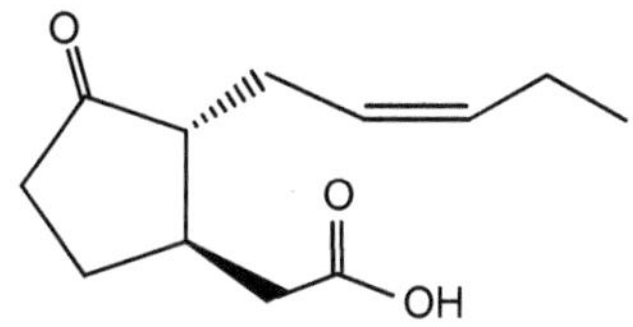

JA (*m/z* 209/59)

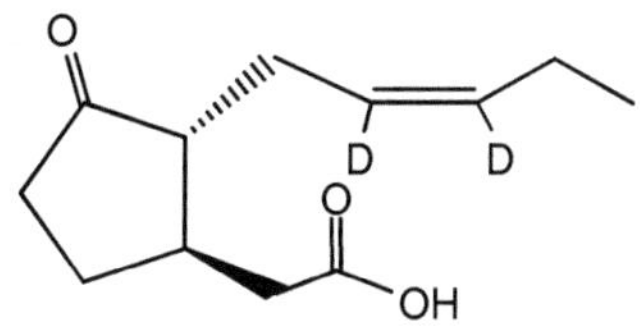

d_2-JA (*m/z* 211/59)

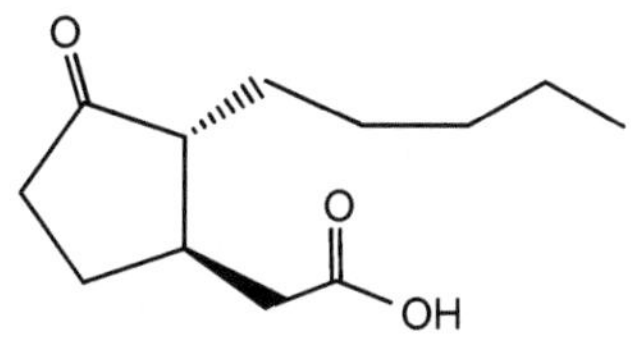

dihydro-JA (*m/z* 211/59)

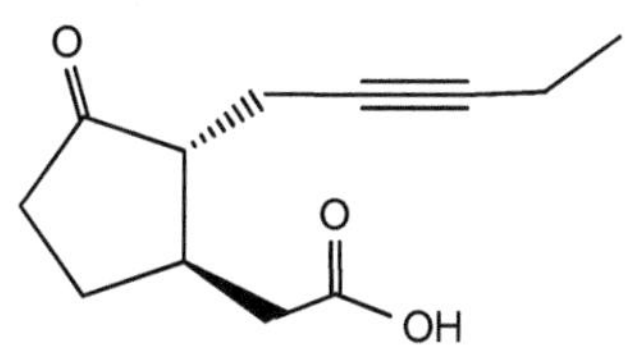

dehydro-JA (*m/z* 207/59)

Fig. 5. Structures of jasmonates with monitoring ions (*m/z* in the parentheses) in MS/MS analysis.

6. To check the complete reduction of the double bond of JA (Fig. 1), the residue is silylated with Sigma-Sil-A and then confirmed by GC-MS analysis.
7. Please make sure that the samples are not thawed, and keep the remaining powdered sample in aliquot of 1.0 or 0.3 g at −80°C.
8. The filter paper (such as Whatman Filter) is used to filter 30 mL of the collected supernatant.
9. Sample caused from bubbling should be noted when acetone is evaporated from the collected solution using a rotary evaporator.
10. The organic phase is extracted three times with diethyl ether (7 mL) from the aqueous phase in a separating funnel.

11. Please control the elution rate according to the manufacturer's instruction accompanying the cartridge.
12. After silylation, the silylated product is immediately injected into GC-MS for analysis.
13. Grind leaves gently with pestle.
14. Dihydro-JA, dehydro-JA, or deuterated-JA (19, 20) can be used as internal standards according to the samples.
15. The pH is critical here and it should be pH 3.0.
16. Please be very careful not to take the interphase/aqueous phase.
17. Jasmonate in the samples is stable at 4°C for many months.
18. When water is still present in the $CHCl_3$ phase, it can be removed by adding anhydrous sodium sulfate.
19. Immediately inject the silylated product into GC-MS.
20. After injection of TMS-derivatized standard to GC-MS, please check the retention time of its GC peak based on its mass spectra in EI mode.
21. Based on their fragmentation pattern in EI mode, ion values of 117, 151, 282 for JA, 117, 155, 286 for D_2DH-JA, 193, 267, 282 for SA, and 193, 252 for PP are selected in SIM mode. Please check the retention time of GC peak of TMS-derivatized standards in both SIM and EI modes.
22. D_2-JA appears at almost same position as JA, but dihydro-JA and dehydro-JA appear before JA. If samples contain impurities that inhibit reappearance in retention times of jasmonates, deuterated jasmonates should be used. D_2-JA and dehydro-JA acid (used as an internal standard) are prepared according to a previous method (19, 21). Analysis parameters should be optimized using synthetic standards.
23. A general ODS-C18 column (2.1 mm × 15 cm) can be used (flow rate of 0.2 mL/min with aqueous 80% MeOH). Formic acid or 0.1% acetic acid can be added to the elutant to sharpen the peaks, but this will decrease the sensitivity for jasmonates.

Acknowledgments

This work was supported by Environment Research and Technology Development Fund (A-0806) of Ministry of Environment, Japan. K.C. was supported at NIES as an Eco-Frontier Fellow (09-Ba086-02). OH appreciates the support of the Korea Research Foundation (2008-313-F00020). GKA and RR greatly appreciate the vision and direction of Prof. V. P. Agrawal (Founder RLABB, Nepal) in their research. GKA appreciates Japan Society for the Promotion of Science (JSPS; ID Number S-10182) for research at NIAS and collaborations therein with ST and RR. RR acknowledges the great support of Professors Yoshihiro Shiraiwa (Chairperson, Faculty of Life and Environmental Sciences, University of Tsukuba) and Seiji Shioda and Dr. Tetsuo Ogawa (Department of Anatomy I, Showa University School of Medicine) in promoting interdisciplinary research and unselfish encouragement.

During the preparation of this manuscript, an advanced method for jasmonates analysis has been reported (22). Moreover, a comprehensive review including recent techniques for jasmonates analysis has also been recently published (23).

References

1. Reymond P, Farmer EE (1998) Jasmonate and salicylate as global signals for defense gene expression. Curr Opin Plant Biol 1:404–411
2. Agrawal GK, Tamogami S, Han O, Iwahashi H, Rakwal R (2004) Rice octadecanoid pathway. Biochem Biophys Res Comm 317:1–15
3. Wasternack C (2007) Jasmonates: an update on biosynthesis, signal transduction and action in plant stress responses, growth and development. Ann Bot 100:681–697
4. Vlot AC, Dempsey DA, Klessig DF (2009) Salicylic acid, a multifaceted hormone to combat disease. Annu Rev Phytopathol 47: 177–206
5. Smith JL, De Moraes CM, Mescher MC (2009) Jasmonate- and salicylate-mediated plant defense responses to insect herbivores, pathogens and parasitic plants. Pest Manag Sci 65: 497–503
6. Vick BA, Zimmerman DC (1984) Biosynthesis of jasmonic acid by several plant species. Plant Physiol 75:458–461
7. Hamberg M, Gardner HW (1992) Oxylipin pathway to jasmonates: biochemistry and biological significance. Biochim Biophys Acta 1165:1–18
8. Coquoz J, Buchala A, Metraux JP (1998) The biosynthesis of salicylic acid in potato plants. Plant Physiol 111:1095–1101
9. Wildermuth MC, Dewdney J, Wu G, Ausubel FM (2001) Isochorismate synthase is required to synthesize salicylic acid for plant defense. Nature 414:562–565
10. Van Poecke RMP, Dicke M (2002) Induced parasitioid attraction by *Arabidopsis thaliana*: involvement of the octadecanoid and the salicylic acid pathway. J Exp Bot 53: 1793–1799
11. Spoel SH, Koornneef A, Claessens SMC, Korzelius JP, Van Pelt JA, Mueller MJ, Buchala AJ, Metraux JP, Brown R, Kazan K, Van Loon LC, Dong X, Pieterse CMJ (2003) NPR1 modulates cross-talk between salicylate- and jasmonate-dependent defense pathways through a novel function in the cytosol. Plant Cell 15:760–770
12. Traw MB, Bergelson J (2003) Interactive effects of jasmonic acid, salicylic acid, and gibberellin on induction of trichomes in Arabidopsis. Plant Physiol 133:1–9
13. Lee A, Cho K, Jang S, Rakwal R, Iwahashi H, Agrawal GK, Shim J, Han O (2004) Inverse correlation between jasmonic acid and salicylic acid during early wound response in rice. Biochem Biophys Res Comm 318:734–738
14. Pena-Cortes H, Albrecht T, Prat S, Weiler EW, Willmitzer L (1993) Aspirin prevents wound-induced gene expression in tomato leaves by blocking jasmonic acid biosynthesis. Planta 191:123–128
15. Sano H, Seo S, Koizumi N, Niki T, Iwamura H, Ohashi Y (1996) Regulation by cytokinins of endogenous levels of jasmonic and salicylic acids in mechanically wounded tobacco plants. Plant Cell Physiol 37:762–769
16. Agrawal GK, Rakwal R (2006) Rice proteomics: a cornerstone for cereal food crop proteomes. Mass Spectrom Rev 25:1–53
17. Kangasjärvi J, Jaspers P, Kollist H (2005) Signalling and cell death in ozone-exposed plants. Plant Cell Environ 28:1021–1036
18. Yang Y, Qi M, Mei C (2004) Endogenous salicylic acid protects rice plants from oxidative damage by aging as well as biotic and abiotic stress. Plant J 40:909–919
19. Tamogami S, Rakwal R, Agrawal G (2008) Interplant communication: airborne methyl

jasmonate is essentially converted into JA and JA-Ile activating jasmonate signaling pathway and VOCs emission. Biochem Biophys Res Commun 376:723–727

20. Rakwal R, Tamogami S, Agrawal G, Iwahashi H (2002) Octadecanoid signaling component "burst" in rice (*Oryza sativa* L.) seedling leaves upon wounding by cut and treatment with fungal elicitior chitosan. Biochem Biophys Res Commun 295:1041–1045
21. Büchi G, Egger B (1971) A new synthesis of cyclopentenones. Methyl jasmonate and jasmine. J Org Chem 36:2021–2023
22. Tamogami S, Agrawal GK, Rakwal R (2010) An in planta technique for cis-/trans-stereochemical analysis of jasmonoyl isoleucine. J Plant Physiol 167:933–937
23. Tamogami S, Rakwal R, Agrawal GK (2011) Jasmonates to Jasmolites in Plants: Past, Present, and Future. Adv Bot Res 60:309–348

Chapter 14

Analysis of Insect-Induced Volatiles from Rice

Nan Zhao, Xiaofeng Zhuang, Gitika Shrivastava, and Feng Chen

Abstract

Plants emit volatile organic compounds (VOCs) from various organs. To understand the biological and ecological roles of plant VOCs as well as their biosynthesis, it is important to determine the quality and quantity of individual plant VOCs. Here we describe three techniques, including solid-phase microextraction, closed-loop stripping system, and an open dynamic system, for sampling volatiles emitted from insect-damaged rice plants. In addition, the protocol for using gas chromatography-mass spectrometry to elucidate the chemical identities of rice volatiles is described.

Key words: Volatile, Headspace sampling, SPME, Closed-loop stripping, Rice, Insect, GC-MS

1. Introduction

Plants synthesize a large array of low-molecular-weight secondary metabolites. Some of these compounds have high vapor pressures and low water solubilities, and therefore can easily volatilize at normal atmosphere. They are collectively called volatile organic compounds (VOCs). Plant VOCs play important roles in many biological/ecological processes (1, 2). Floral VOCs are involved in attracting pollinators. VOCs emitted from plants damaged by insects can function as a direct defense because of their toxicity or as an indirect defense as cues to attract natural enemies of the attacking insects (3). In addition, VOCs emitted from insect-damaged plants may act as intra-plant signals to prime defense responses of nearby healthy plants (4). Plant VOCs are often a complex mixture of compounds derived from multiple biochemical pathways (5). To understand the roles of plant VOCs in specific biological/ecological processes, especially the relevance of individual volatile compounds, it is important to determine their chemical identities and emission rates.

Yinong Yang (ed.), *Rice Protocols*, Methods in Molecular Biology, vol. 956,
DOI 10.1007/978-1-62703-194-3_14, © Springer Science+Business Media, LLC 2013

Chemical analysis of plant VOCs involves volatile sampling, separation, and identification (6). Different methods have been developed and used for volatile isolation. These include solvent extraction, steam distillation, and headspace techniques (6). Both solvent extraction and steam distillation have been widely used in the perfume industry. However, these two methods are liable to produce artifacts. In addition, they are destructive in nature. Therefore, they are not usually preferred for studying biological and ecological functions of plant VOCs. Headspace sampling is a nondestructive technique, which can be used on living plants either in the laboratory or in the field, and sampling can be repeated in time on the same individual. The headspace techniques have been widely used in plant volatile studies in the past 10–15 years (7).

The headspace techniques can be divided into two types: the static and dynamic methods. In the static method, the plants analyzed are enclosed in a container without air circulation. Volatiles are enriched on the absorbing matrix. Solid-phase microextraction (SPME) is one important advance in static headspace analysis. It is based on the absorption and thermal desorption of volatiles from an inert fiber. The SPME method is fast and simple. However, it also has several disadvantages. For instance, accurate quantification of plant VOCs using SPME is often difficult. In addition, repeated injections of the same sample with SPME are impossible. Therefore, the SPME method is best suited for qualitative analysis of plant VOCs at a single time point (7).

In dynamic headspace sampling, a continuous airstream flows through the sample container and the VOCs are trapped on adsorbents. This allows for the collection of VOCs in amounts sufficient for detection or even structure elucidation. Dynamic headspace sampling may be conducted in either closed or open systems, depending on whether the carrier gas circulates within the plant sample container. One best example of the closed dynamic headspace systems is the closed-loop stripping method (8). The closed-loop stripping minimizes trapping of air contaminants and is especially suited for low-emitting plants. The disadvantages of the closed-loop stripping method are related to the accumulation of humidity and other volatiles such as ethylene, which may affect the physiology of the analyzed plants. Therefore, results generated with the closed-stripping method should be compared with those from an open dynamic system to ensure that they are valid (7).

In open dynamic headspace sampling, the carrier gas continuously flows into the container from outside and flows out the container with released plant VOCs, which are trapped at the outlet. To avoid air contaminants, the carrier gas is often purified before flowing into the sample container. The open system avoids problems associated with the closed system such as humidity buildup. Certain open systems may be used to sample volatiles emitted from individual organs of a plant nondestructively. The open dynamic system is the most widely used among all headspace sampling techniques (7).

In addition to sampling techniques, the progress of plant VOCs research has also benefited from advances in instrumentation for chemical structural elucidation, especially gas chromatography-mass spectrometry (GC-MS) (6). In addition to GC-MS, other instruments may be used for chemical identification, such as infrared (IR) or nuclear magnetic resonance (NMR) spectroscopy (9, 10); however they require high amounts of volatiles.

Detailed chemical analysis of plant VOCs will aid in the investigation of volatile biosynthesis, especially in model plant species whose genome has been fully sequenced, such as rice. Association study of gene expression with volatile production can lead to the identification of candidate genes of volatile synthesis. We have previously used an integrated genomics approach to identify genes for synthesis of insect-induced volatiles from rice plants (3). In this study, rice volatiles were sampled using an open dynamic system, which is described in greater detail in this chapter. In addition, we present protocols for using SMPE and closed-loop stripping method to sample insect-induced rice volatiles. Lastly, we describe chemical identification of rice VOCs using GC-MS.

2. Materials

2.1. Rearing Insects

1. General purpose lepidoptera artificial diet (Bio-serv Inc.).
2. Fall armyworm (*Spodoptera frugiperda*) eggs, which can be purchased from commercial sources such as Benzon Research (http://www.benzonresearch.com/insectpurchase.htm).
3. Insect rearing trays (Bio-serv Inc.).
4. Incubator (Fisher scientific).

2.2. Insect Feeding of Rice Plants

1. Rice seeds (*Oryza sativa* spp. japonica CV. Nipponbare).
2. Filter paper.
3. Soil.
4. Glass jars of 200 mL (Sigma).
5. Growth chamber.
6. Fine-haired paint brush.
7. Insect tent with dimensions of L60 × W60 × H60 cm.

2.3. Headspace Sampling Using SPME

1. SPME fiber assembly (Supelco).
2. A 5-L bell jar with lid and Teflon stopper.
3. Test tube holder.

2.4. Headspace Sampling Using Close-Loop Stripping System

1. A 5-L bell jar with lid.
2. Charcoal trap.
3. Furgut circulation pump.
4. Teflon stopper with holes for pump adaptors.
5. Steel column containing a house for volatile trap.
6. Liquid nitrogen.
7. Methylene chloride (Fisher Scientific).
8. Internal standard 1-Octanol (Sigma).
9. Glass pipette.
10. Glass inserts with bottom spring for 2 mL vial (Supelco).
11. 2 mL glass vials (Supelco).

2.5. Headspace Sampling Using an Open Dynamic System

1. Air delivery system composed of charcoal filter and flowmeter (Analytical Research Systems).
2. Pyrex glass volatile collection chamber and removable O-ring snap lid.
3. Teflon tube.
4. SuperQ volatile collection trap (Analytical Research Systems).
5. Glass inserts with bottom spring for 2 mL vial (Supelco).
6. 2 mL glass vials (Supelco).
7. Methylene chloride (Fisher Scientific).
8. Internal standard 1-Octanol (Sigma).

2.6. Chemical Identification Using GC-MS

1. A GC-MS such as Shimadzu 17A gas chromatograph coupled to a Shimadzu QP5050A quadrupole mass selective detector.
2. GC autosampler.
3. Ultrapure helium gas.
4. A capillary column such as Restek SHR5XLB column.
5. Authentic chemical standards.

3. Methods

3.1. Insect Feeding of Rice Plants

1. Germinate rice seeds on moistened filter paper at 30°C for 2 days.
2. Transfer ten germinated rice seeds from filter paper to soil in each glass jar.
3. Grow rice plants in a controlled-climate chamber at 26°C with 16 h of light for 2 weeks.
4. Cut cheese cloth containing fall armyworm eggs into 1–2 cm^2 pieces. Put the pieces in insect rearing trays containing artificial

diet. Incubate the trays in a 28°C incubator until the larvae reach the two-instar stage (see Note 1).

5. Use a fine-haired paint brush to gently transfer two larvae onto each rice plant (see Note 2).
6. Allow insect feeding to proceed overnight before insects are removed from rice plants. The damaged rice plants are ready for volatile profiling.

3.2. Volatile Sampling Using SPME

1. Place one glass jar containing insect-damaged rice plants into a 5-L bell jar. Put on the lid and insert the teflon stopper (Fig. 1a).
2. Align the SPME with the teflon stopper and stabilize SPME to a test tube holder (see Note 3).
3. Push the plunger of the SPME device to expose the SPME fiber to the headspace of rice plants in the bell jar.
4. Continue volatile collection for 1 h (see Note 4).
5. Retract the fiber and remove the SPME from the top of the plant sample container to terminate volatile collection. Proceed with immediate injection of the SPME fiber into the injection port of a GC.

3.3. Volatile Sampling Using Closed-Loop Stripping System

1. Place one glass jar containing insect-damaged rice plants into a 5-L bell jar. Put on the lid and insert the teflon stopper (Fig. 1b).
2. Put a clean charcoal trap (see Note 5) into the steel housing and fasten the steel column onto the circulation pump. Insert the steel columns, which are connected to the circulation pump, into the bell jar through the teflon stopper.
3. Connect the wire out of the circulation pump to a power supply. Turn on the power and start volatile collection (see Note 6).
4. Collect volatiles for 4 h (see Note 7). Turn off the power and remove circulation pump connected to the steel columns from the bell jar. Open the steel housing to retrieve the charcoal trap.
5. Connect the charcoal trap to a glass insert through a short piece of teflon tube whose outer diameter is the same as the inner diameter of the charcoal trap and the glass insert. Take 50 μL of methylene chloride containing an internal standard (see Note 8) using a glass pipette and add it to the charcoal trap (see Note 9). Place the glass insert into liquid nitrogen for a short time (i.e., a few seconds) and then take it out and use fingers to warm the glass insert to drive the organic solvent to flow backwards. Repeat this procedure five times and then leave the glass insert in liquid nitrogen long enough (10 s or longer) to make sure that all the methylene chloride flows into the glass insert.

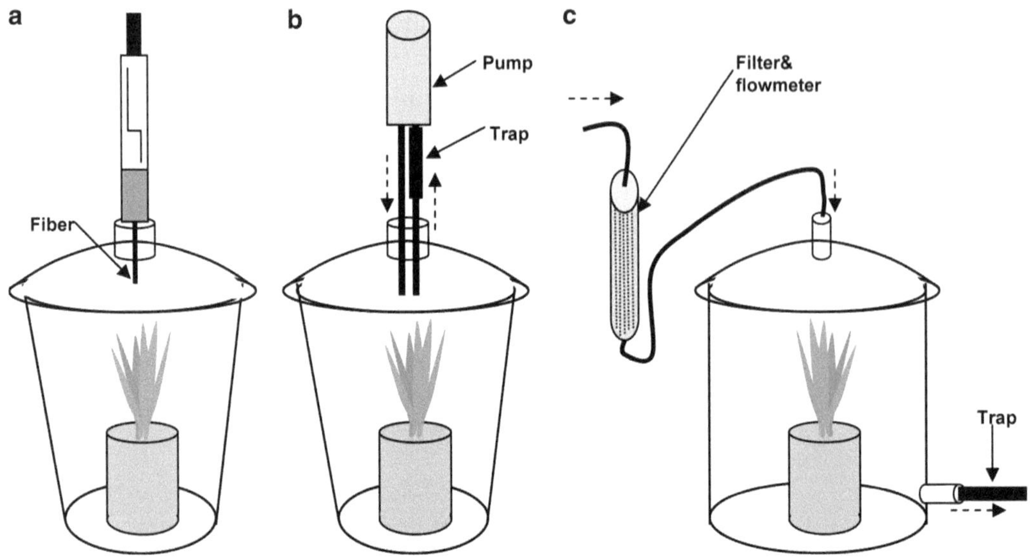

Fig. 1. Plant volatile sampling systems. (**a**) An SPME system. "Fiber" depicts the adsorbent-coated fiber of SPME. (**b**) A closed-loop stripping system. "Pump" and "Trap" depict the circulation pump and the volatile trap, respectively. (**c**) An open dynamic system. "Filter" and "Trap" depict the charcoal filter and the volatile trap, respectively. In both (**b**) and (**c**), the *broken arrows* indicate the directions of airflow.

3.4. Volatile Sampling Using an Open Dynamic System

1. Place one glass jar containing insect-damaged rice plants into a pyrex glass volatile collection chamber. Put on the removable O-ring snap lid (Fig. 1c).
2. Connect the outlet of air source to the inlet of the air delivery system using teflon tube (see Note 10).
3. Connect the outlet of the air delivery system to the inlet of the volatile collection chamber.
4. Place a SuperQ volatile trap at the outlet of the volatile collection chamber.
5. Start the airflow and adjust the flow rate to 1 L/min (see Note 11).
6. Continue volatile collection for 4 h (see Note 12).
7. Remove the volatile trap. Add 50 μL of methylene chloride containing an internal standard to elute the volatiles into a glass insert.

3.5. Chemical Identification of Rice Volatiles Using GC-MS

1. For volatiles collected using SPME, insert the SPME fiber directly into the injection port of the GC.
2. For analysis of volatiles collected using either closed-loop stripping or an open dynamic system, inject 2 μL of the sample into the injection port of the GC using an autosampler.
3. Use helium as the carrier gas at a flow rate of 5 mL/min. Apply a temperature gradient of 5°C/min from 40°C to 240°C. Use

a Restek SHR5XLB column (30 m × 0.25 mm i.d. × 0.25 μm thickness) for separation (see Note 13).

4. Identify the volatiles first by similarity search against commercial chemical libraries such as the NIST library. Then confirm identities of the volatiles by comparing the retention time and mass spectrum with authentic standards if they are available (see Note 14). The quantity of the volatiles can be calculated based on the internal standard.

4. Notes

1. It takes about 1 week for the hatched FAW eggs to reach two-instar larvae stage. Depending on the developmental stage of the rice plant that is going to be analyzed, egg hatching of FAW should be coordinated with rice plant cultivation. So the proper developmental stages of both rice and FAW can be satisfied at the time of feeding experiments.
2. FAW is a nocturnal insect. To mimic natural conditions, place insects on rice plants in late afternoon and allow feeding to proceed overnight.
3. Depending on the chemical properties of the volatiles to be analyzed, different types of SPME fibers should be considered.
4. The collection time will be based on the abundance of the volatiles, which may vary from minutes to hours or even days.
5. To clean the charcoal filter, wash the filter with methylene chloride:methanol (1:3), methylene chloride, and acetone. Each wash should be done three times. Then the filters should be dried in oven (80°C) for 3–4 h or overnight.
6. One important note for operating the circulation pump is to keep the current high but the voltage low. 8–9 V are routinely used in our experiments.
7. The collection time will be based on the abundance of the volatiles, which may vary from minutes to hours or even days.
8. 1-Octonal is routinely used as an internal standard in our analysis of rice volatiles. If the chosen compound is produced by the host plant, then a different compound should be used as an internal standard. The purpose of adding internal standard is for quantification.
9. It is important to handle organic solvent using glass pipettes. Using plastic tips to transfer organic solvents may cause contamination.
10. The air delivery system contains a charcoal filter that will purify air before it flows into the glass volatile collection chamber. This will avoid air contaminates.

11. The flow rate of the air can be adjusted.
12. The collection time will be based on the abundance of the volatiles, which may vary from minutes to hours or even days.
13. Depending on the chemical properties of the volatiles to be analyzed, different types of capillarity columns should be considered.
14. The chemical standards may be purchased from commercial sources such as Sigma or self-prepared if not commercially available.

Acknowledgments

The work described was supported by the Experimental Station of Institute of Agriculture, University of Tennessee.

References

1. Pichersky E, Gershenzon J (2002) The formation and function of plant volatiles: perfumes for pollinator attraction and defense. Curr Opin Plant Biol 5:237–243
2. Dudareva N, Negre F, Nagegowda DA, Orlova I (2006) Plant volatiles: recent advances and future perspectives. Crit Rev Plant Sci 25:417–440
3. Yuan JS, Köllner TG, Wiggins G, Grant J, Degenhardt J, Chen F (2008) Molecular and genomic basis of volatile-mediated indirect defense against insects in rice. Plant J 55:491–503
4. Ton J, D'Alessandro M, Jourdie V, Jakab G, Karlen D, Held M, Mauch-Mani B, Turlings TCJ (2006) Priming by airborne signals boosts direct and indirect resistance in maize. Plant J 49:16–26
5. Dudareva N, Pichersky E, Gershenzon J (2004) Biochemistry of plant volatiles. Plant Physiol 135:1893–1902
6. Knudsen JT, Tollsten L, Bergstrom LG (1993) Floral scents: a checklist of volatile compounds isolated by headspace techniques. Phytochemistry 33:253–280
7. Tholl D, Boland W, Hansel A, Loreto F, Röse US, Schnitzler JP (2006) Practical approaches to plant volatile analysis. Plant J 45:540–560
8. Donath J, Boland W (1995) Biosynthesis of acyclic homoterpenes: enzyme selectivity and absolute configuration of the nerolidol precursor. Phytochemistry 39:785–790
9. Bergstrom J, Bergstriim G (1989) Floral scents of *Bartsia alpina* (Scrophulariaceae): chemical composition and variation between individual plants. Nordic J Bot 9:363–365
10. Borg-Karlson A-K (1987) Chemical basis for the relationship between Opkrys orchids and their pollinators. Chem Scr 27:313–325

Chapter 15

Phenotypic and Physiological Evaluation for Drought and Salinity Stress Responses in Rice

Utlwang Batlang, Niranjan Baisakh, Madana M.R. Ambavaram, and Andy Pereira

Abstract

Drought and salinity stresses seriously affect rice plant growth and yield. The growing need to improve rice cultivars for drought and salt tolerance requires the development of reproducible screening methods that simulate field conditions, and which provide quantitative data for statistical testing and selection of genotypes with differential responses. In addition, the study of molecular responses to drought and salt stress requires controlled conditions for growth and treatments that are reportable and comparable between different laboratories. Drought, also known as soil water deficit, can result from insufficient moisture for a plant to grow adequately and complete its life cycle. Salinity due to excess sodium chloride affects rice at seedling and flowering stages, reducing root and leaf growth. Both these abiotic stresses can lead to major physiological and biochemical changes such as reduced photosynthesis and reprogramming of gene expression. The methods presented in this chapter can be applied for (a) examination of stress responses in rice vegetative and reproductive tissues to identify and characterize molecular and physiological responses; (b) testing of candidate genes by overexpression or knockout studies evaluated for altered stress response phenotypes; and (c) screening of different genotypes such as accessions or segregating populations for their quantitative responses to abiotic stress parameters.

Key words: Drought, Grain yield, Photosynthesis, Salt tolerance, Spikelet fertility, Stomatal conductance, Transgenic plant, Transpiration, Water use efficiency

1. Introduction

Abiotic stresses such as drought and salinity can affect the cellular water potential causing similar mechanistic responses (1). These stresses can cause several similar molecular and physiological responses and the genotypes often provide resistance or tolerance to these stresses simultaneously.

Drought, also known as water deficit, can result from insufficient moisture for a plant to grow adequately and complete its life cycle. Insufficient moisture can be the consequence of a shortage in rainfall,

Yinong Yang (ed.), *Rice Protocols*, Methods in Molecular Biology, vol. 956,
DOI 10.1007/978-1-62703-194-3_15,

coarse-textured soils that retain little water in the root zone, or drying winds (2). Drought stress is one of the most important environmental factors limiting the growth and productivity of economically important plants (3). The effects of drought in plants such as rice are manifested in reduction of biomass, photosynthetic activity, and ultimately the yield. The reduction in yield is mostly due to failure of reproductive development caused by reduced spikelet number (SP) or fertility of spikelets (FS) (4).

In crop production, genetic improvement for drought adaptation is addressed implicitly, or indirectly, by selecting for yield stability over locations and years (5). Despite efforts to improve major crops for drought resistance by traditional breeding, success has been limited. However, advances in molecular approaches hold promise for physiological and breeding research on drought avoidance and tolerance in crop plants, and it is now possible to use these approaches to understand drought sensitivity and improve drought resistance of important crops. Transcriptome analyses and other molecular techniques have been used to identify drought-response mechanisms in rice and other crops. While these identified many candidate genes, it was without correlation to physiological responses. Precise physiological data, which precedes molecular data, is increasingly becoming important for predicting phenotypic differences in breeding programs. Therefore, it is important to develop methods that can allow determination of rice physiological responses to correlate them with molecular data.

Salinity impedes growth and development of crop plants. Salinity tolerance, like responses to any other abiotic stresses, is a complex phenomenon. Rice is naturally salt sensitive and manifests an array of morphological symptoms apart from the physiological and biochemical changes under salt stress. Breeding for salt-tolerant rice to grow in salt-afflicted soils is indispensable to sustainable agriculture. A successful breeding program requires an efficient and reliable technique for identifying salt tolerance in a large number of lines. Lack of reliable large-scale field screening techniques seems to be the biggest problem in genetic improvement of salt tolerance of crop plants. This is confounded by the spatial heterogeneity of soil chemical and physical properties, and seasonal fluctuations in rainfall (6). Screening techniques that can be carried out under controlled environments have therefore often been used. Since detection of salinity-induced injuries is complex, a screening method based on simple selection criterion will be readily acceptable for rapid screening of large number of genotypes.

Rice is considerably tolerant to salinity during germination and maturity. But, it is highly sensitive to salt (electrical conductivity (EC) 5–6 dS/m) at seedling and flowering stages with manifestations of symptoms such as reduced root and leaf length, chlorophyll content, and dry matter (7). We describe here early-vegetative-stage screening and reproductive-stage screening of transgenic vis-à-vis wild-type (WT) rice for salt tolerance.

2. Materials

2.1. Seed Sterilization and Germination

1. Rice seed: Wild-type, transgenic genotypes, plant accessions or populations.
2. Centrifuge tubes (15 mL).
3. Seed germination boxes.
4. Tissue or germination paper.
5. Ethanol (10%).
6. Bleach (NaOCl) (5%).
7. Distilled water.
8. Plant tissue culture chamber.

2.2. Plant Growth

1. Pots (TOP Plastics, Clearwater, MN): 250 mL or 500 mL.
2. Trays: TOP press fill trays (TOP Plastics, Clearwater, MN) to accommodate 18 by 250 mL pots.
3. Soil: A 1:1 mixture of topsoil and compost (made of sphagnum peat moss, vermiculite, and processed bark ash) (Scott-Sierra Horticultural product Co, Marysville, OH).
4. Fertilizers: General purpose fertilizer, 20:20:20 (N:P:K) (Scott-Sierra Horticultural product Co, Marrysville, OH). Weigh 20 g and dissolve in 5 L.
5. Iron and soil acidifier (Planet Natural Garden Supply, Bozeman, MT). Dilute 30 mL iron and soil acidifier in 5 L water.
6. Greenhouse; modulated to 28°C.

2.3. Drought Stress Treatment

1. Balance.
2. Weighing boats to be used as pot bases and the size should accommodate pots used for plant growth.
3. Delta-T soil moisture probe (MLX, Dynamax) to measure volumetric soil moisture content.
4. Pressure plates (Soil Moisture equipments, Santa Barbara, CA) to determine soil water potential.
5. Pulse amplitude modulated fluorometer (OS1-FL, Optiscience).
6. Surgical blades.
7. Analytical balance (HR-60, A & D Co, Tokyo, Japan).
8. Oven (Thermo Fischer Scientific). The temperature range should be up to 105°C.
9. LI-COR 6400 (LI-COR Inc. Lincoln, NE, USA).

2.4. Salt Stress Treatment

1. pH/Electrical conductivity meter.
2. Weighing Balance (1,000-g capacity and 0.001-g readability).
3. Magnetic stirrer and spin bar.

4. KOH and HCl (avoid NaOH for pH adjustment in order to eliminate the possibility of any extra injury due to additional Na in the solution).
5. NaCl (analytical grade).
6. Reagents (analytical grade) for Yoshida's nutrient solution (Table 1).
7. Media storage glass bottles for stock solutions: 5 bottles of 1-L capacity, 1 dark amber glass bottle of 1-L capacity.
8. Forced air (convection) oven.
9. Plastic Petri dish (60 cm × 15 cm).
10. Germination paper.
11. Incubator for seed germination.
12. Volumetric flasks: 100- and 200-mL capacity.
13. Graduated cylinders: 25, 50, and 100 mL capacity.
14. Plastic rectangular trays (black color): 12-L capacity rectangular trays of size.
15. Plastic Beaker: 1,000 mL, 4,000 mL.
16. Styrofoam sheets (commercially available in local stores; 4.25 and 2.5 cm thick for making seedling floats).
17. Storage foam (for plugging).
18. Mixing containers: Plastic buckets, 10-L, 20-L, 50-L capacity.
19. Ruler (meter scale) for measuring length of root, leaf, etc.
20. Chlorophyll meter (SPAD 502 plus, Minolta).
21. Screw-capped scintillation vials.
22. Paper bags.
23. Mortar (9 cm diameter, clear glass) and compatible pestle.
24. Custom-made plastic pots (1.5 cu ft) with holes (3–4 mm) all around.
25. Plastic trays (to fit at least eight pots with at least 25 cm height).
26. Well-ground soil, preferably from paddy field.
27. Fertilizer (Agriform tablets, 20 g).

3. Methods

3.1. Plant Growth for Drought Stress Treatments

In order to conduct controlled experiments for molecular and physiological studies it is important to select uniform plants. Seeds are pre-germinated and grown for approximately 1 week (see Note 1).

3.1.1. Seedling Establishment

1. To break dormancy, heat dry fully filled mature seeds in a forced air oven set to 50°C for 48 h.

Table 1
Elemental composition and preparation of stock solutions for nutrient solution (9)

Element	Reagent (AR grade)	Preparation g/L	mL of stock solution/8 L nutrient solution	Concentration of element (ppm)
Macronutrients				
Stock I (N)	NH_4NO_3	91.4	10	40
Stock II (P)	$NaH_2PO_4{\cdot}2H_2O$	35.6	10	10
Stock III (K)	K_2SO_4	71.4	10	40
Stock IV (Ca)	$CaCl_2{\cdot}2H_2O$	117.35	10	40
Stock V (Mg)	$MgSO_4{\cdot}7H_2O$	324	10	40
Micronutrients				
Stock VI				
Mn	$MnCl_2{\cdot}2H_2O$	1.5	10	0.50
MO	$(NH4)_6{\cdot}Mo_7O_{24}{\cdot}4H_2O$	0.074		0.05
Zn	$ZnSO_4{\cdot}7H_2O$	0.035		0.01
B	H_3BO_3	0.934		0.20
Cu	$CuSO_4{\cdot}5H_2O$	0.031		0.01
Fe	$FeCl_3{\cdot}6H_2O$	7.7		2.00
	Citric acid, monohydrate ($C_6H_8O_7{\cdot}H_2O$)	11.9		0.12

2. Seeds are semi-sterilized to reduce microbial infestations. To do this, fill up sterile 15 mL centrifuge tubes with 10 mL of 10% ethanol and place the required number (e.g., 100) of rice seeds and agitate for 3 min.
3. Remove the ethanol and without rinsing fill up the tubes with 10 mL of 5% bleach. Agitate the tube content for 30 min and remove the bleach and rinse the seeds five times with distilled water.
4. Fill up the tubes with sterile distilled water and allow the seeds to imbibe overnight.
5. Fill up germination boxes with 20 mL distilled water and place tissue/germination paper in the boxes. Allow the paper to moisten and then put semi-sterile seeds on top of it.
6. Cover the boxes with aluminum foil and place them in a plant tissue culture chamber set at 28–30°C for 3 days. The foil is then removed and the plants are grown for another 4 days.
7. The plants can be transplanted into 250 or 500 mL pots and transferred to a growth room with photoperiod of 16-h light or to the greenhouse.

3.1.2. Transplanting and Selection of Plants

Establish and grow the seedlings from Subheading 3.1.1 in 250 or 500 mL pots for vegetative-stage and reproductive-stage screening, respectively. The genotypes can be mutant lines, transgenic lines, or different rice cultivars.

1. Label and weigh a desired number of empty pots and fill each with an equal weight of soil. Fill up the 250 or 500 mL pots with the soil mixture and plant the 1-week-old seedlings.
2. Place ten pots of the 500 mL size or 18 of the 250 mL pots in the trays.
3. Supply irrigation water from surface of the pots for about 2 weeks. This is to allow the soil to compact, and then watering can be done by filling up the tray until the plants reach maturity (see Note 2).
4. If transgenic plants are going to be studied, determine the plants that are positive for the transgene insertion by polymerase chain reaction (PCR) methods.
5. Randomly select the PCR-positive plants or -negative segregants and the WT, and place them in the pot bases (saucers) with predetermined weight.

3.2. Measurement of Soil Moisture Content

3.2.1. Calibration of the Delta-T Soil Moisture Probe

1. To calibrate the *Delta-T Soil Moisture* probe (i.e., Theta), record the soil water content of five water-saturated pots containing rice plants as time point 1. The saturation is achieved by leaving pots in trays with water until the soil water content

is above field capacity. After measuring, weigh each pot and dry them at 105°C for 24 h.

2. Remove the potted plants from trays and allow drainage overnight, and at dawn repeat step 1 above, and this will be measurement for point 2.
3. Allow the plants to evaporate water (dry down) for another day and repeat step 1, and this will be point 3.
4. Repeat step 3 of dry down and measurements as in step 1, to give point 4 (or more), till the plants show leaf rolling.
5. After initiation of leaf rolling, the plants are allowed 3 more days of dry down, till they do not recover at dawn observation. At this stage the plants have reached permanent wilting point. Repeat step 1 and this will be point 5.
6. Multiply the gravimetric moisture content by the bulk density of the soil, and this will represent volumetric soil moisture content. Plot this against theta probe readings to generate a linear curve ($Y = MX + C$), where Y = volumetric soil moisture determined gravimetrically, X = volumetric soil moisture determined theta probe, M = slope of the curve, and C = Y intercept. The equation allows the theta probe measurement to be converted to the gravimetric soil moisture without having to sample the soil every time soil moisture data needs to be taken (see Note 3).

3.2.2. Determination of Soil Water Potential

Laboratory measurements of soil water potential (matric potential) are made on the pressure plate apparatus.

1. Obtain a composite sample of the soil mixture and saturate the soil with water for about 3–4 h in rings about 5 cm in diameter and 0.5 cm thick and place them on a water-permeable ceramic membrane.
2. Subject the samples to pressures equivalent to the following: 0.03, 0.1, 0.5, 1, and 1.5 MPa in a pressure plate until drainage ceases. This can take 3–14 days depending on the magnitude of pressure applied.
3. Remove the soil form the rings and determine volumetric soil moisture as in step 6 of Subheading 3.2.1. The amount of water held under each pressure is equal to the water potential, and the values are negative, that is, soil holding a specific amount of water at 0.03 MPa will have water potential of −0.03 MPa.
4. To construct a water release curve, plot volumetric soil moisture content against water potential of soil (MPa). By applying the Van Genuchten model (1) generate a similar curve.
5. The resulting equation of the curve will be of the form $Y = -1/X$. This equation can be used to estimate soil water potential (Y) at any volumetric water content (X).

3.3. Screening Rice Genotypes for Drought Resistance at Vegetative Stage

3.3.1. Drought Stress Treatment at Vegetative Stage

1. The 1-week-old seedlings from step 6 of Subheading 3.1.1 are grown in 250 mL pots and allowed to grow for at least 8 weeks.
2. Select uniform plants and place on individual pot bases so that one plant will represent an experimental unit.
3. Impose drought stress by withdrawing irrigation on half of the plants and continue to maintain irrigation water at the pot bases for the controls. Such drought stress (DS) and well-watered (WW) units can be used for the different experiments as described in the following sections (see Note 4). The experimental design for the experiment can be in completely randomized design (CRD).

3.3.2. Measurement of Water Use Efficiency by Gravimetric Method

1. Twenty-eight days after planting (DAP), adjust plant–soil experimental units to equal weights (soil + water) by adding water as needed, and mulch with a layer of perlite of fixed weight to minimize evaporative water loss from the soil surface.
2. Transfer the pots from trays and place on weighed pot bases. The 32 pots of each genotype are then divided into two, with 16 for drought stress treatment and 16 for well-watered controls.
3. Drought stress treatment is imposed by withholding irrigation and drying down to approximately 70% field capacity, which takes approximately 3 days. At this time about (30 DAP), shoots on half of the plants (8 DS and 8 WW control treatment) for each genotype/line are harvested and dried at 72°C for 96 h and this biomass is designated as BIO-30.
4. For the remaining plants, gravimetric soil moisture is maintained at 70% (DS) or essentially at field capacity (WW) by replacing water lost through transpiration. This is done at mid-morning and late afternoon. The amount of water added each day for each pot is noted for calculation of cumulative water used (WUc).
5. This watering regime is continued for 14 days. At that point (45 DAP), shoots are harvested from the remaining 8 DS and 8 WW samples, and dried at 72°C for 96 h, and this biomass is designated BIO-45.
6. Cumulative water used (WUc) is calculated from the daily WU, measured as the daily water replaced due to evapotranspiration which is the difference in total plant-pot weight each day, and corrected for the (average) evaporation from soil-filled (and weighed mulch, with water contents as in the plant-pots) pot blank controls without plants. The cumulative $WUc = \Sigma\{\text{daily WU}\}$ over 14 days.
7. Gravimetric water use efficiency (WUEg) is calculated by the formula: $WUEg = ((BIO\text{-}45) - (BIO\text{-}31))/(WUc)$.
8. The experiment is arranged as factorial, with factors being watering regime (drought-exposed or well-watered) and

genotype (experimental genotypes plus WT) arranged in a CRD and replicated eight times.

9. Data collected is subjected to analysis of variance (ANOVA) using the general linear model (Proc GLM) of Statistical Analysis System (SAS). Differences between means are tested by the Least Significance Difference with a 0.05 threshold ($LSD_{0.05}$).

3.3.3. Measurement of Gas Exchange Parameters

1. The experimental plant genotype plants are grown in weighed pots under well-watered/semi-flooded conditions for 8 weeks following procedures as in Subheading 3.3.1.
2. At that point, half of the plants from each genotype are allowed to dry down for 7 days until plants show drought stress symptoms but not leaf rolling.
3. A day before gas exchange measurements, the soil moisture in the pots with drought stress is adjusted to 75% of field capacity.
4. Net photosynthesis (Pn), stomatal conductance (Gs), and transpiration rate (E) are measured on the youngest fully expanded leaves (one per pot) with a portable photosynthesis system LI-COR 6400.
5. The measurements are taken between 10:30 a.m. and noon. An *Arabidopsis* leaf chamber (LI-COR) is used for gas exchange measurements, with an irradiance of 400 μmol/m^2s photosynthetically active radiation (PAR), a temperature of 25–28° C, a CO_2 concentration of 400 μl/L, an airflow rate of 400 μmol/s, and a relative humidity of 55–60%.
6. After the gas exchange measurements, soil moisture content is measured with a Delta-T theta soil moisture probe.
7. Instantaneous water use efficiency (WUEi) is calculated using the formula:

$$\text{WUEi} = (\text{Pn} / \text{E}).$$

8. Experimental design and data analysis are the same as in step 9 of Subheading 3.3.2.

3.4. Determination of Plant Drought Stress Responses

3.4.1. Measurement of Photosynthetic Efficiency

1. Chlorophyll fluorescence (Fv/Fm) and quantum efficiency of photosystem II (PSII) are determined using a pulse amplitude modulated fluorometer.
2. Identify the youngest fully expanded leaf in the plant and a region on it and consistently dark-adapt by covering the region with clips provided with the fluorometer for 15–20 min.
3. Measure chlorophyll fluorescence on the leaf.
4. Measure efficiency of PSII, but use the light-adapted area of the same leaf.

3.4.2. Plant Water Status

The percent relative water content (RWC%) can be determined as described (8).

1. Excise a 10 cm section of the same leaf used for Fv/Fm and PS II, and using a balance determine its fresh weight (FW).
2. Float the leaf sections in deionized water at room temperature for 4–6 h and determine their rehydrated weights referred to as the turgid weight (TW).
3. Dry the leaf sections in an oven at 70°C for 18–24 h and then weigh the samples to get the dry weight (DW).
4. The RWC is calculated using the formula: RWC% = ((FW – DW)/(TW – DW)) × 100, where FW = Fresh weight, DW = Dry weight, and TW = Turgid weight.

3.5. Evaluation of Drought Tolerance at Reproduction Stage

3.5.1. Drought Stress Treatment at Reproductive Stage

1. Plants to be grown to reproductive stage are transplanted in the 500 mL pots. Grow and maintain plants as in Subheading 3.1.1 under well-watered conditions until they reach the R4 or panicle emergence stage (when the panicle emerges on the primary tiller, the secondary tiller will emerge at the booting stage) (see Note 2).
2. At this point, randomly select half of the plants and withhold irrigation to expose them to drought stress (DS treatment). Throughout the experiment irrigate the other half (WW controls).
3. Drought stress in rice is exhibited by leaf rolling. The number of days after leaf rolling can be counted to estimate the extent and severity of drought stress.
4. Different drought stress severities can be applied. A "moderate drought stress" is applied by withholding water until leaf rolling appears, at which point the plants are kept without irrigation for a further day. This normally occurs at around 7 days DS treatment. A "severe drought stress" can be carried out by withholding irrigation for 2 days after leaf rolling is observed. More severe DS is accomplished by maintaining plants without irrigation for 3 or more days of leaf rolling, but this severe treatment generally results in no seed set.
5. Drought-stressed plants are re-watered after the DS treatments and an initial evaluation of DS severity can be made. The number of leaves "fired" (leaves with at least 50% of their lamina exhibiting yellowing or necrosis) is recorded. The percent of leaves fired (LF%) is then calculated as the ratio of leaves fired to total number of leaves on the plant. This can be used to compare genotypes for the similar DS regimes.
6. The DS plants after re-watering are grown to maturity, and the different reproductive parameters can be measured and samples can be taken.

3.5.2. Measurement of Spikelet Fertility and Yield Parameters

1. Allow plant to grow to maturity at which seed production traits can be scored on the primary tiller exhibiting the DS.
2. Excise the panicles at the point of peduncle bifurcation and record the number of filled (contain at least one grain) and empty spikelets. The total of these empty spikelts will be the number of spikelets on the panicle that has experienced drought stress.
3. Remove the grains by hand, dry them in an oven at 37°C for 7 days, and weigh on the balance. The weight of the dried seed is grain yield (GY).
4. The peduncle (the upper internode) length (PL) is then measured to the point of panicle attachment.
5. The main culms (less than their panicles) are harvested at ground level, dried at 70°C for 72 h, and then weighed to determine the shoot biomass (SB).

3.5.3. Assessment of the Effect of Drought on Grain Yield and Components

The effect of drought stress can be assessed by the following parameters:

1. Total number of spikelets (both fertile and sterile) per panicle (SP).
2. Fertile spikelet ratio (FS) (fertile spikelets divided by total spikelets).
3. Grain yield (weight of dried grain).
4. Single grain weight (SGW) (GY divided by the number of grains).
5. Harvest index (HI) is calculated as the ratio of total grain weight (GY) to total above ground dry weight. A grain in this case means whole grain (or caryopsis) with hulls (palea and lemma).
6. The difference in parameters between drought stress and control plants is used to determine the extent of drought stress. Drought-stressed plants will have these parameters reduced significantly. The samples taken from these plants can be used for molecular studies including the differential gene expression between drought-stressed plants and well-watered controls.

3.6. Screening for Salinity Tolerance at the Vegetative Stage

3.6.1. Preparation of the Seedling Floats

1. Cut the rectangular Styrofoam (2.5 cm thickness) to the size that exactly fits midway of the rectangular tray so as to allow it to float over the nutrient solution (8 L), which is about 1 cm smaller than the rectangular tray on all sides (see Note 5).
2. 96 holes (12 × 8) are made with the help of a drill ~1 cm diameter in rows of 8. The purpose of making 96 holes is to make PCR-genotyping compatible with a 96-well PCR plate.
3. Cut the storage foams into pieces (~5 cm).

3.6.2. Preparation of Nutrient Solution

To avoid any nutrient deficiencies and mineral toxicities that would otherwise attribute to salinity stress the following steps are observed (see Note 6).

1. For the macronutrient stock solutions, weigh the required amount of reagent (Table 1) and transfer to a 1,000-mL beaker and do initial mixing with about 750 mL distilled water.
2. Pour the mixture into a 2-L volumetric flask, then add distilled water, and bring up to a 1 L volume. Stir the mixture for 5–10 min using a magnetic stirrer, and then transfer to stock solution to a bottle.
3. For micronutrient stock solution, dissolve separately each reagent in 50 mL distilled water (except for ferric chloride, which must be dissolved in 100 mL distilled water).
4. Mix all solutions together in 750 mL distilled water using a 2.0-L capacity volumetric flask. Add the ferric chloride solution to the mixture just before citric acid and stir the mixture for 15 min using a magnetic stirrer.
5. Finally add 50 mL sulfuric acid to the mixture and make up volume to 1.0 L (the solution looks yellowish brown). Stir for another 5–10 min and store in a dark amber glass bottle.

3.6.3. Seedling Establishment and Salt Treatment

1. Seed should be germinated for uniformity as described in step 1 of Subheading 3.1.1 after breaking dormancy (see Note 7). Place healthy seedlings individually rolled around the coleoptile–radicle juncture with the foam. Insert the rolled seedlings into the holes in the Styrofoam sheets that are floated over the Yoshida's nutrient (9) solution (prepared with deionized water) in the rectangular plastic trays. Care should be taken to avoid mechanical injury to the roots.
2. Let the plants acclimate to these conditions for 1 week to repair the transplanting shock. The plants can be grown in a greenhouse maintained at a temperature regime of 29/21°C day/night, or in a growth room under lighted racks with 14-h light (250 μE/m^2 s):10-h dark, and relative humidity at minimum of 70%.
3. Replace the nutrient solution in trays with salinized solution when the plants are at a two-leaf stage (typically after a week). Add NaCl while stirring up to the desired EC (6 g NaCl/L nutrient solution yields an EC of 12 dS/m). Fill up the trays with this solution high enough (usually 8 L) to touch the bottom of the Styrofoam.
4. Give salinity for first 3 days at EC of 12 dS/m after which increase to EC of 15 dS/m by replenishing the nutrient solution.
5. Maintain the solution volume of the tray(s) to 8 L by adding distilled water every 2 days as there will be loss of solution due to evaporation and transpiration.

6. Monitor and adjust the pH everyday because maintaining the pH (see Note 8) of the culture solution is very critical for making the nutrients available to the plants (see Note 9).

3.6.4. Evaluation of Salt Stress Symptoms

1. Measure the RWC after 1 day of salt stress following the protocol described under Subheading 3.4.2. This gives an indication of how the plants initially tolerate the "physiological drought stress" caused by the salinization (see Note 10). The tolerant lines should have RWC values above 70–80%.
2. Record two observations on the plants after 1 week and 2 weeks of salinization.
3. Measure growth parameters like root and shoot length with a ruler.
4. Count the number of tillers.
5. Use a chlorophyll meter to obtain the SPAD reading for a fully expanded leaf. A non-salinized plant should have a reading of 35–40. Salt-treated plants have readings 20–25 if salinity sensitive, 25–30 if moderately tolerant, and 30–35 if highly tolerant. Transform the SPAD values to chlorophyll content according to reference (10). Alternatively, you could measure the chlorophyll content by acetone/methanol/ethanol extraction method (11).
6. Collect leaf and root samples in paper bags for measurement of dry matter weight and Na+ and K+ content. Dry the samples in an oven at 80°C for 3 days. Grind the tissues to fine powder using mortar and pestle. Digestion of powdered tissues is usually done with HNO_3 followed by measurement in an inductively coupled plasma-mass spectrometry (ICP-MS). Normally 0.5 g of dried tissue powder is sent for analysis to the plant and soil analysis service laboratory.
7. Genotypes with Na/K ratio $\leq$0.170 are considered salt tolerant, <0.250 is moderately tolerant, and >0.250 is susceptible. On an average the ratio is ~0.170 for Pokkali rice, and 0.360 for IR29 rice.
8. Visual scoring of the salt-induced injuries is assigned based on leaf tip bronzing and drying/withering as per the score: 1–3 tolerant; 4–6 moderately tolerant; and 7–10 susceptible.

3.7. Screening for Salinity Tolerance at the Reproductive Stage

The following method provides a screen to compare salinity tolerance of transgenic to WT plants (see Note 11).

3.7.1. Preparation of Pots and Planting

1. Fill the pots with paddy soil (clay loam is the best) and MiracleGrow planting mix (mixed using a 1:2 proportion) 1 cm below the rim of the pot.

2. Place the pots in the plastic tray filled with tap water. This serves as water bath where water level is little higher than the soil level. Extra soil mix will need to be added after a day or 2 as the soil gets settled after absorbing water.
3. Four pre-germinated seeds are placed on the soil surface with their roots gently dipped inside the soil.
4. Place (gently press) one 20 g Agriform tablet per pot once, 2 weeks after seedlings were transplanted and then again after 40 days (maximum tillering). The potting mix provides enough plant food to support growth up to 2 weeks (*w*).
5. Grow the plants under greenhouse conditions as mentioned earlier with agronomic practices as recommended for rice.

3.7.2. Salt Stress Treatment and Replication

1. When the seedlings are 90 days old, remove all water from the tray (water bath). Most of the non-photosensitive varieties come to booting by 90 days after seeding.
2. Allow a 24-h break before siphoning out water that drains out from pots. The water draining capacity of the potting mix–soil mixture is higher than that of the soil.
3. Prepare salinized water solution by dissolving NaCl up to 100 mM with vigorous stirring.
4. Add the salinized water in the water bath so that the water level is just above the soil level in the pot (see Note 13).
5. Each pot with a single entry (WT or transgenic line) is considered as one experimental unit. The four seedlings are subsamples. Either completely randomized (CRD) or randomized complete block (RCB) designs could be used. For RCB, one tray or a set of trays can be taken as a block. We find that if grouping is done as described earlier, CRD is adequate to have statistically sound information.

3.7.3. Scoring for Salt Tolerance

1. Typically at 100 mM salinity level, the plants express phenotypes in about 1 week after salinization. Take note of the symptoms such as leaf rolling and whitening each day thereafter.
2. Score the percent green leaf area (GLA) as it is a convenient and reliable measure of tolerance.
3. Count the number of effective tillers, emergence panicles, etc.
4. Replace the salinized water with fresh tap water after 2 weeks.
5. When the grains turn yellow drain all the water from the tray and stop watering the pots.
6. Harvest the plants individually after 2 weeks and keep in paper bags.
7. Collect the per plant data on the number of spikelets, filled spikelets, and unfilled spikelets.

8. Weigh the total biomass per plant and grain yield per plant (this is considered the best score for tolerance).
9. Analyze the data using suitable statistical analysis such as SAS.

4. Notes

1. Although rice can be directly sown into the soil, such seedlings will be mostly nonuniform and it is difficult to conduct experiments for molecular drought stress studies. Sterilization and pre-germination of seeds therefore allow visual selection of uniform and robust seedling for such studies.
2. After transplanting, the trays should be filled with irrigation water at all times during plant growth phase. The generation-purpose fertilizer should be alternated with irrigation water, followed by the iron and soil acidifier application. This should continue until plants reach the desired stage for drought treatment. Some soils can provide enough iron and maintain the right pH for rice plant growth, but it might be necessary to provide the iron and soil acidifier. However, our experience with commercial plant media shows that this product is absolutely required.
3. At the end of or during drought period experiment, volumetric soil moisture content is determined indirectly by means of Theta probe. Water content as a percentage of dry weight and "C" is not very useful, because a water content representing field capacity in one soil might be below the wilting point in another (5). Percent water content can be converted from weight to volumetric units by multiplying the weight percentage by the bulk density of the soil. The Delta-T theta probe is calibrated by comparison with gravimetric water content measurements from saturated soil water content to permanent wilting point, obtained from conventional oven-dry weights and multiplied by bulk density (12). This allowed indirect determination of soil moisture content and its soil water potential from the probe readings using water release curves following described procedures (13).
4. Samples from these plants can be used for other important drought responses. Water use efficiency and gas exchange measurement are important parameters that can show physiological performance of the plants under drought stress. The underlying gene expression and biochemical studies can further be used to correlate physiology and genomic and biochemical responses underlying the observations. Similar experiments can be conducted and samples used for gene expression and biochemical studies.

5. Black trays are recommended because algae growth is observed in white trays after a few days due to the penetration of the light through the nutrient solution. Algae are not harmful but they tend to increase pH. If substantial algal growth is observed, it is advisable to adjust the pH twice a day. However the best method of algae control will still always be prevention of the problem, so excluding light should be the main emphasis because commonly used algicides could also be very phytotoxic and cause considerable root damage.
6. Prepare fresh stock solutions every 2 months (Table 1). Therefore, the amounts prepared should depend on the number of test entries screened during a 2-month period. Take utmost care while preparing the micronutrient stock solution as most nutrient deficiencies and other toxicities could be traced to improper preparation.
7. Proper breaking of the seed dormancy is very essential for the screening technique. Delays in germination of some entries will likely make these entries more sensitive to salt. If the entries include long-duration genotypes it is safe to keep for 5 days at 50°C.
8. A reliable pre-calibrated pH/EC meter is essential since any deviation of culture solution pH from 5.0 will make some nutrients toxic and others deficient.
9. Early seedlings usually take up more ammonium N than nitrate N leading to drop in the solution pH. The pH will increase when ammonium ions are depleted and more nitrate ions are taken up by the plants. The increase of the pH in the nutrient solution could be used as an indicator that nitrogen source starts to be deficient. Change the nutrient solution every week.
10. Seedling vigor has great advantage at this point since salinization occur at very early seedling stage.
11. In screening transgenic rice lines, we use a row each of IR 29 and Pokkali as test entries, two rows of wild-type plants, and two rows each of two independent transgenic rice lines. We always have an identical set as no-stress control for comparative assessment of effect of salinity on various growth and reproductive parameters.
12. Apply fertilizer to the pot and not to the tray with water.
13. About 2 months after planting, roots may be seen growing into the water bath through the holes in plastic pot. Do not disturb the pots during salinization to avoid any damage to roots. Such damage may eventually interfere with the salt tolerance of the plants.

Acknowledgements

This work was supported by NSF project "Cereal drought stress response and resistance networks" (DBI-0922747) and USDA-CSREES project "Molecular genetic network of water use efficiency in rice" (VAR-2008-01133).

References

1. Van Genuchten MT (1980) A closed-form equation for predicting the hydraulicconductivity of unsaturated soils. Soil Sci Soc Am J 44:892
2. Swindale LD, Bidinger FR (1981) The human consequences of drought and crop research priorities for their alleviation. In: Paleg LG, Aspinal D (eds) The physiology and biochemistry of drought resistance in plants. Academic, Sydney, pp 2–13
3. Tuberosa R, Salvi S (2006) Genomics-based approaches to improve drought tolerance of crops. Trends Plant Sci 11:404–411
4. Ekanayake IJ, De Datta SK, Steponkus PL (1989) Spikelet sterility and flowering response to rice water stress at anthesis. Ann Bot 63:257–264
5. Nguyen HT, Babu RC, Blum A (1997) Breeding for drought resistance in rice: physiology and molecular consideration. Crop Sci 37:1426–1438
6. Munns R, James RA (2003) Screening methods for salinity tolerance: a case study with tetraploid wheat. Plant Soil 253:201–218
7. Pearson GA, Ayers AD, Eberhard DL (1966) Relative salt tolerance of rice during germination and early seedling development. Soil Sci 102:151–156
8. Smart RE, Bingham GE (1974) Rapid estimation of relative water content. Plant Physiol 53:258–260
9. Yoshida S, Forno DA, Cock JH, Gomez KA (1972) Laboratory manual for physiological studies of rice, 2nd edn. International Rice Research Institute, Philippines, pp 53–57
10. Markwell J, Osterman JC, Mitchell JL (1995) Calibration of the Minolta SPAD-502 leaf chlorophyll meter. Photosynth Res 46: 467–472
11. Ritchie RJ (2008) Universal chlorophyll equations for estimating chlorophylls *a*, *b*, *c*, and *d* and total chlorophylls in natural assemblages of photosynthetic organisms using acetone, methanol, or ethanol solvents. Photosynthetica 46:115–126
12. Qiu Y, Fu B, Wang J, Chen L (2001) Soil moisture variation in relation to topography and land use in a hillslope catchment of the Loess Plateau, China. J Hydrol 240:243–263
13. Olson KR (1979) Saran coated method for determining bulk densities, soil moisture values, and linear extensibility. Agronomy Mimeograph. Soil Characterization Laboratory. Cornell University, Ithaca, NY, pp 79–85

Chapter 16

Phenotypic, Physiological, and Molecular Evaluation of Rice Chilling Stress Response at the Vegetative Stage

Benildo G. de los Reyes, Song Joong Yun, Venura Herath, Fuyu Xu, Myoung Ryoul Park, Jeong Il Lee, and Ki Young Kim

Abstract

Rice is a chilling-sensitive plant that is particularly prone to injury during the early stages of seedling development and during flowering. Significant variation exists between subspecies with japonica cultivars generally being less sensitive than most indica cultivars. In most temperate and subtropical countries where rice is grown, crop damage often occurs during the early stages of seedling development due to occasional cold snaps coinciding with the first few weeks after direct seeding in late spring to early summer. Irreversible injuries often result in seedling mortality or if the crop survives a stress episode, plant vigor and resistance to pests and diseases are severely compromised. Recent physiological and molecular studies have shown that oxidative stress is the primary cause of early chilling injuries in rice and the differential responses of indica and japonica cultivars are defined to a large extent by gene expression related to oxidative signaling and defenses. In this chapter, we summarize basic phenotypic, physiological, and molecular procedures that can be adopted for routine evaluation of differential responses between cultivars as well as for functional genomics studies.

Key words: Chilling stress, Oxidative stress, Reactive oxygen scavenging, H_2O_2, Electrolyte leakage, Transcriptome, Regulatory cluster

1. Introduction

Most temperate plants adapt to subfreezing conditions by acclimating at temperatures slightly above 0°C. In many species, maximum acclimation can be achieved in the laboratory by continuous exposure to temperatures between 2°C and 4°C for at least 2 weeks (1). In contrast, plants that are adapted to tropical and subtropical environments do not acclimate and can only maximally endure milder cold stress (chilling). Rice belongs to such a group with an optimum growth temperature around 28°C. Most cultivars are injured irreversibly by exposure to temperatures at or below 15°C particularly

Yinong Yang (ed.), *Rice Protocols*, Methods in Molecular Biology, vol. 956,
DOI 10.1007/978-1-62703-194-3_16, © Springer Science+Business Media, LLC 2013

during emergence and early stages of seedling development. Injury is manifested by mild to severe stress symptoms depending on temperature, duration of exposure, and genotype (2, 3).

Currently, there are no standard criteria and procedures for evaluating the magnitude of injury and differential responses of rice cultivars to chilling stress. In this chapter, we describe a set of procedures (whole plant, physiological, and molecular levels) that have been optimized for rice for some of the relevant methodologies that can be adopted for various applications. Specific emphases are on the evaluation of key physiological parameters associated with chilling-induced oxidative stress. The aim of this chapter is to establish basic parameters and compile a universal set of methodology that can be used by rice geneticists, breeders, physiologists, and molecular biologists when evaluating cultivars for responses related to defenses and for other purposes such as functional genomics.

1.1. Evaluation of Chilling Injury at the Whole Plant Level

Exposure of sensitive indica rice seedlings to a constant day/night temperature between 10°C and 13°C leads to moderate to severe stress symptoms occurring as early as 48 h after exposure (4–7). Moderate symptoms are often characterized by leaf discoloration or yellowing, which lead to severe withering if exposure is continued for at least 2 weeks. Most indica rice cultivars do not recover from such type of injury and even if they survive a stress episode, physiological perturbation often results in growth retardation and reduced tillering (i.e., poor vigor), which severely compromise nutrient uptake ability, competitiveness against weeds, and resistance to diseases and insects (2, 8, 9). In contrast, less sensitive japonica cultivars often show only mild stress symptoms of leaf rolling and most do not exhibit comparable degrees of leaf discoloration. Apparently, most indica cultivars are irreversibly injured even with shorter durations of exposure while many japonica cultivars can withstand similar magnitude of stress for a much longer period. In this section, we describe the most commonly used procedures for chilling tolerance phenotyping based on loss of chlorophyll and leakage of cellular electrolytes.

1.2. Evaluation of Chilling Responses at the Physiological Level

Recent studies have shown that the responses of rice to milder cold stress (10–13°C) are as complex as the process of cold acclimation at lower temperature regimes (around 4°C) in temperate plants in terms of the extent of associated gene expression changes (10–12). Regulatory clusters controlled by *CBF/DREB* transcription factors, which are associated with cold acclimation response, are also activated when rice seedlings are exposed to chilling. This indicates that the *CBF/DREB* regulon also plays an important role in configuring defense mechanisms associated with milder cold stress (5, 13, 14). Moreover, recent studies have shown that chilling-induced oxidative stress is the initial event that triggers progressive physiological injuries and the ability to activate defense responses in a timely manner explains (at least in part) the differential sensitivities of indica and japonica cultivars to chilling (5, 10, 14). Increased activities of scavenging enzymes and other redox regulatory mechanisms provide the first line of direct defenses (15, 16). In this section, we describe several sensitive assays for evaluating cellular H_2O_2 concentration and activities of reactive oxygen scavenging mechanisms in rice seedlings.

1.3. Evaluation of Chilling Responses at the Molecular Level

Evidence directly linking early changes in gene expression in rice to oxidative signaling (probably largely via H_2O_2) has been presented based on the commonalities between direct response to chilling exposure and response to oxidative stress artificially induced by the exogenous application of 4 mM H_2O_2 at ambient temperature (10, 14). These data were further reiterated by the predominance of genes regulated by bZIP-TGA-type transcription factors that are involved in oxidative stress-induced gene expression (17). During the initial 48 h at 10°C, the transcriptome of the japonica cultivar Nipponbare mirrors the typical responses to pathogens, herbivores, and mechanical wounding by virtue of the activities of genes involved in jasmonic acid, salicylic acid, and ethylene signaling, all of which have direct links to H_2O_2 signaling (18–20). Genes involved in redox regulation, cellular detoxification, and membrane repair and biogenesis are among the most dominant defense categories during this time period.

Components of putative regulons that respond directly to primary oxidative signals are generally activated faster in japonica than in the more sensitive indica cultivars. These differences are consistent with earlier findings that the timely execution of immediate but short-term defense mechanisms is important for a species like rice that can only endure moderate and transient cold stress (5, 10). In this section, we describe the optimized real-time PCR procedures for profiling the activities of representative genes that comprise a putative oxidative stress regulatory cluster. These assays serve as molecular indicators of chilling-induced oxidative stress response.

2. Materials

2.1. Evaluation of Chilling Injury

2.1.1. Whole Plant Level Analysis by Loss of Chlorophyll (Cold Air Method)

1. Rice seeds.
2. Whatman No. 3 filter paper (100 cm diameter, round).
3. Sterile disposable petri dishes (100 cm).
4. 70% Ethanol.
5. 5% Sodium hypochlorite.
6. Redi-Earth plug mix (Sun Gro Horticulture, Bellevue, WA) or sterile tap soil.

7. Seedling propagation wedge sheets with 12 columns × 6 rows (3 cm × 3 cm × 5 cm).
8. Seedling propagation flats (52 cm × 26 cm × 5 cm).
9. Plant growth chambers (two) with extended temperature range and humidity control.

2.1.2. Electrolyte Leakage Analysis

1. 50 mL conical polypropylene tubes with cap or equivalent.
2. Deionized water (16 MΩ or higher purity).
3. Electrical conductivity meter.
4. Plant growth chambers with extended temperature range and humidity control.
5. Hot plate for boiling or ultralow (−80°C) freezer.

2.2. Evaluation of Responses at the Physiological Level

(Also applies to Subheading 3.2)

1. Deionized water (sterilized).
2. Amplex Red reagent or 10-acetyl-3,7-dihydrophenoxyazine (Molecular Probes A22188).
3. Dimethyl sulfoxide (DMSO; component of Molecular Probes A22188).
4. Amplex Red reaction buffer (component of Molecular Probes A22188).
5. H_2O_2 solution (30%).
6. Horseradish peroxidase (HRP; component of Molecular Probes A22188).
7. Catalase (CAT; component of Molecular Probes A22180).
8. Tricholoroacteic acid (TCA; 5% w/v).
9. Dowex anion exchange resin (AG1 × 8).
10. 0.25 M Sodium phosphate (pH 7.5).
11. 40 mM Tris–HCl (pH 7.0).
12. 0.5 M Tris–HCl (pH 7.5).
13. Fluorometer or spectrophotometer.

2.3. Evaluation of Responses at the Molecular Level: Quantitative RT-PCR

1. Trizol reagent (Cat. 10296-028; Invitrogen, Carlsbad, CA).
2. 0.25 M Sodium acetate.
3. RNase-free (DEPC-treated) water.
4. Ethanol (absolute).
5. SYBR Green fluorescein dye mix (ABsolute™ QPCR; AB-1219/A or AB-1219/B; ABGene, Rockford, IL).
6. RT primers (Oligo-$dT_{15\text{-}22}$ or random primers).
7. MMLV Reverse Transcriptase (M1701; Promega, Madison, WI).
8. RNase-free DNase RQ1 kit (M6101; Promega, Madison, WI).

9. RNase inhibitor, 40 units/μl (N2611; Promega, Madison, WI).
10. Low-profile 8-tube strips without caps, 0.2 mL (TLS-0851; Bio-Rad, Hercules, CA).
11. Optical flat 8-cap ultraclear strips for 0.2 mL tubes and plates (TCS-0803; Bio-Rad, Hercules, CA).
12. MyiQ Cycler (Bio-Rad, Hercules, CA).

3. Methods

3.1. Evaluation of Chilling Injury

3.1.1. Whole Plant Level Analysis by Loss of Chlorophyll (Cold Air Method)

1. Prior to germination, viable seeds may be selected by specific gravity method using NaCl solution at a density of 1.08–1.13 (optional step).
2. Sterilize 12 healthy seeds by first rinsing with 70% ethanol for 30 s and then with sodium hypochlorite solution for 20 min. Rinse the treated seeds with sterile tap water five to six times.
3. Allow the seeds to soak in fresh sterile tap water for 1 h.
4. Allow the seeds to germinate in a sterile disposable petri dish containing Whatman No. 3 filter paper presoaked with sterile water. Under ideal condition (room temperature, dark) uniform germination can be achieved within 3–4 days after imbibition (see Note 1).
5. An ideal container for seedling propagation is a plastic seedling wedge sheet (6 in a row × 12 in a column = 72 wedges). Place the seedling propagation wedges on a plastic flat and fill the wedges with moderately moist Redi-Earth Plug mix or tap soil.
6. Plant two germinated seeds for each wedge. A 72-wedge flat can be used for 12 cultivars (6 plants per cultivar). Prepare two sets of seedlings for each cultivar (one each for control and chilling conditions).
7. Grow the seedlings until three (V_3)- to five (V_5)-leaf stage in a growth chamber set according to the following parameters: 29°C/21°C day/night temperature, 12-h day length, light intensity of 600–800 μmol/m^2 s, and relative humidity of 70–80%. Keep the seedlings adequately moist during the entire duration of the experiment. Before starting with the treatments, remove one of the two seedlings leaving only one seedling per wedge.
8. At V_3 to V_5 stage, transfer one set of seedlings to a second growth chamber set under chilling stress condition as follows: 10°C constant day/night temperature, 12-h day length, light intensity of 600–800 μmol/m^2 s, and relative humidity of 70–80%. Keep the other set under optimum (control) conditions throughout the duration of the experiment.

9. Monitor the number of plants that show visible signs of chilling injury every 3 days according to the Standard Evaluation Systems developed by the International Rice Research Institute ((21); see Note 2). Rice cultivars that have been well characterized for chilling sensitivity may be used as standards for this analysis including M202, Nipponbare, Quilla 66304 as less sensitive and IR36, IR50, INIAP12 as highly sensitive controls (5, 6, 8).
10. Alternatively, cold sensitivity screening may be conducted at variable day/night temperature regime as follows: 15°C/12°C day/night temperature, 12-h day length, light intensity of 600–800 µmol/m^2 s, and relative humidity of 70–80%. Expose plants to this condition for 5–7 days. Control plants must be kept under the same conditions as the control used for the constant temperature method.
11. Return the cold-treated seedlings to control condition after 5–7 days of chilling stress. Determine the severity of cold injury based on Standard Evaluation Systems developed by the International Rice Research Institute ((21); see Note 3).

3.1.2. Electrolyte Leakage Analysis

1. Plant materials (for control and chilling stress) must be prepared as in Subheading 3.1.1.
2. Cut the third leaf from the base and place in a clean 50 mL conical polypropylene tube.
3. After collecting the leaf samples from all plants, briefly rinse them with deionized water to remove dirt or debris on the leaf surface (see Note 4).
4. Cut the leaves into 1 inch discs and add 20 mL of fresh deionized (>16 MΩ) water into each tube.
5. Make sure that all leaf pieces are covered with deionized water by gentle shaking in a platform shaker (50 rpm) for about 5 min.
6. Leave the samples at room temperature for 5–6 h to allow electrolytes to diffuse in solution. During the incubation period, shake the tubes gently by inverting occasionally to ensure that floating leaf pieces will be soaked in deionized water.
7. During the incubation period, turn the conductivity meter "ON" to warm up before sample measurements begin. Make sure that the probe is clean by rinsing several times with fresh deionized water and blotting it dry with tissue paper. Known solutions representing low and high electrical conductivity must be used to standardize the machine before starting the measurements on the experimental samples. Rinse the electrode with fresh deionized water after taking measurements on the standard.
8. Take electrical conductivity measurements of at least three blank samples (20 mL of >16 MΩ deionized water in 50 mL

conical tubes) at room temperature. The average of these values represents the electrical conductivity of blank sample.

9. To read the conductivity measurements of the samples, remove all the leaf discs in a tube and place them on the bottom side of the tube cap. Measure the electrical conductivity of the solution. The values obtained from these measurements represent the total cellular electrolytes that leaked from the leaf discs as a consequence of their exposure to chilling or control temperature conditions ($EL_{induced}$).
10. Place the leaf discs back into the solution (see Note 5), seal the tube cap, and set aside until measurements of all samples are completed.
11. Place all tubes in a hot water bath and boil for 1 h. During this step, make sure that the tube caps are sealed tight enough to prevent spillage but lose enough to prevent pressure buildup (see Note 6). Allow the samples to cool to room temperature and then centrifuge briefly at low speed to collect the condensation.
12. Remove the leaf discs and measure the electrical conductivity of each sample. The values obtained from these measurements represent the total cellular electrolytes from the leaf tissues (EL_{total}).
13. Calculate the stress-induced electrolyte leakage (EL) with the following formula: $\%\ EL_{stress} = EL_{induced}/EL_{total} \times 100$. Relative injury can also be expressed as the ratio of % EL in stress and control plants.

3.2. Evaluation of Responses at the Physiological Level

These procedures have been optimized for rice using the Amplex Red Assay kit modified according to the manufacturer's (Molecular Probes-Invitrogen, Carlsbad, CA) and previously published protocols on other plant species (22, 23).

3.2.1. Determination of Leaf H_2O_2 Content

1. Prepare a 10 mM stock solution of Amplex Red (AR) in DMSO. Allow AR to dissolve at room temperature.
2. Prepare a 1× solution of AR reaction buffer by combining 4 mL of 5× stock with 16 mL deionized water.
3. Prepare a stock solution (10 U/mL) of HRP by dissolving the content of one vial in 1 mL of 1× AR reaction buffer.
4. Prepare the H_2O_2 assay cocktail by combining 50 μL AR stock solution, 100 μL HRP stock solution, and 4.85 mL AR reaction buffer (1×). This solution is sufficient for 100 assays.
5. With 1× AR reaction buffer, prepare a 20 mM working solution of H_2O_2 from 3% stock (freshly prepared) and use this solution to prepare a dilution series (50 μL volume) for standard curve determination (see Note 7).

6. Place the H_2O_2 dilution (50 μL) series in 1.5 mL dark colored microcentrifuge tubes and combine 50 μL of assay cocktail. Mix by pipetting and then incubate at room temperature in the dark for 30 min.
7. Measure absorbance of the samples at ~560 nm at least three times after the 30-min incubation period. Alternatively, fluorescence of the samples may be measured at 530–560 nm excitation and ~580 nm emission. Use the average of three measurements to plot the H_2O_2 standard curve.
8. After plotting the standard curve, prepare the tissue samples (from seedlings at control and chilling conditions) by cutting two ~0.5 in. leaf discs and placing them in pre-weighed 1.5 mL dark colored microcentrifuge tubes containing 0.1 mL 5% TCA.
9. Grind the leaf tissues with a sterilized tissue homogenizer. Weigh the tubes to determine the amount of tissue sample.
10. Centrifuge the mixture at 12,000 × *g* for 15 min at 4°C and collect the supernate.
11. Add 10 mg of Dowex anion exchange resin to the supernate and then incubate at room temperature for 1 min (see Note 8).
12. Centrifuge at 10,000 × *g* for 2 min and transfer the clear and colorless supernate in a fresh tube. Adjust the volume to 0.2 mL with TCA (5%).
13. Dilute a 10 μL aliquot of the purified tissue extract in 40 μL 0.25 M sodium phosphate (pH 7.5).
14. In a dark colored microcentrifuge tube, combine the purified tissue extract with 50 μL assay cocktail. Mix by pipetting and incubate at room temperature in the dark for 30 min.
15. Measure absorbance or fluorescence of samples as in step 7 of Subheading 3.2.1.
16. Determine the H_2O_2 amount per mg leaf tissue using the standard curve.

3.2.2. Determination of Peroxidase Activity in Leaf Tissues

1. With the HRP stock solution (from **step 3** of Subheading 3.2.1) prepare a peroxidase (POX) dilution series in 1× AR reaction buffer for standard curve determination (50 μL total volume).
2. Prepare the POX assay cocktail by combining 50 μL AR stock solution (from step 1 of Subheading 3.2.1), 500 μL H_2O_2 working solution (from step 5 of Subheading 3.2.1), and 4.5 μL of 1× AR reaction buffer (from step 2 of Subheading 3.2.1). This solution is enough for 100 reaction assays.
3. In dark colored 1.5 mL microcentrifuge tubes, combine 50 μL of HRP dilution series and 50 μL of POX assay cocktail. Mix the solution by pipetting.

4. Incubate the mixture at room temperature in the dark for 30 min.
5. Measure absorbance of the samples at ~560 nm at least three times after the 30-min incubation period. Alternatively, fluorescence of the samples may be measured at 530–560 nm excitation and ~580 nm emission. Use the average of three measurements to plot the POX activity standard curve (see Note 9).
6. To prepare the tissue samples, harvest ~0.1 g of leaves from control and chilling-stressed seedlings.
7. Homogenize the tissues in 1 mL 40 mM Tris–HCl (pH 7.0) by grinding in a chilled porcelain mortar and pestle.
8. Transfer the tissue homogenate to 1.5 mL microcentrifuge tube and centrifuge at 12,000 × *g* for 20 min at 4°C.
9. Determine total protein content of soluble extract (supernate) by Bradford method.
10. Dilute an appropriate volume of leaf extract containing about 50 μg soluble protein in 0.5 M Tris–HCl (pH 7.5) to a total volume of 50 μL and place in dark colored 1.5 mL microcentrifuge tube.
11. Add 50 μL of the POX assay cocktail to the soluble protein extract, mix by pipetting, and incubate the mixture at room temperature in the dark for 30 min.
12. Measure absorbance or fluorescence of samples as in **step 5** of Subheading 3.2.2. POX activity is directly proportional to absorbance/fluorescence.
13. Determine total POX activity per mg protein using the standard curve.

3.2.3. Determination of Catalase Activity in Leaf Tissues

1. This assay makes use of similar set of reagents from steps 1–3 of Subheading 3.2.1.
2. Prepare the CAT stock solution (1 unit/μL) by dissolving the contents of 1 vial in 100 μL of distilled water. Unused solution may be stored at -20°C.
3. Prepare a CAT dilution series in 1× AR reaction buffer for standard curve determination (25 μL total volume).
4. Prepare a 40 μM working solution of H_2O_2 by diluting 10 μL of 20 mM solution from step 3.2.1 in a total volume of 5 mL with 1× AR reaction buffer.
5. Mix 50 μL of AR (from step 1 in Subheading 3.2.1) and 20 μL of HRP (from step 3 in Subheading 3.2) in a total volume of 5 mL (with 1× AR reaction buffer) to make the CAT assay cocktail.
6. Combine 25 μL of CAT standard curve samples or 25 μL of soluble protein extracts (from step 9 of Subheading 3.2.2) in individual 1.5 mL microcentrifuge tubes (dark colored).

7. Add 25 μL of H_2O_2 (from step 5 of Subheading 3.2.1) to each tube containing the CAT standard curve samples or soluble protein extracts. Incubate mixture at room temperature for 30 min (see Note 10).
8. Add 50 μL of CAT assay cocktail to each tube and incubate at 37°C in the dark for 30 min or longer.
9. Measure absorbance or fluorescence of samples as in step 5 of Subheading 3.2.2. CAT activity is inversely proportional to absorbance/fluorescence.

3.3. Evaluation of Responses at the Molecular Level: Quantitative RT-PCR

1. Harvest leaf tissues from rice seedlings (V_3 to V_5 stage) exposed to 10°C for 0, 2, 6, 12, 24, 48, and 96 h and freeze immediately in liquid nitrogen (see Note 11).
2. Extract total RNA from leaf tissues of control and chilling-stressed plants by standard method with Trizol reagent.
3. Set up DNase treatment of total RNA samples by combining the following: total RNA (1 μg/μL) in water or TE buffer (5 μL); 10× DNAse reaction buffer (5 μL); and DNase, 1 unit/μL (5 μL).
4. Nuclease-free water (to 50 μL final volume): Reaction volume may be scaled up depending on the amount of RNA needed for a given number of genes to be assayed. Mix by pipetting and incubate at 37°C for 30 min.
5. Terminate the reaction by adding 5 μl of DNase stop solution.
6. Inactivate DNase by incubation at 65°C for 10 min.
7. Centrifuge at 10,000 ×*g* for 2 min at 4°C to pellet the precipitated reagents from the DNase stop reaction and transfer the supernatant (DNA-free RNA) to a fresh tube.
8. Precipitate the DNA-free RNA sample by adding 3 volumes of nuclease-free water, 2.5 volumes of absolute ethanol, and 0.1 volume of 0.25 M Sodium Acetate.
9. Incubate at –20°C for 1 h and centrifuge at 13,000 ×*g* for 1 h at 4°C.
10. Dissolve RNA in 40 μL RNase-free water.
11. Estimate the concentration of pure RNA and perform quality assessment by running 100 ng aliquot in formaldehyde agarose gel.
12. Perform cDNA synthesis (20 μL reaction) by combining reagents as follows: DNase-treated total RNA (1 μg); nuclease-free water to 11 μl; and oligo-dT primers (500 ng/μL) and random primers (400 ng/μL) mixture in 3:1 ratio.
13. Mix by pipetting and incubate at 70°C for 5 min to denature secondary structures. Cool immediately in ice and centrifuge briefly at 10,000 ×*g* to collect condensation.

14. Add the following reagents for a 20 μL volume first-strand cDNA synthesis cocktail: 5× first-strand cDNA synthesis buffer (4 μL); dNTPs, 500 μM (2 μL); RNase inhibitor, 40 unit/μL (1 μL); and MMLV Reverse Transcriptase (1 μL).
15. Mix by pipetting and incubate at 42°C for 60 min.
16. Inactivate MMLV Reverse Transcriptase by incubation at 70°C for 15 min. The cDNA samples may be stored in small aliquots at −20°C. Two to three independent cDNA synthesis reactions on pooled biological replicates are recommended for robust technical replication.
17. Set up 20 μL RT-PCR cocktail for each sample by combining the following: cDNA from 10 ng of total RNA (2 μL); 2× SYBR Green master mix (10 μL); gene-specific primer-F (Table 1; see Note 12), 40 ng/μL(2 μL); gene-specific primer-R (Table 1; see Note 12), 40 ng/μL (2 μL); and nuclease-free water (4 μL).
18. Mix by pipetting and spin whole strip briefly to collect droplets at the bottom of the tubes and to remove trapped air bubbles.
19. Perform amplification reaction in the MyiQ™ Single-Color Real-Time PCR Detection System (Bio-Rad) under the following cycling parameters: initial denaturation at 95°C (15 min); cycle denaturation at 95°C (30 s, 40 cycles); primer annealing at 57°C (30 s, 40 cycles); and elongation at 72°C (30 s, 40 cycles).
20. Calculate relative expression by the 2-delta-delta-C_T method (24) as follows and compare temporal expression profiles between samples:

$$R = 2^{\Delta \text{Ct.target (control} - \text{sample)}} / 2^{\Delta \text{Ct.reference (control} - \text{sample)}}.$$

4. Notes

1. Complete germination is defined as the stage when the coleoptiles are at least 1 mm elongated.
2. Scoring of injury symptoms at constant day/night temperature is based on the following scale: 1 = seedlings are normal and green; 3 = tips of most leaves are pale green; 5 = seedlings are yellowing and partially withered; 7 = seedlings are brown and withered; and 9 = seedlings are dead. Mild symptom (score of 3) is often visible in sensitive cultivars as early as 1 week after chilling exposure.
3. Scoring of injury symptoms at variable day/night temperature is based on the following scale: 1 = seedlings are normal and green; 3 = tips of most leaves turned to pale green; 5 = about

Table 1
Gene-specific primers for RT-PCR analysis of representative chilling stress response "indicator" genes

Locus ID[a]	Gene name	Primer sequence	Amplicon (bp)
Normalization control			
Os10g36650	Actin-2	CATGAAGATCAAGGTTGTCGCTCC (F) CCAGATTCTTCATACTCAGCCCTTG (R)	126
Transcription factors			
Os08g43090 (a)	*RF2b* (bZIP)	GGAGCAACAAGCACAGCTAA (F) AGTTTCAGACGCTGGACCTC (R)	69
Os06g41100 (a)	*TGA10* (bZIP)	CACCTCTGGCATGATCCAA (F) TCCTTCAGGGTTCCTTCTTG (R)	67
Os08g38020 (a)	*bZIP148* (bZIP)	GGATTATGCTTTGGGATTGC (F) GCAGCAGTTGGGTTATTATGG (R)	75
Os01g09760 (a)	*DIRP* (MYB)	GCCACGCTTCTTGCAATC (F) GTCGCAGCATTAAGCAAGC (R)	94
Os02g41510 (a)	*OsMyb4* (MYB)	CGGAGGAGGAGTTCCAGAT (F) CTCATCTCCATCCCGGAGT (R)	88
Os09g35020 (a)	*DREB1D/CBF4*	CGAAATGGACTACGACACGTACTA (F) GATGTCAGCGCCTTCGTC (R)	109
Os08g43210 (b)	*DREB1B/CBF1*	GAGGCATCACCGTTCGAG (F) CGCTAAGCTCGCGTAGTACA (R)	69
Targets of transcription factors			
Os03g03910 (a)	Catalase-1	CCGTATGGAACAACAACAACTC (F) GATACGCTCCCTGTCGAAGT (R)	107
Os05g14260 (a)	Peroxidase-29	AGCTGGTTCGACAACCTCTACTA (F) GAGAAGGCCCGGAAGAAG (R)	146
Os02g43790 (a)	Ethylene-responsive protein	TCACTCTCTGCAGAATAGAAGAAGTT (F) TGCCTGGAACTCGTCACAT (R)	75
Os06g13460 (a)	Jasmonate O-methyl transferase	GAAGGATGAGGATGTATACAATGGA (F) AGGATTCCAACTTTTCCTTGTTC (R)	113
Os10g36270 (a)	NBS-LRR protein	CATGGAGGTTGTAGGATGTGG (F) TTCTGAAATCTCCAGTCTTTCTAGTG (R)	85

(a)—induced by chilling and exogenous H_2O_2 (oxidative mediated response)
(b)—induced by chilling but not by exogenous H_2O_2 (oxidative independent response)
[a]Locus number based on MSU (TIGR) annotation

1/3 of the leaf area is yellow; 7 = 2/3 of leaf area of most leaves is brown; and 9 = most leaves are dead.

4. The EL method tends to produce noisy data and much of the noise originates from inadequately cleaned (soiled) tissue samples. It is absolutely important that leaf tissues are clean for consistent readings of electrical conductivity. Leaf tissues should be cleaned by first rinsing them in deionized water and then very gently blotting them dry on sterile wipes. Vigorous wiping could cause tissue abrasion and this must be avoided. Air drying is not an acceptable alternative because further tissue

dehydration could affect electrical conductivity measurements. Use of lint-free wipes in all steps is also recommended.

5. It is important that all tissues are saved in order to establish accurate measurements of the total cellular electrolytes (EL_{total}) per unit weight of the tissue samples.
6. Loose tube caps can be held in place by wrapping the edges with parafilm or plastic wrap to prevent sample overflow during boiling. Sample overflow causes inaccuracies in total cellular electrolyte (EL_{total}) measurements, which severely affect the ratio between induced leakage and total leakage ($EL_{induced}/EL_{total}$).
7. For measuring leaf H_2O_2 concentrations of rice, a range from 0 to 2 μM at 0.1–0.25 μM interval is recommended for more accurate determinations of values from experimental samples. Sample measurements may also be performed in 1/10 or 1/100 dilution if values are too high relative to the scale of the standard curve.
8. Passing the sample extract through Dowex anion exchange resin helps remove colored impurities. Leaf extracts could be yellowish to brownish in color depending on the age and condition of the tissues and fluorescence/absorbance values often fluctuate due to these impurities. Sample solutions should be colorless after this step. Two to three changes of resin may be performed until a colorless solution is obtained.
9. For measuring POX activity of rice leaves, a standard curve covering a range between 0 and 2 μunits/μL is recommended.
10. For measuring CAT activity of rice leaves, a standard curve covering a range between 0 and 4 μunits/μL is recommended.
11. The purpose of this experiment is to assess differential expression (temporal) between rice cultivars for representative genes involved in early response mechanisms. Chilling-induced expression in intolerant indica cultivars is often characterized by 6–12-h delay (and often less robust) relative to the tolerant japonica cultivars. Nipponbare and INIAP12 are ideal standards for tolerant and intolerant cultivars, respectively.
12. Whenever it was possible, real-time PCR primers were designed to span an intron–exon junction at the 3′ end of the coding region. Before proceeding with the real-time PCR experiment, trial amplification by conventional PCR procedure is highly recommended to assess possible genomic DNA contamination on the cDNA samples. Two bands will be visible in agarose gel if the cDNA sample has significant contamination with genomic DNA (genomic = larger band with intron; cDNA = smaller band without intron). If cDNA is significantly contaminated with genomic DNA, go back to the original RNA sample and treat with DNase for longer periods (at least for another 2 h) before repeating the cDNA synthesis.

Acknowledgement

This work was supported by a grant from USDA-CSREES-NRI, Plant Genome Research (2006-35604-1669), to BGDR and by a grant from Korea Research Foundation (KRF-2006-352-F00002) and BioGreen 21 Program-Rural Development Administration (20080401034024), Republic of Korea, to SJY and MRP.

References

1. Pearce RS (1999) Molecular analysis of acclimation to cold. Plant Growth Reg 29:47–76
2. Li CC, Rutger JN (1980) Inheritance of cool-temperature seedling vigor in rice and its relationship with other agronomic characters. Crop Sci 20:295–298
3. Adair CR (1968) Testing rice seedlings for cold tolerance. Crop Sci 8:264–265
4. Jeong EG, Kim DS, Lee JI, Kim SL, Kim KJ, Yea JD, Son JR (2006) Effects of cold water irrigation on quality properties of rice. Kor J Crop Sci 51S:119–124
5. Morsy MR, Almutairi AM, Gibbons J, Yun SJ, De los Reyes BG (2005) The *OsLti6* genes encoding low-molecular-weight membrane proteins are differentially expressed in rice cultivars with contrasting sensitivity to low temperature. Gene 344:171–180
6. Andaya VC, Mackill DJ (2003) Mapping of QTLs associated with cold tolerance during the vegetative stage of rice. J Expt Bot 54:2579–2585
7. Jeong EG, Yea JD, Baek MK, Moon HP, Choi HC, Yoon KM, Ahn SN (2000) Estimation of critical temperature for traits related to cold tolerance in rice. Kor J Breed 32:363–368
8. Andaya VC, Tai TH (2006) Fine mapping of the qCTS4 locus associated with seedlings cold tolerance in rice (*Oryza sativa* L.). Mol Breed 20:349–358
9. Peterson ML, Jones DB, Rutger JN (1978) Cool temperature screening of rice lines for seedling vigor. II Riso 27:269–274
10. Cheng C, Yun KY, Ressom H, Mohanty B, Bajic VB, Jia Y, Yun SJ, De los Reyes BG (2007) An early response regulatory cluster induced by low temperature and hydrogen peroxide in seedlings of chilling-tolerant japonica rice. BMC Genomics 8:175
11. Fowler S, Thomashow MF (2002) Arabidopsis transcriptome profiling indicates that multiple regulatory pathways are activated during cold acclimation in addition to the *CBF* cold response pathway. Plant Cell 14:1675–1690
12. Rabbani MA, Maruyama K, Abe H, Khan MA, Katsura K, Ito Y, Yoshiwara K, Seki M, Shinozaki K, Yamaguchi-Shinozaki K (2003) Monitoring expression profiles of rice genes under cold, drought, and high salinity stresses and abscisic acid application using cDNA microarray and RNA gel-blot analyses. Plant Physiol 133:1755–1767
13. Dubouzet JG, Sakuma Y, Ito Y, Kasuga M, Dubouzet EG, Miura S, Seki M, Shinozaki K, Yamaguchi-Shinozaki Y (2003) *OsDREB* genes in rice, *Oryza sativa* L., encode transcription activators that function in drought-, high salt-, and cold-responsive gene expression. Plant J 33:751–763
14. Yun KY, Park MR, Mohanty B, Herath V, Xu F, Mauleon R, Wijaya E, Bajic VB, Bruskiewich R, De los Reyes BG (2010) Transcriptional regulatory network triggered by oxidative signals configures the early response mechanisms of japonica rice to chilling stress. BMC Plant Biol 10:16
15. D'Atreaux B, Toledano MB (2007) ROS as signaling molecules: mechanisms that generate specificity in ROS homeostasis. Nat Rev 8:813–824
16. Foyer CH, Noctor G (2005) Oxidant and antioxidant signaling in plants: a re-evaluation of the concept of oxidative stress in a physiological context. Plant Cell Environm 28: 1056–1071
17. Orozco-Cardenas ML, Narvaez-Vasquez J, Ryan CA (2001) Hydrogen peroxide acts as a secondary messenger for the induction of defense genes in tomato plants in response to wounding, systemin, and methyl jasmonate. Plant Cell 13:179–191
18. Johnson C, Boden E, Arias J (2003) Salicylic acid and NPR1 induce the recruitment of trans-activating TGA factors to a defense gene promoter in Arabidopsis. Plant Cell 15: 1846–1858
19. Neill S, Desikan R, Hancock J (2002) Hydrogen peroxide signaling. Curr Opin Plant Biol 5: 388–395

20. Memelink J, Verpoorte R, Kijne JW (2001) ORCAnization of jasmonate-responsive gene expression in alkaloid metabolism. Trends Plant Sci 6:212–219
21. International Rice Research Institute (2002) Standard evaluation system for rice. IRRI, Philippines, p 55
22. Zhou M, Diwu Z, Panchuk-Voloshina N, Haugland RP (1997) A stable nonfluorescent derivative of rosorufin for the fluorometric determination of trace hydrogen peroxide: applications in detecting the activity of phagocyte NADPH oxidase and other oxidases. Anal Biochem 253:102–168
23. De los Reyes BG, McGrath JM (2003) Cultivar-specific seedling vigor and expression of a putative oxalate oxidase germin-like protein in sugar beet (*Beta vulgaris* L.). Theor Appl Genet 107:54–61
24. Livak KJ, Schmittgen TD (2001) Analysis of relative gene expression data using real-time quantitative PCR and the 2(-Delta Delta C(T)) method. Methods 25: 402–408

Chapter 17

Analysis of Rice Root Hair Morphology Using Cryo-Scanning Electron Microscopy

Zhiming Yu, Hanmin Chen, Yanli Tong, and Ping Wu

Abstract

Root hairs are highly polarized long tubular outgrowths from the surface of epidermal trichoblast cells. Root hair development is a simple process that has facilitated for the study of cell fate determination and tissue differentiation in higher plants. Root hair patterning types in dicot and monocot plants are different. Rice is a monocot model plant with type II root hair pattern. The method to examine root hair fine structure and cell shape in rice may help elucidate the mechanism of cell initiation and differentiation in monocot plants. Compared with the critical-point-drying SEM, the Cryo-SEM method has great advantage, as the Cryo-SEM can well maintain the delicate structure of root hairs in their natural situation. Here we provide the methodology developed to investigate several rice mutants with impaired root hair cells using Cryo-SEM.

Key words: *Oryza sativa* L., Root hair, Cryo-SEM

1. Introduction

Plant root hairs are important organs for the uptake of nutrients and water from the soil and serve as sites of interaction with soil microorganisms (1). Root hairs are also very important for acquiring phosphorus in crops (2) and *Arabidopsis* (3). Many plant species exhibit increased length and density of root hairs in response to phosphorus deficiency (4, 5), which may be a strategy for enhancing phosphorus acquisition.

Three types of root hair pattern have been suggested (6). *Arabidopsis* has the striped pattern (Type 3) of root hairs. In this type, root hairs are formed from specific epidermal cells (trichoblasts, H cells) that are arranged in rows located over the intercellular spaces between two underlying cortical cells, whereas cells located outside a single cortical cell normally give rise to non-hair cells (atrichoblasts, N cells) (7–11).

Yinong Yang (ed.), *Rice Protocols*, Methods in Molecular Biology, vol. 956,
DOI 10.1007/978-1-62703-194-3_17, © Springer Science+Business Media, LLC 2013

In contrast to *Arabidopsis*, Type 2 patterns depend on asymmetrical cell division which is found in rice (*Oryza sativa* L.), Barley (*Hordeum vulgare*), and wheat (HH). Hair-forming cells (trichoblasts) in Type 2 roots derive from a late, unequal transverse cell division in the epidermal stem cell (6). Type 3 patterns that show a random occurrence of root hairs from epidermal cells are found in maize (Zea mays). Several root hair mutants have been identified in barley (12, 13), rice (14), and maize (15), but only a few genes have been cloned. These include the β-expansin gene in barley (16), *OsCSLD1* in rice (17), and *RTH1* and *RTH3* genes in maize. *RTH1* encodes an SEC3-like protein (18) whereas *RTH3* encodes a COBRA-like protein that is unique to monocots and required for root hair elongation and normal grain yield in maize (19). However, our understanding of the processes of root hair development in monocot crops is still limited. To increase our knowledge on the mechanisms of root hair patterning, it is important to finely examine the morphological traits of crop roots.

2. Materials

2.1. Seedling Root Preparation

1. Rice root hair mutants isolated from an MES-generated rice mutant library (Nipponbare, Japonica).
2. 75 % EtOH.
3. Sodium hypochlorite.
4. MS salts.
5. Phytagel (Sigma).

2.2. Cryo-SEM

1. Liquid nitrogen.
2. Nitrogen (99.99 % purification).
3. Nitrocellulose paper.
4. HITACHI S-3000N SEM.
5. GATAN ALTO 2100.

3. Methods

1. Seeds separated from the hull were surface-sterilized for 2 min in 75 % EtOH and washed three times using sterile distilled water. Then, seeds were sterilized for 30 min in 10 % sodium hypochlorite and thoroughly rinsed in sterile distilled water.
2. Seeds were germinated in the sterile distilled water in a continuous 37°C dark growth cabinet.

3. Germinated seeds were transferred to sterile absorbing paper, and then sown in 0.5× MS solid medium (0.5 % phytagel). The plates were sealed with 3 M Micropore Surgical Tape and placed vertically (to allow the roots grow along the surface) in growth chambers at temperature regime of 30/22°C (day/night) and 70 % humidity under a 12-h photoperiod (15,000 lux).
4. After 3 days of growth, cut the root tips (about 1 cm) using sharp blade. Quickly place cut roots on moist nitrocellulose paper (by immersing in distilled water and drying on absorbing paper) that is pasted on a stub by carbon-rich conductive glue. (Glue is conductive to allow discharge of electrons.)
5. Pour liquid nitrogen into the slush chamber and press the SLUSH/VENT button. After about 2 min the liquid nitrogen will change to solid. Press the SLUSH/VENT button again to vent the nitrogen into the chamber (see Note 1) (20).
6. Remove the lid and quickly plunge the sample holder into the slushy nitrogen to rapidly cool down the sample. Keeping the sample holder in the slush, slide the VTD down the rod until it sits on the O-ring in the little recess on top of the slushing chamber (20).
7. Press the SLUSH/VENT button to activate slush chamber pumping. The sample holder retracts fully into the VTD by pulling the rod upwards before the liquid nitrogen turns solid, thereby pushing the trap door valve into place on the VTD (see Note 2).
8. Press the SLUSH/VENT button once again to vent the slush pot. The sample is now held under vacuum and is ready for transfer into the prep chamber. Place the VTD onto the GATE VALVE airlock and press the LOAD PUMP button to pump it.
9. When at least six bars are illuminated in the Gate Valve airlock vacuum gauge, open the valve fully and push the sample holder onto the dovetail on the sample stage in the preparation chamber (see Note 3).
10. Set the temperature of the prep-chamber stage at −95°C. Press the HEAT button to start warm-up. For rice root hairs, 180 s is commonly enough, but the required duration also depends on root hairs' degree of humidity.
11. Press the HEAT button again to stop heat and recover the temperature to ≤140°C. Set the sputter timer (120 s should suffice for root hair requirements) by holding down the TIMER button whilst using the RAISE and LOWER buttons to add and subtract time (see Note 4).
12. The Argon admit solenoid valve will close automatically after sputtering has finished and the vacuum level of the pre-chamber will start to recover (see Note 5).

13. Once the ball valve is opened the sample may be introduced into the SEM chamber.
14. With the sample inside the SEM chamber and the ball valve closed, the beam may be run up. Bring the sample to a suitable working distance.
15. The HITACH SEM is mainly operated by the mouse, keyboard, and rotary knobs. Mouse is used for most of the operations including opening/closing of dialog, execution of processes, and settings. Keyboard operation is used to input characters and comments. Rotary knobs may be adjusted including the MAGNIFICATION, CONTRAST, BRIGHTNESS, and FOCUS knobs (21).
16. Use the MAGNIFICATION knob to change magnification and adjust contrast and brightness of image by turning CONTRAST and BRIGHTNESS clockwise or anticlockwise. Using FOCUS knob adjust the focus roughly at low magnification, then increase the magnification, and finely adjust the focus (21).
17. Click H.R. Capture and scan (see Note 6). The scan is made at the selected scan speed and the image data is acquired in the SEM memory. The data is transferred to the PC memory at the selected resolution. Input a file name and click the SAVE button (see Figs. 1 and 2).
18. Turn off filament and open the BALL VALVE when imaging is complete.
19. Haul the rod backwards and out of the SEM cold stage. Shove the airlock valve closed and push the LOAD PUMP button twice to vent the airlock to remove the VTD.
20. Place the VTD on the slushing pot and engage the trap-door valve onto the control lever.

4. Notes

1. Adjust the nitrogen pressure to 7.5 psi and ventilate the cryo-pipeline for at least 5 min before running the cryo-SEM system.
2. Ensure that only the sample holder is immersed when slushing the sample. Never let the rod enter the liquid nitrogen as cooling it could freeze the O-ring in the VTD and cause an unnecessary thermal load.
3. When the prep-chamber vacuum is more than eight bars, the liquid nitrogen is filled into the Alto prep-chamber. It will take about 30 min for the prep-chamber to cool down to under −140°C.

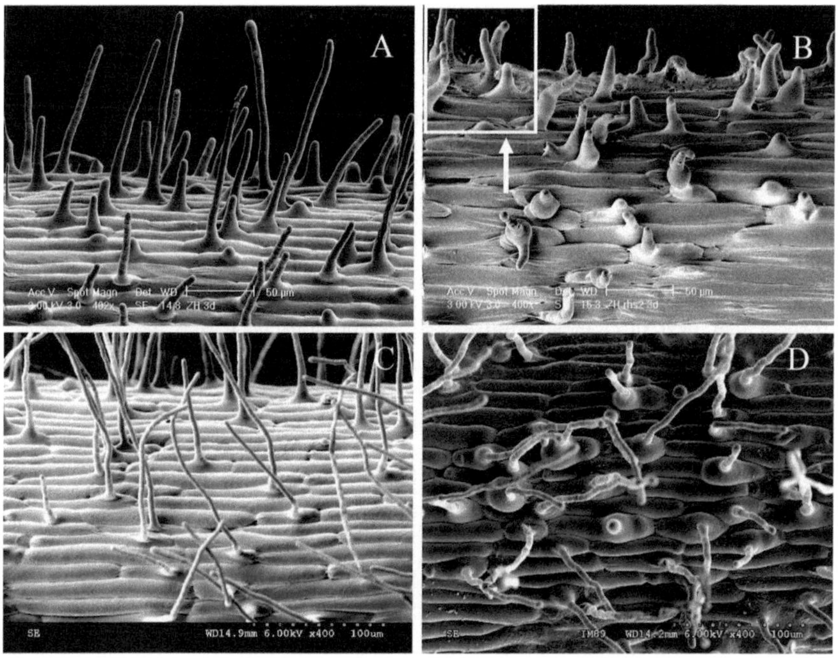

Fig. 1. Root hairs of the wild-type (**a**, **c**) vs. abnormal root hair cells of the mutant (**b**, **d**) (400×).

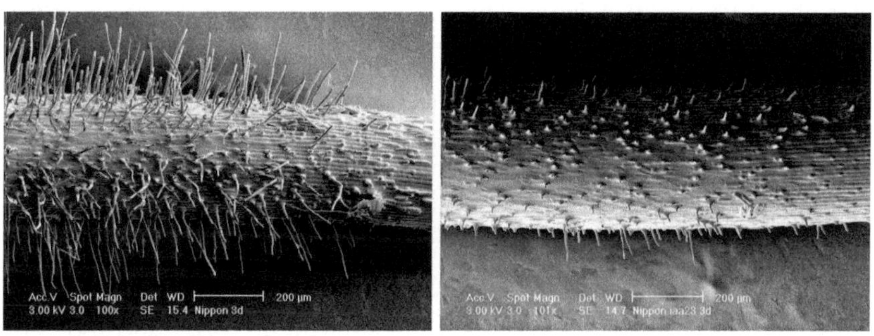

Fig. 2. Root hairs of the wild-type (left) vs. short root hairs of the mutant (right) (100×).

4. Before coating the sample with gold, it should be observed on the SEM cold stage to discern whether the root hair was sufficiently sublimated. If root hair was not sublimated well, repeat the sublimation step.
5. Open the Argon cylinder valve and carefully adjust the pressure to 12 psi. Adjust the voltage dial to set the current to 10 mA for gold. If the sputter head switches off during the setup, simply RESET the timer and press the SPUTTER button again.
6. The image may be distorted during the high-resolution capture process and this is not an abnormality. It will return to the original form when the process ends. If this does not happen, you can start the capture process again.

Acknowledgments

This work was supported by the Key Basic Research Special Foundation of China (2005CB120901).

References

1. Gilroy S, Jones DL (2000) Through form to function: root hair development and nutrient uptake. Trends Plant Sci 5:56–60
2. Gahoonia TS, Care D, Nielsen NE (1997) Root hairs and acquisition of phosphorus by wheat and barley cultivars. Plant Soil 191: 181–188
3. Bates TR, Lynch JP (2000) The efficiency of *Arabidopsis thaliana* (Brassicaceae) root hairs in phosphorus acquisition. Am J Bot 87: 964–970
4. Bates TR, Lynch JP (2000) Plant growth and phosphorus accumulation of wild type and two root hair mutants of *Arabidopsis thaliana* (*Brassicaceae*). Am J Bot 87:958–963
5. Zhou J, Jiao FC, Wu ZC, Li YY, Wang XM, He XW, Zhong WQ, Wu P (2008) OsPHR2 is involved in phosphate-starvation signaling and excessive phosphate accumulation in shoots of plants. Plant Physiol 146:1673–1686
6. Clowes FAL (2000) Pattern in root meristem development in angiosperms. New Phytol 146: 83–94
7. Dolan L, Costa S (2001) Evolution and genetics of root hair stripes in the root epidermis. J Exp Bot 52:413–417
8. Kwak SH, Schiefelbein J (2007) The role of the SCRAMBLED receptor-like kinase in patterning the *Arabidopsis* root epidermis. Dev Biol 302:118–131
9. Lee MM, Schiefelbein J (1999) WEREWOLF, a MYB-related protein in *Arabidopsis*, is a position-dependent regulator of epidermal cell patterning. Cell 99:473–483
10. Schiefelbein JW (2000) Constructing a plant cell. The genetic control of root hair development. Plant Physiol 124:1525–1531
11. Dolan L, Costa S (2001) Evolution and genetics of root hair stripes in the root epidermis. J Exp Bot 52:413–417
12. Engvild KC, Rasmussen SK (2004) Root hair mutants of barley. Barley Genet Newsl 34: 13–15
13. Gahoonia TS, Nielsen NE, Joshi PA, Jahoor A (2001) A root hairless barley mutant for elucidating genetic of root hairs and phosphorus uptake. Plant Soil 235:211–219
14. Suzuki N, Taketa S, Ichii M (2003) Morphological and physiological characteristics of a root-hairless mutant in rice (*Oryza sativa* L.). Plant Soil 255:9–17
15. Wen TJ, Schnable PS (1994) Analyses of mutants of three genes that are influence root hair development in *Zea mays* (Gramineae) suggest that root hairs are dispensable. Am J Bot 81:833–842
16. Kwasniewski M, Szarejko I (2006) Molecular cloning and characterization of beta-expansin gene related to root hair formation in barley. Plant Physiol 141:1149–1158
17. Kim CM, Park SH, Je BI, Park SH, Park SJ, Piao HL, Eun MY, Dolan L, Han CD (2007) OsCSLD1, a cellulose synthase-like D1 gene, is required for root hair morphogenesis in rice. Plant Physiol 143:1220–1230
18. Wen TJ, Hochholdinger F, Sauer M, Bruce W, Schnable PS (2005) The roothairless1 gene of maize encodes a homolog of sec3, which is involved in polar exocytosis. Plant Physiol 138:1637–1643
19. Hochholdinger F, Wen TJ, Zimmermann R, Chimot-Marolle P, da Costa e Silva O, Bruce W, Lamkey KR, Wienand U, Schnable PS (2008) The maize (*Zea mays* L.) roothairless3 gene encodes a putative GPI-anchored, monocot-specific, COBRA-like protein that significantly affects grain yield. Plant J 54:888–898
20. Alto 2100 operators handbook. (2005) GATAN
21. Instruction manual of Model S-3000N scanning electron microscope (2006) HITACHI

Chapter 18

Inoculation and Virulence Assay for Bacterial Blight and Bacterial Leaf Streak of Rice

Bing Yang and Adam Bogdanove

Abstract

Xanthomonas oryzae pv. oryzae (Xoo) and *Xanthomonas oryzae* pv. oryzicola (Xoc) cause bacterial blight and bacterial leaf streak in rice, respectively. Despite being very closely related, the pathogens colonize different tissues and cause distinct diseases. The diseases are economically important and also serve as model systems for studying plant–bacterial interactions. Here we describe protocols for Xoo and Xoc inoculation and disease scoring methods that are appropriate to their different modes of infection. These methods are routinely used to evaluate pathogen virulence or host responses under controlled environmental conditions.

Key words: *Xanthomonas oryzae*, Inoculation, Virulence assay, Bacterial blight, Bacterial leaf streak, Plant–bacterial interactions

1. Introduction

Bacterial blight and bacterial leaf streak of rice are caused by *Xanthomonas oryzae* pv. oryzae (Xoo) and *X. oryzae* pv. oryzicola (Xoc), respectively. Bacterial blight is a vascular disease starting with infection of rice leaves through wounds and hydathodes; Xoo multiplies and spreads within the xylem, causing long, grey to white, opaque necrotic lesions that typically spread from the tip of the leaf (1). In contrast, bacterial leaf streak is a nonvascular disease; Xoc enters leaves through stomata and colonizes the intercellular spaces of mesophyll parenchyma, causing discrete, yellowish, translucent, water-soaked lesions between the veins that may expand and coalesce, become necrotic, and eventually resemble bacterial blight superficially (2). Several artificial inoculation methods for Xoo and Xoc have been used and are briefly summarized (1), including needle-pricking, spraying, and clipping of leaves, or dipping of non-leaf parts of rice with bacterial suspension. Our laboratories have found that leaf clip inoculation for Xoo and syringe-infiltration for Xoc are

Yinong Yang (ed.), *Rice Protocols*, Methods in Molecular Biology, vol. 956,
DOI 10.1007/978-1-62703-194-3_18, © Springer Science+Business Media, LLC 2013

effective, simple means of inoculation appropriate to the natural modes of infection of these two pathogens. These choice methods give consistent and reliable results for quantitative evaluation of bacterial virulence and for assessment of plant responses to infection. Syringe infiltration is also a useful method for rapid qualitative evaluation of rice resistance to Xoo.

In the leaf clip assay, originally developed by Kauffman et al. (3), leaf tips are cut off with Xoo-infested scissors, thus depositing the inoculum directly in the exposed veins across the whole cut edge near the tip. This method guarantees a near 100% chance of infection. The bacteria multiply and move down the inoculated leaf against the transpirational water flow, generating a lengthening lesion. The lesion length (i.e., the distance from the tip to the leading edge of grayish to chlorotic tissue) can be taken as a measure of the progression of blight disease. As an example, we describe the use of leaf clip inoculation to evaluate the virulence contribution by *pthXo1*, a type III effector gene and a major virulence factor in Xoo strain PXO99^{A} (4).

In the syringe infiltration method, bacterial cells in suspension are introduced through stomata directly into the interveinal intercellular spaces of leaves by delivery through a blunt (needleless) syringe pressed against the underside of the leaf. This method similarly guarantees a near 100% chance of infection. In susceptible plants, the water-soaked lesions that develop following infiltration spread in both directions, rapidly for Xoc, and less rapidly and to a lesser extent for Xoo. Due to gene-for-gene interactions (5), resistant plants do not become water soaked. Rather the infiltrated spot turns brown and desiccates within 2–3 days and bacterial spread is restricted. As examples of this method, we provide an assay comparing the virulence of an Xoc mutant to that of the wild-type strain, and present a demonstration of avirulence gene function in Xoo.

2. Materials

1. Tryptone Sucrose (TS) medium (1 L): Tryptone, 10 g; sucrose, 10 g; glutamic acid, 1 g; pH, 5.7 (see Note 1). Solid medium is made by supplementing with 15 g agar/L.
2. Cephalexin (Cx): Stock, 10 mg/mL in water; working concentration, 10 mg/L.
3. Spectinomycin (Sp): Stock, 100 mg/mL in water; working concentration, 100 mg/L.
4. Kanamycin (Km): Stock, 50 mg/mL in water; working concentration, 50 mg/L.
5. Plant materials: Rice variety: IRBB7 (near-isogenic line of IR24 containing resistance gene *Xa7*), IR24 (recurrent parental line of IRBB7).

6. *Xanthomonas oryzae* pv. oryzae: Strains PXO99^A (Philippine Xoo race 6, J. Leach), ME2 (PXO99^A mutant with *pthXo1* knockout) and PXO99^A/*avrXa7* (PXO99^A containing avirulence gene *avrXa7*); *Xanthomonas oryzae* pv. oryzicola: strains BLS303 (a Philippines isolate, C. Vera-Cruz) and BLS303-38C10 (BLS303 mutant (6)).
7. Soil: Baccto soil (Hummert International).
8. Fertilizer (Miracle-Grow for acid loving plants or similar 30-10-10 fertilizer), used at 185 mg/L concentration.
9. Iron chelate micronutrient (Becker-Underwood Sprint 330 or a similar 10% chelated iron), used at 600 mg/L concentration.

3. Methods

3.1. Generation of Xoo Transformants by Electroporation

3.1.1. Xoo Competent Cell Preparation

1. Initiate 5 mL cultures in TS medium with a few single colonies of PXO99^A or ME2 and incubate overnight at 30°C with agitation (~220 rpm).
2. Inoculate 500 mL of TS with the 5 mL overnight culture and incubate at 30°C with agitation (~220 rpm) until $OD_{600}=0.8–1.0$.
3. Place the culture on ice for 30 min and transfer it into autoclaved, prechilled 250 mL centrifuge tubes.
4. Centrifuge the culture for 10 min at 4,500×*g* at 4°C and then resuspend the pellet very gently by hand (do not vortex) in 100 mL prechilled autoclaved water; set the resuspension on ice for 15 min.
5. Centrifuge the cells for 10 min at 4,500×*g* at 4°C and then resuspend the pellet very gently by hand (do not vortex) in 100 mL prechilled autoclaved 10% glycerol; set the resuspension on ice for 15 min.
6. Centrifuge the cells for 10 min at 4,500×*g* at 4°C and then resuspend the pellet very gently by hand (do not vortex) in 10 mL prechilled autoclaved 10% glycerol.
7. Aliquot the final preparation in prechilled microcentrifuge tubes, 75 μl each. Use immediately or flash freeze the cells by placing the tubes in an ethanol/dry ice bath and store them at −80°C.

3.1.2. Xoo Transformation

1. Use 75 μl of PXO99^A (for *avrXa7*) or ME2 (for *pthXo1*) mixed with 0.5–1.0 μg of respective plasmid DNA in ≤5 μl of TE or water for electroporation in a 2 mm cuvette with pulse settings of 2.5 kV, 25 μF, and 200 Ω.
2. Immediately following electroporation, add 1 mL of TS medium without antibiotics and incubate at 30°C for 1.5–2 h.

3. Plate the electroporated cells on TS agar supplemented with spectinomycin and incubate at 30°C for appearance of single colonies.
4. Confirm the transformants by extracting plasmid DNA and retransform *E. coli* in expectation of a reasonable number of colonies resistant to spectinomycin.

3.2. Leaf Clip Inoculation for Xoo

1. Plant growth conditions: Plants are grown in a growth chamber under approximately 1,000 μmol/m^2 s light density provided by metal halide lamps and high-pressure sodium lamps, at 30°C for a 12-h light period and 28°C for a 12-h dark period, with 75% relative humidity. Prior to sowing in soil, rice seeds are incubated in water at 37°C overnight. Five rice plants are grown per 3 in. square pot in a tray containing standing water. At 2 weeks and weekly thereafter, the water is replaced with 30-10-10 fertilizer at 185 mg/L and 10% iron chelate solution at 600 mg/L.
2. Bacterial inoculum preparation: The bacterial stocks of PXO99^A, ME2, and ME2(*pthXo1*) preserved in −80°C freezer are streaked out on TS agar plates supplemented with appropriate antibiotics and incubated at 28°C for 2–4 days (see Note 2). The cells are scraped off from the plates and resuspended in sterilized distilled water. The concentration of this suspension is adjusted to OD_{600} = ~0.5.
3. The two youngest fully expanded leaves of 4–5-week-old plants (see Note 3) are clipped about 2 cm from the tip with scissors that have been immersed in bacterial suspension immediately prior to each clipping (Fig. 1a). Five plants are used for inoculation of each strain. The inoculated leaves are monitored for the appearance of curling starting at the tip and for the progress of lesions. Usually tip curling and lesion development start at 5 days after inoculation of PXO99^A.
4. Measurement: The lesion length (the distance from the tip to the leading edge of the grayish symptom) is measured for individual inoculated leaves 14 days after inoculation. The mean lesion length of ten leaves is used for each treatment. Treatments are compared by using the Tukey test for analysis after ANOVA. A P value of 0.05 is used to determine significance (Fig. 1b). After lesion lengths are measured, bacterial populations can be measured as a function of colony-forming units (cfu) recovered from the inoculated leaves. The leaves from each treatment are pooled and cut into small (~1 cm) pieces, and grounded in water (1 mL per leaf) using a sterilized mortar and pestle. The resulting suspension is diluted serially and plated in triplicate on TS agar with appropriate antibiotics. Counts are taken from appropriate dilutions on the three plates and converted into cfu per leaf.

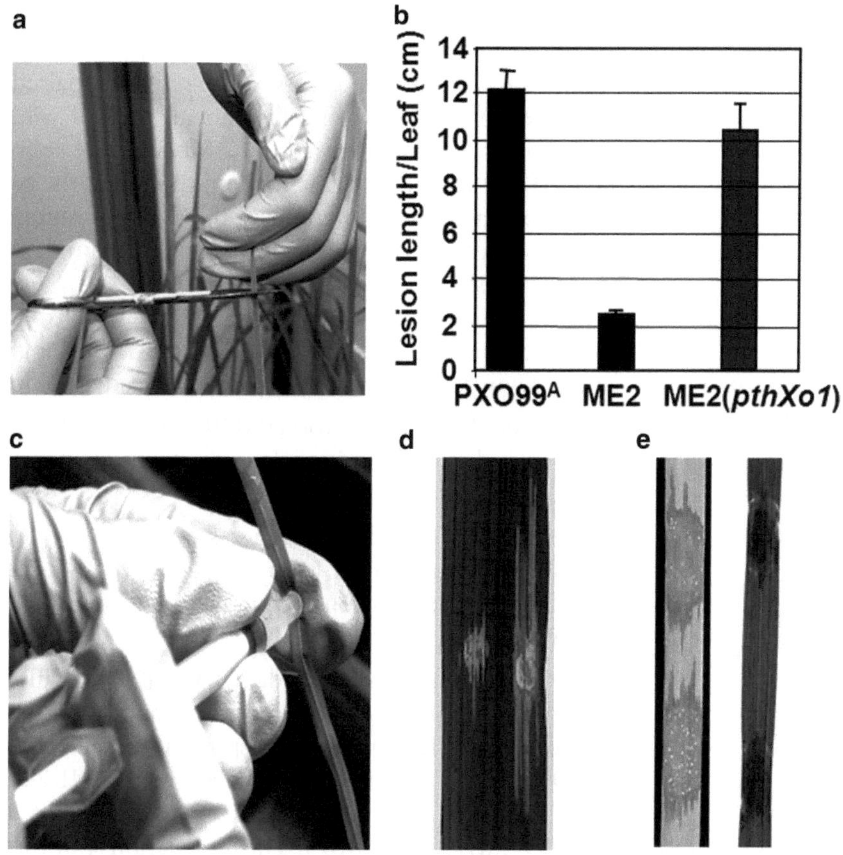

Fig. 1. Methods for inoculation and quantification of rice bacterial blight and bacterial leaf streak diseases. (**a**) Leaf clip inoculation. A fully expanded rice leaf is cut about 2 cm from the tip with scissors that are coated with bacterial suspension. (**b**) Results from a virulence assay by leaf clipping inoculation. Lesion lengths 14 days after inoculation on 4-week-old IR24 plants are shown for the strains indicated. (**c**) Syringe infiltration. A blunt-ended syringe is used to force the bacterial inoculum into the intercellular spaces of the leaf tissue. (**d**) Quantitative virulence assay comparing an Xoc mutant strain to the wild type on an 8-week-old IR24 plant. The mutant strain (38C10) was inoculated on the left side of the midrib of a leaf, and the wild type (WT) was inoculated directly opposite by syringe infiltration. The photo was taken 10 days after inoculation (image copyright © American Society for Microbiology, Appl Environ Microbiol 73: 8023–8027, 2007, reprinted with permission). (**e**) Phenotypic symptoms of susceptible reaction and resistant response of rice to Xoo. Photos were taken 3 days after syringe inoculation of 3-week-old IR24 leaf (*left*) with PXO99^A and same-age IRBB7 (*right leaf*) with PXO99^A/*avrXa7*.

3.3. Infiltration Inoculation

3.3.1. Xoc Inoculation by Syringe

1. Prepare bacterial inoculum and grow plants (typically to 4–8 weeks old) as described for Xoo inoculation;
2. Infiltrate bacterial suspensions into leaves from the underside by pressing the mouth of the syringe to the leaf. Support syringe with a finger propped behind it and gently depress the plunger (Fig. 1c). Take care not to crush the leaf. Also, be sure not to overlap the midrib as this will raise the syringe and break the necessary seal. Gently wipe the leaf with a paper towel to remove liquid from the surface of the leaf. Hold the leaf briefly up to the light to confirm infiltration which will appear as water-soaking (see Note 4).

3. For comparison of mutant strains to wild type, inoculate the mutant on one side of the midrib and the wild type on the other (Fig. 1d). In a typical assay, five replicate, paired inoculations are done. Perform similar inoculations of wild type vs. wild type to control for variation. Measure the lesion length after 10 days, calculate the lesion length ratio of the two inoculations on each leaf, and compare the ratios of mutant to wild type vs. wild type to wild type by using a paired, two-tailed Student *t* test.

3.3.2. Xoo Inoculation by Syringe

1. Grow rice plants (IR24 and IRBB7) as described for clip inoculation. Two- to three-week-old plants are ready for inoculation.
2. The bacterial inocula of Xoo ($PXO99^A$ and $PXO99^A$/*avrXa7*) are prepared as for clip inoculation with the concentration of $OD_{600} = \sim 0.5$.
3. Strains are inoculated by syringe as described for Xoc above. Inoculate each strain on both IR24 and IRBB7. Five plants with three to five spots on each fully expanded leaf are used for each treatment.
4. Monitor the symptoms on a daily basis after inoculation. The water-soaking on the inoculated leaves appears 36–48 h after inoculation at the infiltration spot in susceptible reactions ($PXO99^A$-IR24 or -IRBB7; $PXO99^A$/*avrXa7*-IR24), but the water-soaking is much weaker in resistant reaction ($PXO99^A$/*avrXa7*-IRBB7). The inoculation spot turns brown and bacterial spread is restricted in the resistant reaction 2–3 days after infiltration, while the symptoms in susceptible plants progress with much stronger water-soaking extending in both directions and eventually bacterial exudate appearing at the infiltration spot (Fig. 1e).

4. Notes

1. Other media are also suitable for culturing *X. oryzae*, including peptone–sucrose medium (per liter: peptone, 10 g; sucrose, 10 g; glutamic acid, 1 g; pH, 5.7), nutrient broth (per liter: beef extract, 3 g; peptone, 5 g; sucrose, 10 g), and glucose yeast extract (per liter: glucose, 20 g; yeast extract, 10 g).
2. We grow inoculum directly from the working stock for each inoculation experiment. There are reports that repeated subculture of bacteria could result in partial or severe loss of bacterial virulence.
3. Plants of different stages may be used for different purposes, such as measuring plant disease resistance that is developmentally regulated.

4. For easiest syringe infiltration, maintain rice plants at high humidity and inoculate at the beginning of the light cycle as this keeps stomata open.

References

1. Mew TW (1984) Scanning electron microscopy of virulent and avirulent strains of *Xanthomonas campestris* pv. *oryzae* on rice leaves. Phytopathology 74:635–641
2. Ou SH (1985) Rice diseases. Commonwealth Agricultural Bureau, Kew, Surrey
3. Kauffman HE, Reddy APK, Hsieh SPY, Merca SD (1973) An improved technique for evaluating resistance of rice varieties to *Xanthomonas oryzae* (bacterial blight). Plant Dis Rep 57:537–541
4. Yang B, White FF (2004) Diverse members of the AvrBs3/PthA family of type III effectors are major virulence determinants in bacterial blight disease of rice. Mol Plant Microbe Interact 17:1192–1200
5. Keen NT (1990) Gene-for-gene complementarity in plant-pathogen interactions. Ann Rev Genet 24:447–463
6. Wang L, Makino S, Subedee A, Bogdanove AJ (2007) Novel candidate virulence factors in rice pathogen *Xanthomonas oryzae* pv. oryzicola revealed by mutational analysis. Appl Environ Microbiol 73: 8023–8027

Chapter 19

Inoculation and Scoring Methods for Rice Sheath Blight Disease

Yulin Jia, Guangjie Liu, Dong-Soo Park, and Yinong Yang

Abstract

Sheath blight disease of rice caused by the soilborne fungal pathogen *Rhizoctonia solani* has been a major disease of rice with a serious threat to stable rice production worldwide. Although various cultural practices have been used to manage the disease, it is advantageous and important to screen rice germplasm and identify resistant rice cultivars for more effective disease control. Recent advances in methods for the fungal inoculation and disease evaluation have enabled a better measurement of host resistance by minimizing confounding factors from plant architectures and environmental conditions. This chapter introduces five such methods: (1) detached leaf method; (2) micro-chamber method; (3) mist-chamber method; (4) parafilm sachet method; and (5) aluminum foil method. These methods are useful for screening and evaluating disease reactions of rice germplasm and facilitating the genetic mapping of disease resistance genes.

Key words: Controlled inoculation, Disease reaction, Sheath blight, *Rhizoctonia solani*

1. Introduction

Phenotype reflects a combined effect of the expression of the genes of an organism and the influence of environmental factors in addition to possible interactions between them. Phenotyping is one of the most critical components for rice genetic studies and modern rice breeding. This is particularly true for a phenotype that is governed by multiple genetic loci and environmental factors, such as resistance to sheath blight disease caused by the soilborne fungal pathogen *Rhizoctonia solani* IG-1A (1). In nature, *R. solani* survives as sclerotia on plant residue in soil. After a permanent flood is established, *R. solani* sclerotia float and attach to the bottom of rice plants. The fungus grows from sclerotia and moves up along the rice sheath and infects the sheath after a certain period of vegetative growth. Sheath blight disease has been known to cause significant crop losses annually ever since the large-scale deployment

Yinong Yang (ed.), *Rice Protocols*, Methods in Molecular Biology, vol. 956,
DOI 10.1007/978-1-62703-194-3_19, © Springer Science+Business Media, LLC 2013

of semidwarf rice varieties grown under intensified cultural management. To date, sheath blight disease often rivals rice blast to become one of the most damaging rice diseases worldwide.

Over the years, classic phenotyping in rice paddy has allowed the identification of some quantitative resistance loci (QTLs). Some of these QTLs have been utilized in genetic studies and rice breeding programs worldwide (2). However, field evaluation can only be performed in rice-growing areas with longer growing season, larger space, and higher cost, and is strongly influenced by environmental conditions and plant architectures. Rapid and accurate phenotyping of sheath blight disease under highly controlled conditions should facilitate germplasm screening, QTL discovery, and evaluation of rice cultivars developed by traditional and molecular breeding. During the past decade, significant progress has been made in improving the method for sheath blight inoculation and evaluation under controlled conditions in a growth chamber or a greenhouse. These methods include (1) detached leaf (3); (2) micro-chamber (4); (3) mist-chamber (5); (4) parafilm sachet; and (5) aluminum foil (6). To date, these new methods have been successfully applied for gene expression studies and QTL mapping. The basic principles for these new methods were designed to reduce confounding plant and environmental factors in order to determine genetic contribution to overall plant responses after the infection. In comparison with classic field evaluation, these new methods have the following advantages: (1) independent of weather, these methods were performed under controlled lab and greenhouse conditions; (2) shorter time, the total time needed for one experiment is 2–12 weeks, versus 4–6 months of field evaluation; (3) less space, each experiment can be performed in relatively smaller areas; and (4) fewer seeds for each test, tests can be performed repeatedly from one single plant using detached leaf method. Thus far, these new methods have been effectively used for gene expression analysis after *R. solani* infection using message RNAs isolated from detached leaf method (5). Using micro-chamber and mist-chamber methods, one major and several new sheath blight-resistant QTLs were successfully identified in a rice cultivar (5). It is anticipated that these new methods can also be modified for studying other necrotrophic plant diseases.

2. Materials

1. Rice seeds of test entries, susceptible check cultivar "Lemont" and resistant check cultivar "Jasmine 85."
2. Sterilized soils.
3. Sclerotia of a *R. solani* isolate.
4. 2- or 3-L empty soft drink bottles.

5. Square Petri dish (24.3 cm×24.3 cm×1.8 cm) (PGC Scientific).
6. 12×12 cm plastic nursery pots with holes at the bottom.
7. 56×28×6 cm plastic trays.
8. Parafilm®.
9. Aluminum foil.
10. 3MM Whatman filter paper (22.5 cm×22.5 cm).
11. Petri dish (10.0 cm×1.5 cm).
12. Ruler.
13. 1-ml Eppendorf pipette tip.
14. Electronic digital caliper (ISO 17025).
15. Scissors.
16. Autoclave oven.
17. Microscopic slides.
18. Potato dextrose agar (PDA).
19. Potato dextrose broth (PDB).
20. Petri dishes (9 cm in diameter).
21. 200 μL and 1,000 μL pippetter tips.
22. 1,000 μL pipette tip.
23. Plastic box container.
24. Plastic chambers: 4 m×3 m×1.90 m with a 500 pot capacity (for mist-chamber method).
25. Fertilizer equivalent to NPK (180:60:60 kg/ha).

3. Methods

3.1. Detached Leaf Inoculation Method

1. Remove the second youngest leaf from rice plants at the 4-leaf to heading stage, and leaves are cut into 16-cm-long segments with scissors (see Note 1).
2. Immediately place these detached leaves into square Petri dishes, each containing a 3 MM Whatman filter paper moistened with sterilized water (see Note 2).
3. Excise mycelial disks (0.7 cm diameter) with a 1-ml Eppendorf pipette tip using a 3-day-old mycelial culture grown on 20 ml PDA containing tetracycline (0.005%w/v) in a Petri dish (100×15 mm).
4. Remove a 0.7 cm diameter PDA plug from similar periphery of the culture by using a sterilized toothpick (Fig. 1a).
5. Place a PDA plug near the middle of the abaxial surface of leaf segments (Fig. 1b). As a control, leaf segments from the same cultivar are also inoculated with a PDA plug without mycelia.

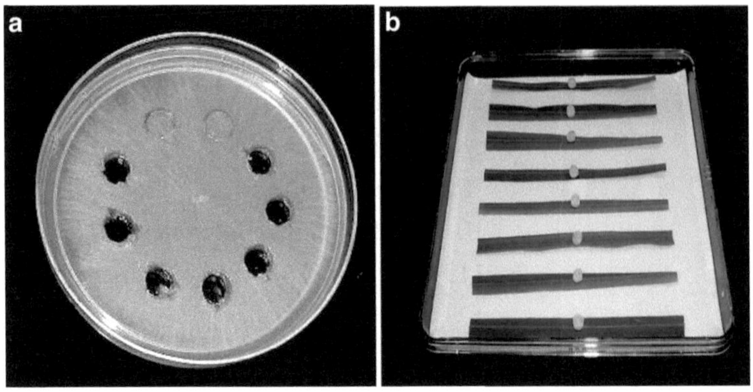

Fig. 1. Schematic presentation of detached leaf method. (**a**) Mycelia grown on a Petri Dish showing the position of plug removal; (**b**) detached leaves inoculated with the fungus on mycelia plug.

6. Incubate Petri dishes containing inoculated leaf segments on a laboratory bench at 21–24°C under continuous fluorescent light (10–20 μE/m^{2}s) for 72 h.
7. Measure the length of disease lesions from the inoculation site to both ends of disease lesions with a ruler. An average of measurements can be used to determine the final disease reaction of each cultivar or transgenic line.
8. Disease reactions of cultivars to *R. solani* isolates are determined based on the lesion length as resistant (R, <3.0 cm), moderately resistant (MR, 3.0–5.0 cm), moderately susceptible (MS, 5.1–6.5 cm), and susceptible (S, >6.5 cm).

3.2. Micro-Chamber Method

1. Germinate seeds on moist filter paper in Petri dishes for 3 days at 30°C to obtain uniform growth of rice seedlings.
2. Transfer healthy and fully germinated seedlings to a 15-cm-diameter pot using sterilized soil, and grow them in a greenhouse until they reach two- to three-leaf stage.
3. Remove mycelial disks approximately 0.7 cm diameter with a 1-ml Eppendorf pipette tip using a 3-day-old mycelial culture grown on PDA medium containing tetracycline (0.005%w/v). To maintain uniformity, the mycelial disks must be excised from similar radii in Petri dishes.
4. Place the mycelia disks onto the stem base of each seedling in a pot using sterile toothpicks. Make sure that mycelia attach to the stem of seedlings.
5. Immediately after inoculation, cover each pot with a clean and clear 2- or 3-L plastic soft drink bottle, with the bottom of the bottle removed.
6. Push soft drink bottle slightly into the soils to maintain humidity and enhance infection. Leave the bottles uncapped (Fig. 2).

Fig. 2. Presentation of micro-chamber method. Pictures showing inoculated rice cultivars, Jasmine 85 and Lemont in soft drink bottles ready for evaluation.

7. Place the inoculated pot on the tray and add water to the tray so that the seedlings can receive water from the bottom of the pot. Fill water up to one-fourth height of the tray.
8. When plants grow to the top of the bottle (approximately 5–10 days after inoculation in the summer), carefully remove the bottles. On the following day, measure disease lesion length on the culm and the full height of the culm of each seedling and calculate disease lesion length in proportion to the culm length and convert to percentage (see Note 3).
9. Disease reaction is evaluated as a relative value to the susceptible and resistant controls. A visual rating is recommended for verification.

3.3. Mist-Chamber Method

1. Grow *R. solani* on PDA medium for 3 days at 30°C.
2. Grow rice plants in pots containing steam-sterilized soil. Pots are also steam sterilized after evaluation. Sterilized soil and pots can be reused several times.
3. Inoculate 50-day-old rice plants (flag leaf well developed) by placing an agar plug (0.5 cm diameter) with mycelia at the bottom or the lower part of the main tiller of each rice plant.
4. Grow 5 plants per pot (15 cm diameter) and inoculate each plant in the pot. Inoculate a minimum of three replications for a total of 15 plants per rice line/cultivar.
5. After inoculation, incubate rice plants for two time periods in a plastic growth chamber (Fig. 3) at a high relative humidity by

Fig. 3. Presentation of mist-chamber method. Picture showing two humidifiers in a chamber with inoculated plants in a mist-chamber.

turning on four humidifiers 1 h in the morning (10–11 am) and 1 h in the afternoon (2–3 pm). After the first incubation period for 10 days, rice plants are moved to a greenhouse bench for 5 days (see Note 4, or just leave the plastic door open without moving the plants out).

6. Move the plants back for a second incubation period of 10 days before evaluation. The plastic doors of the chamber are kept closed during the two incubation periods; however, they should be partially open if temperature reaches 32°C. It might be necessary to adjust the time humidifiers are on and the doors partially open to maintain a high relative humidity and a temperature between 27 and 32°C inside the chamber. Plants should be watered if necessary to avoid drying out.

7. After the second incubation period, the plants are removed to a greenhouse bench and the inoculated tillers are evaluated for sheath blight reaction. To measure the reaction, a modified scale based on the percentage of area affected is used (Fig. 4): 30% maximum for flag leaf area with sheath blight symptoms; 15% for each of the second and third leaves; 10% for each of the fourth and fifth leaves; and 20% maximum for the stem area affected. The line/cultivar reaction is the average for the 15 plants inoculated.

8. For each inoculation/replication, a set of controls with known field sheath blight reaction (resistant, intermediate, and susceptible) is included to determine disease reaction.

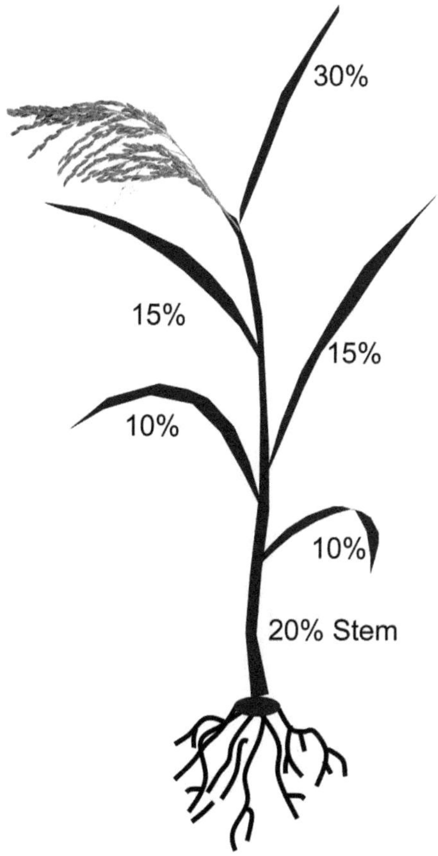

Fig. 4. Schematic presentation of rating standard for mist-chamber method. Percentages of disease lesions were estimated based on lesion length and position of each leaf in a plant.

3.4. Parafilm Sachet Method

1. Sow three rice seeds in a 12 cm × 12 cm plastic nursery pot filled with sterilized soils: Eight pots are lined in the plastic tray (56 cm × 28 cm × 6 cm) containing water at the depth of 4 cm.
2. At 3 weeks after sowing, rice plants are thinned to a single plant per pot. Plants grown to as early as the four-leaf stage can be used.
3. Cut parafilm into the 10 cm × 10 cm pieces.
4. Parafilm is half folded into 5 cm × 10 cm and then folded again to 2.5 cm × 2.5 cm to make a crease.
5. Unfold parafilm piece and lay on the top of a 1,000 μL pippetter tip support. At the center right half of the 10 cm × 10 cm pieces, a 200 μL pipette tip is gently pressed down through a hole on the support, making a 5-mm-depth indentation to accommodate an inoculum disk.
6. Grow *R. solani* by placing sclerotia onto PDA medium and then transfer newly grown mycelia to fresh medium to produce active mycelia.

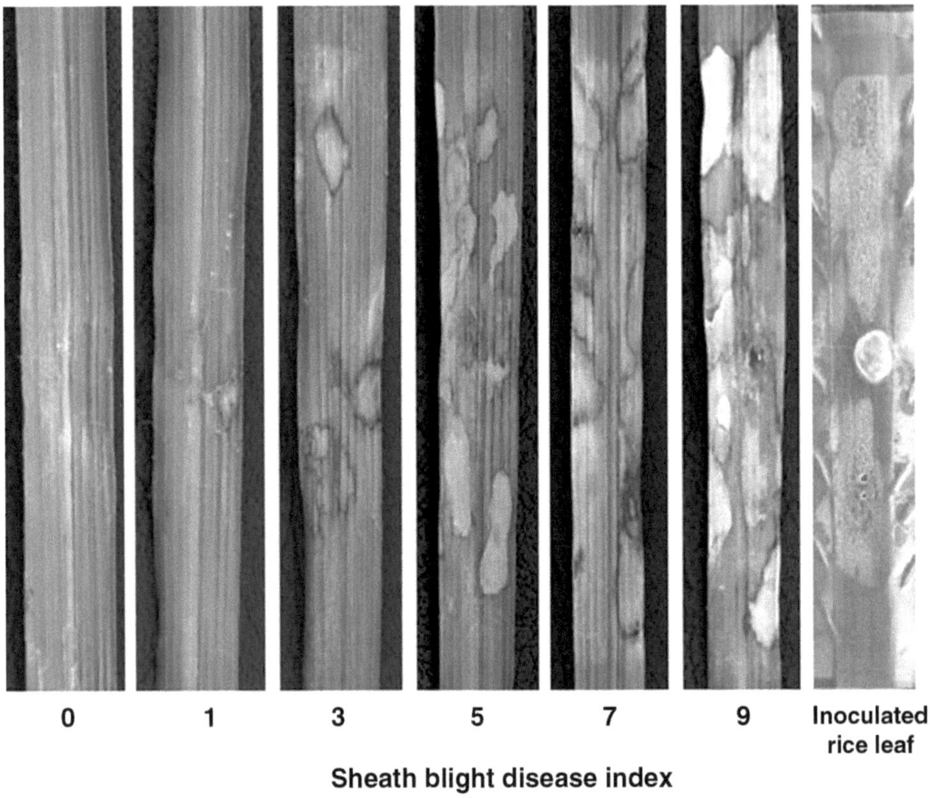

Fig. 5. Rating standard of sachet method. (**a**) 0 indicates no infection, 1 indicates that infection restricts at sites of inoculation, 3 indicates that restricted infection occurs at the site and within 2 cm of the site of inoculation, 5 indicates unrestricted infection within 4–5 cm of the total infected areas, 7 indicates that expanded infection is greater than 5 cm length of the inoculated area, and 9 indicates that the complete destruction of infection is greater than 6 cm length of the inoculated areas; an inoculated leaf with a PAD plug covered by parafilm 3 days after inoculation.

7. Pick a round disk approximately 0.7 cm diameter from similar periphery of the culture using a pipette tip.
8. Place one inoculum disk in an indentation of a Parafilm sachet.
9. Enclose at the middle of the second youngest leaf of plant. The inoculated plants are lined in the plastic trays laid on benches in a shaded greenhouse.
10. Evaluate disease reaction at 7–10 days after inoculation by first removing the Parafilm sachets from rice leaves, and then determine the percentage of the diseased area under the coverage of a sachet (10 cm), visually using the diseased area evaluation chart (Fig. 5). Disease scores are converted using the evaluation standard (Table 1).

3.5. Aluminum Foil Method

1. Inoculate a piece of mycelia or sclerotia on PDA and culture at room temperature (22–24°C) under continuous light for 3 days (Fig. 6a).

Table 1
The evaluation standard for disease rating based on the percentage of the diseased areas caused by *R. solani*

Disease score	Percentage of the diseased area (%)
0	0
1	<10.0
3	10.1–30.0
5	30.1–50.0
7	50.1–70.0
9	>70.1

2. Inoculate a piece of fresh mycelia from PDA into 200 ml of PDB, and culture on a shaker (130–140 rpm) at 26–28°C under darkness (Fig. 6b).
3. After 7 days of liquid culture, fungal mycelia are harvested and cut into small pieces (about 0.5 cm in diameter, see Note 5, Fig. 6c).
4. Using forceps, inoculate mycelial balls beneath the leaf sheath (Fig. 6d) of rice plants (late tillering stage, about 2 months old), then immediately cover with aluminum foil (Fig. 6e), and place rice plants into a micro-chamber in a greenhouse.
5. Sheath blight lesions typically appear 3 days later (Fig. 6f). The aluminum foil can then be removed (Fig. 6g).
6. *R. solani*-infected plants are left in a micro-chamber for about 18 days to allow disease development (the period may vary depending on environmental conditions). Disease evaluation (Fig. 6h, i) is generally conducted when the control cultivar (Nipponbare, Fig. 6h) exhibits an index of 70% disease susceptibility.
7. To evaluate disease resistance, lesion length and degree of disease severity in each sheath of inoculated plants are recorded. Disease severity values are assigned as follows: a value of 0 represents no lesion; 1 represents the appearance of water-soaked lesion; 2 represents the appearance of necrotic lesion; 3 represents less than 50% necrosis on the leaf cross section; 4 represents more than 50% necrosis on the leaf cross section; and 5 represents necrosis across the entire leaf section resulting in leaf death.
8. The index of disease susceptibility was calculated as follows: Susceptibility Index $= (5n5 + 4n4 + 3n3 + 2n2 + 1n1 + 0n0)/5\,N \times 100$, where $n1$–5 is the number of leaves in each degree (1–5),

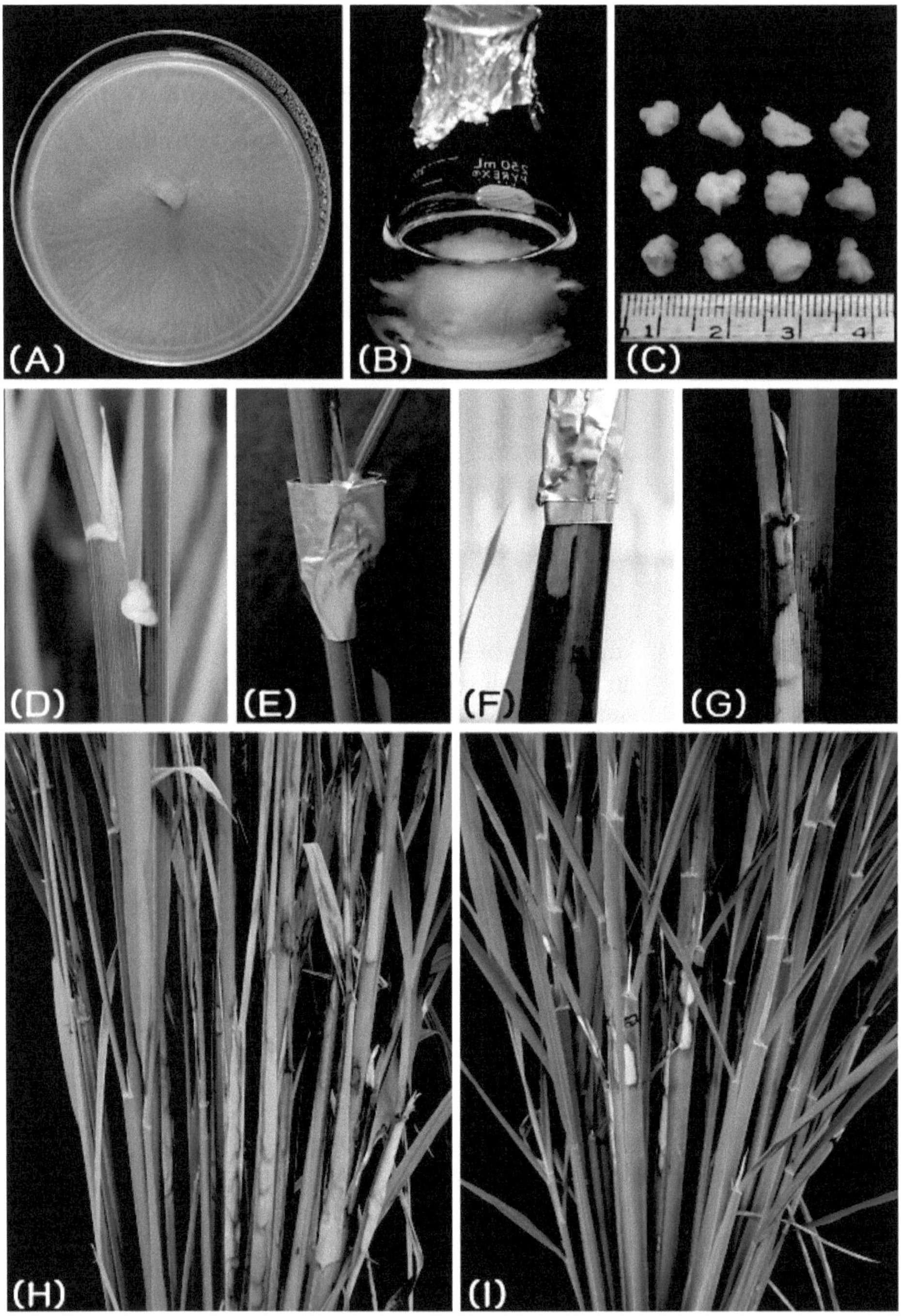

Fig. 6. Inoculation of *R. solani* on rice plants using the aluminum foil method. (**a**) 3 days after inoculation of *R. solani* on potato dextrose agar. (**b**) 7 days after inoculation in potato dextrose broth. (**c**) Mycelia balls harvested 7 days after inoculation. (**d**) Inoculation of mycelia ball beneath leaf sheath. (**e**) Inoculum and stem covered with aluminum foil. (**f**, **g**) 3 days after inoculation with and without aluminum foil. (**h**) Susceptible Nipponbare cultivar and (**i**) resistant Jasmine 85 cultivar showing a different level of disease development.

and N is the number of total leaves investigated in a culm. All the lesions from a culm are added together to give a total lesion length for each culm. An individual culm is considered a replication and 25 replications are used for each cultivar. The analysis of variance of the susceptibility index and the lesion length is performed using SAS or other statistical software.

4. Notes

1. Repeated tests are required to obtain accurate measurement. It is recommended that three seedlings with four replications are needed for determining disease reaction of each cultivar/line. It is also recommended that known susceptible (for example cv. Lemont) and resistant (cv. Jasmine 85) cultivars are used as controls.
2. In order to prevent leaf rollover, avoid excessive exposure to light and maintain adequate moisture by adding sterilized water for detached leaf method.
3. Seedlings may be removed from the soil surface and lesions can be measured using an electronic digital caliper (ISO 17025) or a ruler scale (mm). If further observation is required, measure the seedlings in the pot.
4. The 5-day intermediate period on the greenhouse bench allows a plant recovery and expression of resistance for better differentiation between susceptible and resistant plants during the second incubation period.
5. In order to prepare a large number of mycelial balls for the aluminum method, 96-well plates can be used to make small mycelia balls (approximately 0.5 cm in diameter) uniformly and quickly.

Acknowledgements

The authors thank Ellen McWhirter for proofreading, Anna McClung of USDA-ARS Dale Bumpers National Rice Research Center, and Fernando Correa-Victoria of RiceTec for technical assistance. This work was supported in part by the USDA Cooperative State Research, Education and Extension Service—National Research Initiative—Applied Plant Genomics Program (USDA/CSREES grant 2004-35317-14867) and by USDA-ARS National Program NP301 project "Response of Diverse Rice Germplasm to Biotic and Abiotic Stresses."

References

1. Wamishe Y, Jia Y, Singh P, Cartwright RD (2007) Identification of field isolates of *Rhizoctonia solani* to detect quantitative resistance in rice under greenhouse conditions. Front Agric China 1:361–367
2. Jia Y, Liu G, Costanzo S, Lee S, Dai Y (2009) Current progress on genetic interactions of rice with rice blast and sheath blight fungi. Front Agric China 3:231–239
3. Venu RC, Jia Y, Gowda M, Jia MH, Jantasuriyarat C, Stahlberg E, Li H, Rhineheart A, Boddhireddy P, Singh P, Rutger JN, Kudrna D, Wing R, Nelson JC, Wang G (2007) RL-SAGE and microarray analysis of the rice transcriptome after *Rhizoctonia solani* infection. Mol Genet Genomics 278:421–431
4. Jia Y, Correa-Victoria F, McClung A, Zhu L, Liu G, Wamishe Y, Xie J, Marchetti MAM, Pinson SR, Rutger JN, Correll JC (2007) Rapid determination of rice cultivar responses to the sheath blight pathogen *Rhizoctonia solani* using a micro-chamber screening method. Plant Dis 91:485–491
5. Liu G, Jia Y, Correa-Victoria F, Prado GA, Yeater KM, McClung A, Correll JC (2009) Mapping quantitative trait loci responsible for resistance to sheath blight in rice. Phytopathology 99:1078–1084
6. Park D-S, Sayler R, Hong Y-G, Nam M-H, Yang Y (2007) A method for inoculation and evaluation of rice sheath blight disease. Plant Dis 92:25–29

Chapter 20

Molecular Approaches to Improve Rice Abiotic Stress Tolerance

Junya Mizoi and Kazuko Yamaguchi-Shinozaki

Abstract

Abiotic stress is a major factor limiting productivity of rice crops in large areas of the world. Because plants cannot avoid abiotic stress by moving, they have acquired various mechanisms for stress tolerance in the course of their evolution. Enhancing or introducing such mechanisms in rice is one effective way to develop stress-tolerant cultivars. Based on physiological studies on stress responses, recent progress in plant molecular biology has enabled discovery of many genes involved in stress tolerance. These genes include regulatory genes, which regulate stress response (e.g., transcription factors and protein kinases), and functional genes, which protect the cell (e.g., enzymes for generating protective metabolites and proteins). Both kinds of genes are used to increase stress tolerance in rice. In addition, several quantitative trait loci (QTLs) associated with higher stress tolerance have been cloned, contributing to the discovery of significantly important genes for stress tolerance.

Key words: Abiotic stress, Stress tolerance, Transgenic rice, Transcription factor, Protein kinase, Metabolic engineering, Stress-inducible promoter

1. Introduction

Abiotic stresses (such as high salinity, drought and flood, high and low temperatures) are major factors that reduce the agricultural productivity of rice (*Oryza sativa*). Improving the abiotic stress tolerance of rice will increase agricultural productivity and extend cultivatable areas, thus supporting an increasing population. There is, therefore, a strong demand for abiotic stress-tolerant rice cultivars.

Abiotic stress tolerance is a very complex trait. It is difficult to develop significantly stress-tolerant rice cultivars by conventional breeding because of the many factors that influence stress tolerance and because of difficulties in reproducing stress strength and duration in the field. Genetic engineering, however, appears to be an effective way to develop rice cultivars with improved stress tolerance.

Yinong Yang (ed.), *Rice Protocols*, Methods in Molecular Biology, vol. 956,
DOI 10.1007/978-1-62703-194-3_20,

Genes which have been shown to be involved in stress tolerance can be introduced by transgenic methods or by introgression of a tolerant allele.

There are three possible strategies to improve stress resistance:

1. To change the life cycle of crops, enabling them to *escape* from stress, e.g., by growing and flowering only during more favorable seasons;
2. To modify the architecture of organs or tissues to *avoid* development of stressful internal environments in plants, such as extremely high osmotic potentials and salt concentrations in tissues;
3. To improve tissue and cell viability in order to *tolerate* stressful internal environments.

Physiological and molecular biological studies on plant abiotic stress tolerance have been predominantly undertaken using the third strategy above, by investigating cellular responses. In this review, we will therefore focus on the application of that strategy.

Genes involved in stress tolerance can be categorized into regulatory genes and functional genes (Fig. 1). Regulatory genes such as protein kinases and transcription factors can regulate multiple cellular processes via signal transduction and gene expression. Functional genes such as enzymes and protective proteins can enhance specific features of stress tolerance. As we describe below, both types of genes have been reported to enhance the stress tolerance of rice.

2. Regulatory Genes

During the course of plant adaptation to abiotic stress, stress signals are received by specific receptors and transduced by signaling components such as protein kinases and phosphatases. These signals activate transcription factors, which enhance transcription of downstream stress-tolerance genes (Fig. 1). Modifying the expression of regulatory genes is an efficient strategy to improve stress tolerance in plants because it mimics or enhances stress signals and enables regulation of many downstream genes at once.

2.1. Transcription Factors

Transcription factors regulate expression of downstream genes by binding to *cis*-acting elements in their promoters. Overrepresentation of transcription factor genes in plant genomes (1, 2) and high expansion rates of transcription factor gene families during evolution of plants (3) suggest the important role of transcription factors in the adaptation of plants to various environments. In fact, several transcription factors are responsible for stress-related quantitative trait loci (QTLs), indicating the significance of transcription factors

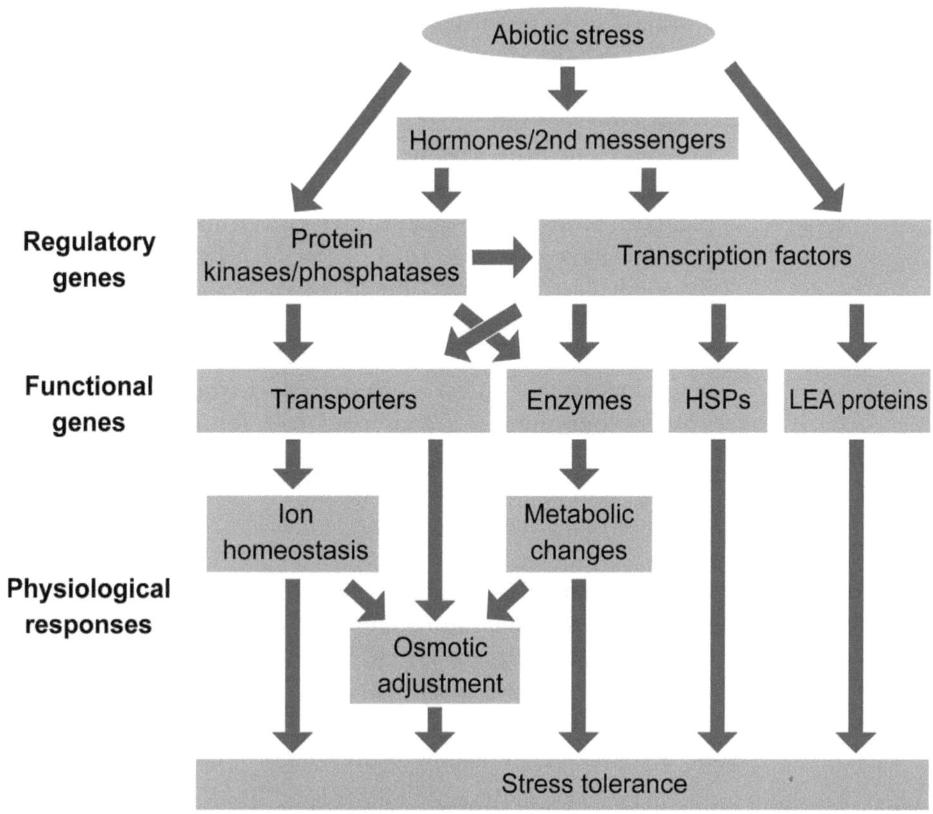

Fig. 1. Schematic model for molecular responses to abiotic stress in plants. When plants are exposed to abiotic stress, stress signals activate regulatory genes, which can be dependent or independent of hormones or second messengers. In turn, regulatory genes act on many functional genes. Expression of functional genes can lead to protection of cells, either directly or by causing a physiological response in the cells. Plants acquire abiotic stress tolerance via either mechanism. Abbreviations: *HSP* heat shock protein, *LEA* late embryogenesis abundant.

as key regulators of stress tolerance in natural environments. The DREB1/CBF family of transcription factors, for example, is associated with QTLs for freeze-tolerance in Arabidopsis (*Arabidopsis thaliana*) (4) and in grass species such as barley (*Hordeum vulgare*) (5), wheat (*Triticum aestivum*) (6), and diploid wheat (*Triticum monococcum*) (7). In addition, the ethylene response factor (ERF)-like transcription factor Sub1A has been identified for submergence tolerance in rice (8, 9) by cloning of a major QTL.

2.1.1. DREB1/CBF Transcription Factors

The *DREB1/CBF* family of genes was first identified in Arabidopsis by isolation of proteins that bind to the dehydration-responsive element/C-repeat (DRE/CRT), a *cis*-acting element found in promoters of dehydration- or cold-inducible genes (10, 11). In *Arabidopsis*, two families of APETALA2 (AP2)/ERF-type transcription factors, DREB1/CBF and DREB2, were identified as DRE/CRT binding proteins (10, 11). *DREB1/CBF* genes are immediately induced by cold stress and induce expression of cold-inducible genes, while *DREB2* genes are induced by drought, salt, and heat stress (10–12). Transgenic Arabidopsis strains expressing *DREB1A/CBF3* under the control of constitutive or stress-inducible promoters exhibit remarkably enhanced freezing tolerance as well as drought and salt tolerance (11, 13). *DREB1/CBF* genes have been isolated from various plant species, including grass species, and have been shown to enhance abiotic stress tolerance in other plants (14, 15). Rice plants that constitutively express *DREB1* genes from *Arabidopsis* (16, 17), rice (16, 18, 19), and barley (20) have enhanced drought, salt and cold tolerance. These plants show enhanced expression of a number of stress-inducible genes (16, 17, 20), and accumulate metabolites such as proline and sugars at a level comparable to stress-treated wild-type plants (16). These observations suggest a conserved role for *DREB1* genes in the acquisition of abiotic stress tolerance between plant species, and indicate the potential usefulness of *DREB1* genes for improving abiotic stress tolerance of rice.

2.1.2. DREB2 Transcription Factors

DREB2 genes have been isolated from grass species. Most were found to respond to high salinity and drought conditions (14), but because the function of DREB2 transcription factors is tightly regulated by alternative splicing in grass species (21) and by posttranslational regulation at least in *Arabidopsis* (22, 23), simple expression of *DREB2* genes is not promising for crop improvement. In *Arabidopsis*, artificial modification of the *DREB2A* gene was necessary for expression of downstream genes and enhancement of stress tolerance in transgenic plants (23). However, one report indicated that transgenic rice seedlings overexpressing the rice *OsDREB2B* gene showed a higher survival rate than the non-transgenic control after drought stress (18), although the molecular mechanisms of the drought tolerance were not clear.

2.1.3. Other AP2/ERF Transcription Factors

The *HARDY* (*HRD*) gene of *Arabidopsis* is a close relative to the *DREB1* genes. *Arabidopsis* plants overexpressing *HRD* show strong highly branched roots, thick mesophyll tissue, high water use efficiency, and increased tolerance to drought and salt. Overexpression of *HRD* in rice also results in altered shoot tissue structures, higher water use efficiency, and improved drought resistance (24). Rice may have *HRD* homologs that cause these responses during growth under water-deficit conditions.

Submergence 1A (Sub1A), a transcription factor that belongs to the ERF subfamily of AP2/ERF transcription factors, was identified by cloning of *Sub1*, a major QTL for submergence tolerance (8, 9). The *Sub1A-1* allele from a submergence-tolerant *indica* cultivar, FR13A, confers submergence tolerance to an intolerant *japonica* cultivar, Liaogeng, when ectopically expressed as a transgene. Introgression of the FR13A *Sub1* haplotype into an intolerant *indica* variety, Swarna, by marker-assisted selection also confers submergence tolerance.

2.1.4. bZIP Transcription Factors

The plant hormone abscisic acid (ABA) plays an important role in the physiological response of plants to abiotic stress and developmental roles in seed maturation and dormancy (25). Many drought- and high-salinity-inducible genes in rice respond to ABA (26), suggesting an important role in gene expression under these conditions. One of the major *cis*-acting elements that regulate gene expression in response to ABA is the ABA-responsive element (ABRE). Basic leucine zipper (bZIP) transcription factors TRAB1 and AREB/ABFs were isolated as ABRE binding proteins from rice (27) and Arabidopsis (28, 29), respectively. These proteins belong to the same subfamily of bZIP proteins and have conserved amino-terminal domains. They are activated by phosphorylation in response to ABA (30, 31). Overexpression of a constitutive active form or wild-type AREB/ABFs enhances drought tolerance of transgenic *Arabidopsis* (32, 33) and rice (17). Recently, overexpression of OsbZIP23, a close relative to TRAB1, was reported to enhance ABA sensitivity and expression of stress-inducible genes (34). This correlated with improved drought and salt tolerance of transgenic rice. Furthermore, a null mutant of *OsbZIP23* showed decreased tolerance to drought and salinity, suggesting a major role of this gene in ABA-dependent acquisition of drought and salt tolerance (34).

2.1.5. NAC Transcription Factors

NAM, ATAF, and CUC (NAC) proteins were first identified as a group of transcription factors that function in plant development. Several members of the NAC family have also been shown to be involved in abiotic stress responses in *Arabidopsis* (35, 36). In rice, *OsNAC6* is induced by ABA and abiotic stresses such as cold, drought, and high salinity, as well as methyl jasmonate, wounding, and blast disease (37). Expression of *OsNAC6* under the control of both constitutive and stress-inducible promoters results in improved drought and salinity tolerance in transgenic rice (37). While constitutive expression causes growth retardation, expression from stress-inducible promoters such as *OsNAC6* and *LIP9* promoters significantly reduces negative effects on growth, thus demonstrating the utility of stress-inducible promoters. Other stress-inducible *NAC* genes, *SNAC1* and *SNAC2*, have also been reported to enhance water-stress tolerance in transgenic rice (38, 39). Notably, transgenic rice overexpressing *SNAC1* exhibit improved drought tolerance at reproductive stage in the field (38), suggesting the potential utility of stress-inducible *NAC* transcription factors in developing abiotic stress-tolerant rice cultivars.

2.1.6. Other Transcription Factors

OsWRKY11, a heat- and drought-inducible member of the WRKY transcription factor family, enhances tolerance of rice to drought, heat, and even a combination of drought and heat, when overexpressed under the control of the heat-inducible *HSP101* promoter (40).

2.2. Protein Kinases

Protein kinases are involved in various stages of plant development and adaptation. The SnRK2 family of protein kinases is known to be expressed or activated in response to water deficiency and ABA in many species (41), including rice (42). Analysis of SnRK2 mutants in *Arabidopsis* showed the roles of SnRK2s in stomata regulation (43) and gene expression during drought (44, 45). One of the important roles of SnRK2s is activation of AREB/ABF transcription factors in *Arabidopsis* (31) and rice (46) via ABA-dependent phosphorylation. Overexpression of the SnRK2 gene *SAPK4* also increases salt tolerance in transgenic rice (47).

Two protein kinase families, calcium-dependent protein kinase (CDPK) and calcineurin B-like proteins-interacting (CBL-interacting) protein kinase (CIPK), mediate calcium-dependent signal transduction in response to abiotic stress (41). In rice, overexpression of *OsCDPK7* conferred cold, drought, and salt tolerance (48), whereas overexpression of the *OsCDPK13* gene confers only cold tolerance (49). CIPKs are activated by interacting with CBL calcium sensors; the function of several CBL-CIPK signaling pathways has been thoroughly studied in the salt-stress response of *Arabidopsis* (50). From a comprehensive study of the rice CIPK family, 20 out of 30 CIPK members were shown to be stress-inducible, and overexpression of *OsCIPK03*, *OsCIPK12*, and *OsCIPK15* resulted in enhanced cold, drought, and salt tolerance in transgenic rice, respectively (51).

MAP kinase cascades are also known to be activated by abiotic stress (41). *OsMAPK5*, which was identified as a blast-infection-inducible gene, is also activated by abiotic stress (52). Transgenic rice that overexpress *OsMAPK5* exhibit reduced tolerance to biotic stress, but enhance tolerance to cold, salt, and drought. Conversely, suppression of *OsMAPK5* expression results in reduced tolerance to these abiotic stresses.

3. Functional Genes

Adaptation of plants to abiotic stress involves cellular physiological responses such as accumulation of specific metabolites, synthesis of protective proteins, and the alteration to the lipid composition of membranes (Fig. 1). One way to specifically induce such physiological responses is through the use of functional genes that encode enzymes or protective proteins directly involved in these processes. This approach is effective in targeting important known processes arising from abiotic stress conditions.

3.1. Enzymes for Synthesis of Soluble Metabolites

Accumulation of low-molecular-weight organic compounds that are highly soluble in water and nontoxic at high concentrations (i.e., "compatible solutes") in response to abiotic stress is an important physiological survival mechanism for plants. Plants accumulate a variety of molecules such as sugars, amino acids, and betaines depending on the species of plant (53). In addition to compatible solutes, several metabolites accumulate at relatively low concentrations in response to abiotic stress.

3.1.1. Proline

Proline is an important compatible solute in rice. Δ^1-Pyrroline-5-carboxylate synthetase (P5CS) and proline dehydrogenase (ProDH) are key enzymes for proline synthesis and catabolism, respectively. Rice has two *P5CS* genes which are induced by ABA, cold, high salinity, and osmotic stress, and rice accumulates proline in response to salt or osmotic stress (54–56). Knockout of the *OsP5CS2* gene in rice results in slow growth under salt or cold conditions (56), suggesting the importance of *OsP5CS2* in salt and cold tolerance. In addition, expression of a mothbean (*Vigna aconitifolia*) P5CS enhances tolerance to water and salt stress in transgenic rice (57). In *Arabidopsis*, suppression of the *ProDH* gene enhances salt and freezing tolerance (58).

3.1.2. Glycine Betaine

Glycine betaine (GB) is a quaternary amine, synthesized in response to abiotic stress, and found ubiquitously in higher plants (59). Several plant species have been genetically engineered to accumulate GB in order to increase abiotic stress tolerance (60). In plants, GB is synthesized from choline via two sequential enzymatic reactions catalyzed by choline monooxygenase (CMO) and betaine aldehyde dehydrogenase. In *Arthrobacter*, GB is synthesized directly from choline by choline oxidase. Rice does not accumulate GB because it lacks a functional CMO gene (61). However, introduction of a spinach CMO gene (61) or a choline oxidase gene (62–64) into rice results in accumulation of GB and enhanced tolerance to salt or cold. It has also been shown that targeting choline oxidase to chloroplasts is effective for protecting rice plants from photoinhibition under salt or cold conditions (62).

3.1.3. Polyamines

Polyamines are also implicated in protective function against abiotic stress. Expression of polyamine biosynthetic genes such as an arginine decarboxylase from jimsonweed (*Datura stramonium*) (65) and *S*-adenosylmethionine decarboxylase from tritordeum (66) in transgenic rice increases polyamine levels and confers drought and salt tolerance, respectively.

3.1.4. Sugars and Polysaccharides

Fructans are fructose polymers based on sucrose and are found in vacuoles of various plants. In addition to their role in carbon storage, fructans may have a role in abiotic stress tolerance (67). Rice lacks enzymes for fructan biosynthesis and does not accumulate fructans (68). However, introduction of wheat fructan-synthesizing enzymes into rice results in accumulation of fructans and improved survival after cold stress (69).

Although the function of trehalose as an osmoprotectant is implicated in stress tolerance of a variety of organisms, it seems that trehalose plays a role in regulation of sugar metabolism and development in angiosperms (70). However, trehalose may also have some role in abiotic stress response, because a slight accumulation of trehalose is found in rice exposed to drought, salt, and cold (71, 72). Enhancement of trehalose metabolism improves salt and drought tolerance of transgenic rice (71, 73, 74), although enhanced expression of stress-inducible genes suggests indirect effects of trehalose biosynthesis in abiotic stress tolerance of rice (73, 74).

Transgenic plants (other than rice) have been engineered to accumulate several other sugars, such as mannitol, d-ononitol, and sorbitol (53). Oligosaccharides in the raffinose family, such as galactinol and raffinose, may also enhance abiotic stress tolerance (75). Further evaluation will be needed to determine which carbohydrates are suitable for improving the abiotic stress tolerance of rice.

3.2. Enzymes for Membrane Lipid Biosynthesis

The level of unsaturated fatty acids in membrane lipids directly affects the behavior of the membrane at different temperatures. High-unsaturation levels of membrane lipids contribute to cold tolerance, and low unsaturation levels contribute to heat tolerance (76). Thus, engineering unsaturation levels is a direct way to change optimal growth temperature.

The level of *cis*-unsaturated fatty acids in plastidial phospholipid phosphatidylglycerol (PG) is largely determined by substrate specificity of glycerol-3-phosphate acyltransferase (GPAT) and is an important factor that affects cold tolerance in plants (76). In rice, a comparative analysis of cultivars has shown a relationship between the levels of unsaturated fatty acids in PG and reduced photoinhibition at a cold temperature (77). The cold-sensitive tobacco plant has been engineered to be cold tolerant by transformation with a plastid GPAT from the cold-tolerant plant, *Arabidopsis* (78). Similarly, transgenic rice overexpressing GPAT cDNAs from *Arabidopsis* or spinach exhibit increased levels of *cis*-unsaturated fatty acids in PG and higher cold tolerance (79, 80).

The level of unsaturated fatty acids in bulk membrane lipids also has a strong relationship with temperature adaptation (76). The conversion from saturated to unsaturated fatty acids is catalyzed by fatty acid desaturases in the endoplasmic reticulum and plastids. Overexpression, co-suppression, or mutation of fatty acid desaturase genes can modulate levels of unsaturated membrane lipids and, as a result, change the cold or heat tolerance of several plants, including rice (81, 82).

3.3. Protective Proteins

In addition to metabolites, plants accumulate protective proteins in response to abiotic stress. Late embryogenesis abundant (LEA) proteins are associated with water stress, whereas heat shock proteins (HSPs) are associated with heat stress.

3.3.1. LEA Proteins

LEA proteins include several families of very hydrophilic proteins that accumulate during seed maturation and abiotic stress conditions such as drought and cold (83). Most meet the definition of hydrophilins, a group of hydrophilic proteins distributed across all domains of organisms, including eubacteria and archaea. Although their actual mode of function is unclear, they are considered to protect proteins and membranes in dehydrated cells, and have been shown to enhance dehydration tolerance of transgenic plants.

The rice genome has 35 LEA genes (84), several of which were identified as abiotic stress-inducible genes (26, 85, 86). Overexpression of barley *HVA1* in rice conferred tolerance to water deficit and salt stress (87, 88). The *OsLEA3-1* gene was also expressed in transgenic rice by constitutive promoters (*CaMV35S* and *Actin1*) or a stress-inducible promoter (*OsLEA3-1*) (89). After screening the transgenic lines for single-copy insertions, overexpression of a transgene, and better performance under drought conditions, homozygous lines were tested for drought resistance in the field. Transgenic lines utilizing *CaMV35S* or *OsLEA3-1* promoters, but not the *Actin1* promoter, showed significantly improved grain yield under drought conditions, demonstrating the critical importance of promoter selection.

3.3.2. HSPs

HSPs include five families (Hsp100/ClpB, Hsp90/HtpG, Hsp70/DnaK, Hsp60/GroEL, and small HSP) and are thought to act as molecular chaperones (90). Many plant HSPs are induced by heat stress and their role in heat tolerance has been shown in transgenic or mutant plants. Some HSPs have also been found to have a developmental function (90). Overexpression of an *HSP101* gene from *Arabidopsis* enhances heat tolerance of transgenic rice (91). In addition, both transgenic rice overexpressing a rice small HSP gene and wild-type rice pre-exposed to heat stress show enhanced drought tolerance (92).

3.4. Transporters

Cation transporters contribute to maintaining cellular ion homeostasis under salt stress. Several members of transporters located in the plasma membrane have been characterized and found to prevent Na^+ accumulation in tissues. SKC1/OsHKT1;5 is a Na^+ transporter in the high-affinity K^+ transporter (HKT) family. SKC1/OsHKT1;5 is located in the plasma membrane and was identified by positional cloning of a QTL whose allele from a salt-tolerant *indica* variety was associated with higher K^+ content in shoots under salt stress. It is proposed that *SKC1* contributes to salt tolerance by unloading Na^+ into root xylem parenchyma cells from xylem sap (93). In *Arabidopsis*, the plasma membrane Na^+/H^+ antiporter SALT OVERLY SENSITIVE1 (SOS1) is proposed to control long-distance Na^+ transport from root to shoot (94). SOS1 improves salt tolerance of transgenic *Arabidopsis* when overexpressed under the control of the *CaMV35S* promoter. A functional homolog of SOS1 (OsSOS1) was isolated from rice, suggesting conservation of the same system in rice (95).

Compartmentalization of Na^+ into the vacuole by vacuolar Na^+/H^+ antiporters (NHXs) lowers cytosolic Na^+ concentration and enables cells to use Na^+ for osmotic adjustment (96). Functional analysis of AtNHX1 using both overexpression and mutation analysis has confirmed these roles of NHXs in *Arabidopsis* (97, 98). Overexpression of vacuolar NHXs from rice (OsNHX1) or an Amaranthaceae halophyte *Atriplex gmelini* (AgNHX1) improved salt tolerance of transgenic lowland rice (99, 100) as well as upland rice (101). Although the use of transporters does provide a direct approach for enhancing salt tolerance, precise promoters should be used for transgene expression so that ion translocation and homeostasis are not perturbed.

4. Perspectives

As outlined above, many transgenic plants have been reported to show improved stress tolerance. Field trials, however, are only at the initial stages of development, and most of the implicated genes need further validation. In most studies, stress tolerance is tested at seedling stages only, although plants are clearly susceptible to stress at their reproductive stage. Causal relationships between transgenes and stress-tolerant phenotypes should be confirmed by comparison of tolerance between homozygous and transgene-free segregates, and by evaluating the effects of the transgenes, e.g., expression of downstream genes or accumulation of proteins or metabolites. Finally, stress tolerance must be tested in the field.

It appears that optimization of promoters for transgene expression is critical, because expression of genes involved in stress tolerance often causes negative effects on plant growth by negative regulation. Several studies showed that use of appropriate stress-inducible promoters reduce such negative effects (37, 89). In particular, using the genes' own promoters, which confer appropriate tissue specificity, seems to be effective. It is noted that, in several QTLs, strong gene expression is associated with stress-tolerant alleles, suggesting the importance of promoters (4, 7, 9).

Owing to developments in DNA sequencing technologies, the $1,000 genome era is expected to come within a few years. Accumulation of whole genome sequences from stress-tolerant rice cultivars and other *Oryza* species will accelerate QTL cloning and marker development, as well as the discovery of new alleles of stress tolerance genes. Molecular characterization of these new alleles and genes should provide a good resource for developing stress-tolerant rice cultivars.

References

1. Riechmann JL, Heard J, Martin G et al (2000) Arabidopsis transcription factors: genome-wide comparative analysis among eukaryotes. Science 290:2105–2110
2. Goff SA, Ricke D, Lan T-H et al (2002) A draft sequence of the rice genome (*Oryza sativa* L. Ssp. *japonica*). Science 296:92–100
3. Shiu S-H, Shih M-C, Li W-H (2005) Transcription factor families have much higher expansion rates in plants than in animals. Plant Physiol 139:18–26
4. Alonso-Blanco C, Gomez-Mena C, Llorente F, Koornneef M, Salinas J, Martínez-Zapater JM (2005) Genetic and molecular analyses of natural variation indicate *CBF2* as a candidate gene for underlying a freezing tolerance quantitative trait locus in Arabidopsis. Plant Physiol 139:1304–1312
5. Francia E, Barabaschi D, Tondelli A et al (2007) Fine mapping of a hv *CBF* gene cluster at the frost resistance locus *Fr-H2* in barley. Theor Appl Genet 115:1083–1091
6. Vágújfalvi A, Galiba G, Cattivelli L, Dubcovsky J (2003) The cold-regulated transcriptional activator *Cbf3* is linked to the frost-tolerance locus *Fr-A2* on wheat chromosome 5A. Mol Gen Genomics 269:60–67
7. Knox AK, Li C, Vágújfalvi A, Galiba G, Stockinger EJ, Dubcovsky J (2008) Identification of candidate *CBF* genes for the frost tolerance locus *Fr-A^{m}2* in *Triticum monococcum*. Plant Mol Biol 67:257–270
8. Fukao T, Xu K, Ronald PC, Bailey-Serres J (2006) A variable cluster of ethylene response factor-like genes regulates metabolic and developmental acclimation responses to submergence in rice. Plant Cell 18:2021–2034
9. Xu K, Xu X, Fukao T et al (2006) *Sub1A* is an ethylene-response-factor-like gene that confers submergence tolerance to rice. Nature 442:705–708
10. Jaglo-Ottosen KR, Gilmour SJ, Zarka DG, Schabenberger O, Thomashow MF (1998) *Arabidopsis CBF1* overexpression induces *COR* genes and enhances freezing tolerance. Science 280:104–106
11. Liu Q, Kasuga M, Sakuma Y et al (1998) Two transcription factors, DREB1 and DREB2, with an EREBP/AP2 DNA binding domain separate two cellular signal transduction pathways in drought- and low-temperature-responsive gene expression, respectively, in Arabidopsis. Plant Cell 10:1391–1406
12. Sakuma Y, Maruyama K, Qin F, Osakabe Y, Shinozaki K, Yamaguchi-Shinozaki K (2006) Dual function of an *Arabidopsis* transcription factor DREB2A in water-stress-responsive and heat-stress-responsive gene expression. Proc Natl Acad Sci USA 103:18822–18827
13. Kasuga M, Liu Q, Miura S, Yamaguchi-Shinozaki K, Shinozaki K (1999) Improving plant drought, salt, and freezing tolerance by gene transfer of a single stress-inducible transcription factor. Nat Biotechnol 17:287–291
14. Nakashima K, Ito Y, Yamaguchi-Shinozaki K (2009) Transcriptional regulatory networks in response to abiotic stresses in Arabidopsis and grasses. Plant Physiol 149:88–95
15. Yamaguchi-Shinozaki K, Shinozaki K (2006) Transcriptional regulatory networks in cellular responses and tolerance to dehydration and cold stresses. Annu Rev Plant Biol 57:781–803
16. Ito Y, Katsura K, Maruyama K et al (2006) Functional analysis of rice DREB1/CBF-type transcription factors involved in cold-responsive gene expression in transgenic rice. Plant Cell Physiol 47:141–153
17. Oh S-J, Song SI, Kim YS et al (2005) Arabidopsis CBF3/DREB1A and ABF3 in transgenic rice increased tolerance to abiotic stress without stunting growth. Plant Physiol 138:341–351
18. Chen J-Q, Meng X-P, Zhang Y, Xia M, Wang X-P (2008) Over-expression of OsDREB genes lead to enhanced drought tolerance in rice. Biotechnol Lett 30:2191–2198
19. Wang Q, Guan Y, Wu Y, Chen H, Chen F, Chu C (2008) Overexpression of a rice *OsDREB1F* gene increases salt, drought, and low temperature tolerance in both *Arabidopsis* and rice. Plant Mol Biol 67:589–602
20. Oh S-J, Kwon C-W, Choi D-W, Song SI, Kim J-K (2007) Expression of barley HvCBF4 enhances tolerance to abiotic stress in transgenic rice. Plant Biotechnol J 5:646–656
21. Qin F, Kakimoto M, Sakuma Y et al (2007) Regulation and functional analysis of *ZmDREB2A* in response to drought and heat stresses in *Zea mays* L. Plant J 50:54–69
22. Qin F, Sakuma Y, Tran L-SP et al (2008) *Arabidopsis* DREB2A-interacting proteins function as RING E3 ligases and negatively regulate plant drought stress-responsive gene expression. Plant Cell 20:1693–1707
23. Sakuma Y, Maruyama K, Osakabe Y et al (2006) Functional analysis of an *Arabidopsis* transcription factor, DREB2A, involved in drought-responsive gene expression. Plant Cell 18:1292–1309
24. Karaba A, Dixit S, Greco R et al (2007) Improvement of water use efficiency in rice by

expression of *HARDY*, an *Arabidopsis* drought and salt tolerance gene. Proc Natl Acad Sci USA 104:15270–15275
25. Finkelstein RR, Gampala SSL, Rock CD (2002) Abscisic acid signaling in seeds and seedlings. Plant Cell 14:S15–S45
26. Rabbani MA, Maruyama K, Abe H et al (2003) Monitoring expression profiles of rice genes under cold, drought, and high-salinity stresses and abscisic acid application using cDNA microarray and RNA gel-blot analyses. Plant Physiol 133:1755–1767
27. Hobo T, Kowyama Y, Hattori T (1999) A bZIP factor, TRAB1, interacts with VP1 and mediates abscisic acid-induced transcription. Proc Natl Acad Sci USA 96:15348–15353
28. Choi H, Hong J, Ha J, Kang J, Kim SY (2000) ABFs, a family of ABA-responsive element binding factors. J Biol Chem 275: 1723–1730
29. Uno Y, Furihata T, Abe H, Yoshida R, Shinozaki K, Yamaguchi-Shinozaki K (2000) Arabidopsis basic leucine zipper transcription factors involved in an abscisic acid-dependent signal transduction pathway under drought and high-salinity conditions. Proc Natl Acad Sci USA 97:11632–11637
30. Kagaya Y, Hobo T, Murata M, Ban A, Hattori T (2002) Abscisic acid-induced transcription is mediated by phosphorylation of an abscisic acid response element binding factor, TRAB1. Plant Cell 14:3177–3189
31. Furihata T, Maruyama K, Fujita Y et al (2006) Abscisic acid-dependent multisite phosphorylation regulates the activity of a transcription activator AREB1. Proc Natl Acad Sci USA 103:1988–1993
32. Kim S, Kang JY, Cho D-I, Park J-H, Kim SY (2004) ABF2, an ABRE-binding bZIP factor, is an essential component of glucose signaling and its overexpression affects multiple stress tolerance. Plant J 40:75–87
33. Fujita Y, Fujita M, Satoh R et al (2005) AREB1 is a transcription activator of novel ABRE-dependent ABA signaling that enhances drought stress tolerance in *Arabidopsis*. Plant Cell 17:3470–3488
34. Xiang Y, Tang N, Du H, Ye H, Xiong L (2008) Characterization of OsbZIP23 as a key player of the basic leucine zipper transcription factor family for conferring abscisic acid sensitivity and salinity and drought tolerance in rice. Plant Physiol 148:1938–1952
35. Fujita M, Fujita Y, Maruyama K et al (2004) A dehydration-induced NAC protein, RD26, is involved in a novel ABA-dependent stress-signaling pathway. Plant J 39:863–876
36. Tran L-SP, Nakashima K, Sakuma Y et al (2004) Isolation and functional analysis of Arabidopsis stress-inducible NAC transcription factors that bind to a drought-responsive *cis*-element in the *early responsive to dehydration stress 1* promoter. Plant Cell 16:2481–2498
37. Nakashima K, Tran L-SP, Van Nguyen D et al (2007) Functional analysis of a NAC-type transcription factor OsNAC6 involved in abiotic and biotic stress-responsive gene expression in rice. Plant J 51:617–630
38. Hu H, Dai M, Yao J et al (2006) Overexpressing a NAM, ATAF, and CUC (NAC) transcription factor enhances drought resistance and salt tolerance in rice. Proc Natl Acad Sci USA 103:12987–12992
39. Hu H, You J, Fang Y, Zhu X, Qi Z, Xiong L (2008) Characterization of transcription factor gene *SNAC2* conferring cold and salt tolerance in rice. Plant Mol Biol 67:169–181
40. Wu X, Shiroto Y, Kishitani S, Ito Y, Toriyama K (2009) Enhanced heat and drought tolerance in transgenic rice seedlings overexpressing *OsWRKY11* under the control of *HSP101* promoter. Plant Cell Rep 28:21–30
41. Boudsocq M, Laurière C (2005) Osmotic signaling in plants. Multiple pathways mediated by emerging kinase families. Plant Physiol 138:1185–1194
42. Kobayashi Y, Yamamoto S, Minami H, Kagaya Y, Hattori T (2004) Differential activation of the rice sucrose nonfermenting1-related protein kinase2 family by hyperosmotic stress and abscisic acid. Plant Cell 16:1163–1177
43. Mustilli A-C, Merlot S, Vavasseur A, Fenzi F, Giraudat J (2002) Arabidopsis OST1 protein kinase mediates the regulation of stomatal aperture by abscisic acid and acts upstream of reactive oxygen species production. Plant Cell 14:3089–3099
44. Yoshida R, Hobo T, Ichimura K et al (2002) ABA-activated SnRK2 protein kinase is required for dehydration stress signaling in *Arabidopsis*. Plant Cell Physiol 43:1473–1483
45. Umezawa T, Yoshida R, Maruyama K, Yamaguchi-Shinozaki K, Shinozaki K (2004) SRK2C, a SNF1-related protein kinase 2, improves drought tolerance by controlling stress-responsive gene expression in *Arabidopsis thaliana*. Proc Natl Acad Sci USA 101:17306–17311
46. Kobayashi Y, Murata M, Minami H et al (2005) Abscisic acid-activated SNRK2 protein kinases function in the gene-regulation pathway of ABA signal transduction by phosphorylating ABA response element-binding factors. Plant J 44:939–949

47. Diédhiou CJ, Popova OV, Dietz K-J, Golldack D (2008) The SNF1-type serine-threonine protein kinase *SAPK4* regulates stress-responsive gene expression in rice. BMC Plant Biol 8:49
48. Saijo Y, Hata S, Kyozuka J, Shimamoto K, Izui K (2000) Over-expression of a single Ca^{2+}-dependent protein kinase confers both cold and salt/drought tolerance on rice plants. Plant J 23:319–327
49. Abbasi F, Onodera H, Toki S, Tanaka H, Komatsu S (2004) OsCDPK13, a calcium-dependent protein kinase gene from rice, is induced by cold and gibberellin in rice leaf sheath. Plant Mol Biol 55:541–552
50. Gong D, Guo Y, Schumaker KS, Zhu J-K (2004) The SOS3 family of calcium sensors and SOS2 family of protein kinases in Arabidopsis. Plant Physiol 134:919–926
51. Xiang Y, Huang Y, Xiong L (2007) Characterization of stress-responsive CIPK genes in rice for stress tolerance improvement. Plant Physiol 144:1416–1428
52. Xiong L, Yang Y (2003) Disease resistance and abiotic stress tolerance in rice are inversely modulated by an abscisic acid-inducible mitogen-activated protein kinase. Plant Cell 15:745–759
53. Chen THH, Murata N (2002) Enhancement of tolerance of abiotic stress by metabolic engineering of betaines and other compatible solutes. Curr Opin Plant Biol 5:250–257
54. Igarashi Y, Yoshiba Y, Sanada Y, Yamaguchi-Shinozaki K, Wada K, Shinozaki K (1997) Characterization of the gene for Δ^1-pyrroline-5-carboxylate synthetase and correlation between the expression of the gene and salt tolerance in *Oryza sativa* L. Plant Mol Biol 33:857–865
55. Hien DT, Jacobs M, Angenon G et al (2003) Proline accumulation and Δ^1-pyrroline-5-carboxylate synthetase gene properties in three rice cultivars differing in salinity and drought tolerance. Plant Sci 165:1059–1068
56. Hur J, Jung K-H, Lee C-H, An G (2004) Stress-inducible *OsP5CS2* gene is essential for salt and cold tolerance in rice. Plant Sci 167:417–426
57. Zhu B, Su J, Chang M, Verma DPS, Fan Y-L, Wu R (1998) Overexpression of a Δ^1-pyrroline-5-carboxylate synthetase gene and analysis of tolerance to water- and salt-stress in transgenic rice. Plant Sci 139:41–48
58. Nanjo T, Kobayashi M, Yoshiba Y, Kakubari Y, Yamaguchi-Shinozaki K, Shinozaki K (1999) Antisense suppression of proline degradation improves tolerance to freezing and salinity in Arabidopsis thaliana. FEBS Lett 461:205–210
59. Rhodes D, Hanson AD (1993) Quaternary ammonium and tertiary sulfonium compounds in higher plants. Annu Rev Plant Physiol Plant Mol Biol 44:357
60. Chen THH, Murata N (2008) Glycinebetaine: an effective protectant against abiotic stress in plants. Trends Plant Sci 13:499–505
61. Shirasawa K, Takabe T, Takabe T, Kishitani S (2006) Accumulation of glycinebetaine in rice plants that overexpress choline monooxygenase from spinach and evaluation of their tolerance to abiotic stress. Ann Bot 98:565–571
62. Sakamoto A, Murata A, Murata N (1998) Metabolic engineering of rice leading to biosynthesis of glycinebetaine and tolerance to salt and cold. Plant Mol Biol 38:1011–1019
63. Mohanty A, Kathuria H, Ferjani A et al (2002) Transgenics of an elite *indica* rice variety Pusa Basmati 1 harbouring the *codA* gene are highly tolerant to salt stress. Theor Appl Genet 106:51–57
64. Su J, Hirji R, Zhang L, He C, Selvaraj G, Wu R (2006) Evaluation of the stress-inducible production of choline oxidase in transgenic rice as a strategy for producing the stress-protectant glycine betaine. J Exp Bot 57:1129–1135
65. Capell T, Bassie L, Christou P (2004) Modulation of the polyamine biosynthetic pathway in transgenic rice confers tolerance to drought stress. Proc Natl Acad Sci USA 101:9909–9914
66. Roy M, Wu R (2002) Overexpression of S-adenosylmethionine decarboxylase gene in rice increases polyamine level and enhances sodium chloride-stress tolerance. Plant Sci 163:987–992
67. Vijn I, Smeekens S (1999) Fructan: more than a reserve carbohydrate? Plant Physiol 120:351–360
68. Ji X, Van den Ende W, Schroeven L et al (2007) The rice genome encodes two vacuolar invertases with fructan exohydrolase activity but lacks the related fructan biosynthesis genes of the Pooideae. New Phytol 173:50–62
69. Kawakami A, Sato Y, Yoshida M (2008) Genetic engineering of rice capable of synthesizing fructans and enhancing chilling tolerance. J Exp Bot 59:793–802
70. Paul MJ, Primavesi LF, Jhurreea D, Zhang Y (2008) Trehalose metabolism and signaling. Annu Rev Plant Biol 59:417–441
71. Garg AK, Kim J-K, Owens TG et al (2002) Trehalose accumulation in rice plants confers high tolerance levels to different abiotic stresses. Proc Natl Acad Sci USA 99:15898–15903
72. Pramanik MHR, Imai R (2005) Functional identification of a trehalose 6-phosphate

phosphatase gene that is involved in transient induction of trehalose biosynthesis during chilling stress in rice. Plant Mol Biol 58: 751–762
73. Jang I-C, Oh S-J, Seo J-S et al (2003) Expression of a bifunctional fusion of the *Escherichia coli* genes for trehalose-6-phosphate synthase and trehalose-6-phosphate phosphatase in transgenic rice plants increases trehalose accumulation and abiotic stress tolerance without stunting growth. Plant Physiol 131:516–524
74. Ge L-F, Chao D-Y, Shi M, Zhu M-Z, Gao J-P, Lin H-X (2008) Overexpression of the trehalose-6-phosphate phosphatase gene *OsTPP1* confers stress tolerance in rice and results in the activation of stress responsive genes. Planta 228:191–201
75. Taji T, Ohsumi C, Iuchi S et al (2002) Important roles of drought- and cold-inducible genes for galactinol synthase in stress tolerance in *Arabidopsis thaliana*. Plant J 29:417–426
76. Nishida I, Murata N (1996) Chilling sensitivity in plants and cyanobacteria: the crucial contribution of membrane lipids. Annu Rev Plant Physiol Plant Mol Biol 47:541–568
77. Zhu S-Q, Zhao H, Liang J-S, Ji B-H, Jiao D-M (2008) Relationships between phosphatidylglycerol molecular species of thylakoid membrane lipids and sensitivities to chilling-induced photoinhibition in rice. J Integr Plant Biol 50:194–202
78. Murata N, Ishizaki-Nishizawa O, Higashi S, Hayashi H, Tasaka Y, Nishida I (1992) Genetically engineered alteration in the chilling sensitivity of plants. Nature 356: 710–713
79. Ariizumi T, Kishitani S, Inatsugi R, Nishida I, Murata N, Toriyama K (2002) An increase in unsaturation of fatty acids in phosphatidylglycerol from leaves improves the rates of photosynthesis and growth at low temperatures in transgenic rice seedlings. Plant Cell Physiol 43:751–758
80. Yokoi S, Higashi S-I, Kishitani S, Murata N, Toriyama K (1998) Introduction of the cDNA for shape *Arabidopsis* glycerol-3-phosphate acyltransferase (GPAT) confers unsaturation of fatty acids and chilling tolerance of photosynthesis on rice. Mol Breed 4:269–275
81. Upchurch RG (2008) Fatty acid unsaturation, mobilization, and regulation in the response of plants to stress. Biotechnol Lett 30:967–977
82. Matsuda O, Iba K (2005) Trienoic fatty acids and stress responses in higher plants. Plant Biotechnol 22:423–430
83. Battaglia M, Olvera-Carrillo Y, Garciarrubio A, Campos F, Covarrubias AA (2008) The enigmatic LEA proteins and other hydrophilins. Plant Physiol 148:6–24
84. Hundertmark M, Hincha DK (2008) LEA (late embryogenesis abundant) proteins and their encoding genes in *Arabidopsis thaliana*. BMC Genomics 9:118
85. Moons A, Bauw G, Prinsen E, Van Montagu M, Van Der Straeten D (1995) Molecular and physiological responses to abscisic acid and salts in roots of salt-sensitive and salt-tolerant indica rice varieties. Plant Physiol 107:177–186
86. Aguan K, Sugawara K, Suzuki N, Kusano T (1991) Isolation of genes for low-temperature-induced proteins in rice by a simple subtractive method. Plant Cell Physiol 32: 1285–1289
87. Xu D, Duan X, Wang B, Hong B, Ho T-HD, Wu R (1996) Expression of a late embryogenesis abundant protein gene, *HVA1*, from barley confers tolerance to water deficit and salt stress in transgenic rice. Plant Physiol 110:249–257
88. Rohila JS, Jain RK, Wu R (2002) Genetic improvement of basmati rice for salt and drought tolerance by regulated expression of a barley *Hva1* cDNA. Plant Sci 163:525–532
89. Xiao B, Huang Y, Tang N, Xiong L (2007) Over-expression of a LEA gene in rice improves drought resistance under the field conditions. Theor Appl Genet 115:35–46
90. Kotak S, Larkindale J, Lee U, von Koskull-Döing P, Vierling E, Scharf K-D (2007) Complexity of the heat stress response in plants. Curr Opin Plant Biol 10:310–316
91. Katiyar-Agarwal S, Agarwal M, Grover A (2003) Heat-tolerant basmati rice engineered by over-expression of *hsp101*. Plant Mol Biol 51:677–686
92. Sato Y, Yokoya S (2008) Enhanced tolerance to drought stress in transgenic rice plants overexpressing a small heat-shock protein, sHSP17.7. Plant Cell Rep 27:329–334
93. Ren Z-H, Gao J-P, Li L-G et al (2005) A rice quantitative trait locus for salt tolerance encodes a sodium transporter. Nat Genet 37:1141–1146
94. Shi H, Quintero FJ, Pardo JM, Zhu J-K (2002) The putative plasma membrane Na^+/H^+ antiporter SOS1 controls long-distance Na^+ transport in plants. Plant Cell 14: 465–477
95. Martínez-Atienza J, Jiang X, Garciadeblas B et al (2007) Conservation of the salt overly sensitive pathway in rice. Plant Physiol 143: 1001–1012

96. Munns R, Tester M (2008) Mechanisms of salinity tolerance. Annu Rev Plant Biol 59: 651–681
97. Apse MP, Aharon GS, Snedden WA, Blumwald E (1999) Salt tolerance conferred by overexpression of a vacuolar Na^+/H^+ antiport in *Arabidopsis*. Science 285:1256–1258
98. Apse MP, Sottosanto JB, Blumwald E (2003) Vacuolar cation/H^+ exchange, ion homeostasis, and leaf development are altered in a T-DNA insertional mutant of *AtNHX1*, the *Arabidopsis* vacuolar Na^+/H^+ antiporter. Plant J 36:229–239
99. Ohta M, Hayashi Y, Nakashima A et al (2002) Introduction of a Na^+/H^+ antiporter gene from *Atriplex gmelini* confers salt tolerance to rice. FEBS Lett 532:279–282
100. Fukuda A, Nakamura A, Tagiri A et al (2004) Function, intracellular localization and the importance in salt tolerance of a vacuolar Na^+/H^+ antiporter from rice. Plant Cell Physiol 45:146–159
101. Chen H, An R, Tang J-H et al (2007) Overexpression of a vacuolar Na^+/H^+ antiporter gene improves salt tolerance in an upland rice. Mol Breed 19:215–225

Chapter 21

Molecular Strategies to Improve Rice Disease Resistance

Emily E. Helliwell and Yinong Yang

Abstract

Rice diseases such as blast (*Magnaporthe oryzae*), sheath blight (*Rhizoctonia solani*) and bacterial blight (*Xanthomonas oryzae* pv *oryzae*) are a major obstacle to achieving optimal yields. To complement conventional breeding method, molecular and transgenic method represents an increasingly important approach for genetic improvement of disease resistance and reduction of pesticide usage. During the past two decades, a wide variety of genes and mechanisms involved in rice defense response have been identified and elucidated. These include components of pathogen recognition, signal transduction, downstream defense-related proteins, and crosstalk among different signaling pathways. In addition, various molecular strategies including use of specialized promoters, modification of target protein structures have been studied and proposed to improve the effectiveness of transgenes. While genetically improving rice for enhanced disease resistance, it is important to consider potential effects of the transgene on rice yield, tolerance to abiotic stresses, and defense against other pathogens.

Key words: Biotic stress, Disease resistance, Defense signal transduction, Transgenic rice

1. Introduction

In order to keep pace with increasing human population growth, rice yields will need to double by 2050 (1). A major obstacle in achieving optimal yields is the occurrence of rice diseases caused by numerous fungal, bacterial, and viral pathogens. Among the three most devastating diseases are rice blast, sheath blight, and bacterial blight. Rice blast is caused by the fungal ascomycete *Magnaporthe oryzae*, which is found in the majority of rice-producing regions in the world (2). Rice blast epidemics often result in between 10% and 30% yield loss, which on a lower estimate, would be enough rice to feed 60 million people for 1 year (1). Sheath blight is caused by the basidiomycete *Rhizoctonia solani* and is the second-most devastating disease of rice behind rice blast. Yield losses from sheath blight are typically between 10% and 25% (3). The third devastating

Yinong Yang (ed.), *Rice Protocols*, Methods in Molecular Biology, vol. 956,
DOI 10.1007/978-1-62703-194-3_21, © Springer Science+Business Media, LLC 2013

disease of rice is bacterial blight, caused by *Xanthomonas oryzae* pv. *oryzae*. Yield losses due to bacterial blight are variable, heavily dependent on the cultivar used and the environment. In Japan, yield losses ranged typically between 20% and 30% after distribution of high-yielding dwarf varieties (4). Among tropical climates, yield losses up to 75% were reported in Indonesia, India, and the Philippines (3, 5).

In modern agricultural practices, there are three major strategies used to enhance disease resistance in crops (6). The first strategy involves improvement of cultural practices. This strategy is concerned with altering the environment in such a way that it is not conducive to disease development. Some examples of this include proper spacing of plants to reduce canopy thickness, soil aeration and crop rotation to reduce soil-borne pathogen populations, and pesticides to protect against infection and disease development. Although a combination of these strategies has proven to be effective in preventing yield loss, there are drawbacks present in cost-effectiveness, as well as potential impact on the environment and human health due to overuse of chemicals.

The second strategy involves the improvement of crops through conventional or molecular marker-assisted breeding of disease resistant cultivars. This is done through crossing and/or backcrossing a cultivar containing genetic resistance into a cultivar with superior agronomic traits. The breeding of disease resistant crops often reduces the use of pesticides and the need for cultural control, however, there are drawbacks as well. Several generations of crossing, selfing, and testing plants for resistance are required to create a superior, redundant cultivar, which may translate into many years of breeding efforts. In addition, traditional breeding is often negatively affected by linkage drag, which refers to the transfer of loci conferring potentially undesired agronomic traits due to its close linkage with resistance loci.

The third crop improvement strategy is through the direct transformation of resistance-related genes into elite cultivars. Research conducted within the last decade has greatly improved our understanding of the mechanisms involved in the interactions between host and pathogen that result in resistance and susceptibility. Along with this, techniques for gene transfer such as the *Agrobacterium*-mediated transformation are becoming more efficient and reproducible. The combination of these two advances has made gene transfer a readily available technique that is less time-consuming than conventional breeding, and avoid the problem of linkage drag. The use of molecular techniques such as the *Agrobacterium*-mediated transformation allows the stable transfer of a transgene shown to enhance resistance in a different variety or species, as well as to use a different promoter on a resistance-related gene to alter the intensity of gene expression. On the other hand,

RNA interference techniques can be used to silence a signaling component that negatively regulates resistance (7). The end result is a stronger defense response, which manifests as a reduction in disease development.

To evade host defense responses, successful pathogens either avoid recognition by host receptors, attenuate or inhibit downstream defense signaling, or hijack host signaling mechanisms to favor establishment of disease. There are different types of disease resistance in terms of host-pathogen specificity and durability of resistance. Broad-spectrum resistance (BSR) is classified as resistance against two or more pathogen species, and/or against several races within one pathogen species. Durable resistance (DR) is defined as resistance that sustains its effectiveness for many years after deployment in an environment with the pathogen present (8). Generally, the most effective forms of resistance are conferred by race-specific resistance (*R*) genes, which correspond only to a specific pathogen avirulence (*Avr*) gene. This specificity means that the resistance is not broad-spectrum in nature. Additionally, resistance conferred by *R* genes is prone to breakdown due to changes (e.g., deletion and mutation of *Avr* genes) in the genetic structure of pathogen populations, and is therefore not a durable form of resistance. There are a few examples of known rice *R* genes that confer durable, broad spectrum resistance, such as *pi21* (9) and *Pi9* (10). However, these are the exception and not the rule. Additionally, plants may not have a form of race-specific resistance to particular pathogens, such as in the case of *Rhizoctonia solani*. To date, no *R* genes corresponding to *R. solani* have been identified. Despite their transient resistance, *R* genes are favored in conventional and marker-based breeding due to their effectiveness, and straightforward transfer during the breeding process. Genes that contribute to basal resistance theoretically will be more durable and protect against a wider range of pathogen races and species than genes that mediate race-specific resistance (11). However, it has been shown to be difficult to characterize sources of basal resistance due to its polygenic, additive nature (11, 12).

There are various molecular and biotechnological approaches to genetically improve rice crop for effective, durable and/or broad-spectrum resistance to major diseases. Such molecular manipulation and genetic modification may be carried out with rice genes and components involved in pathogen recognition (e.g., R proteins and pattern recognition receptors), host signaling (e.g., reactive oxygen species, hormones, protein kinases, transcription factors, etc.), or defense responses (e.g., pathogenesis-related proteins and other defense genes). This review will discuss the various factors involved in host signaling, and their efficacy as transgenes for the purpose of improving genetic resistance to rice disease.

2. Improving Host Recognition of Pathogens

In order for a potential host to launch a defense response against attacking pathogens, the host must first be able to recognize a non-self entity. Plants have numerous receptors and sensors that recognize and depending on the nature of the receptor, directly or indirectly interact with a variety of microbial components. Generally, plant immunity is divided into two levels, based on the recognition of specific microbial components (13). The first level of defense is often triggered by host recognition of conserved microbial components, known as pathogen-associated molecular patterns (PAMPs), or microbial-associated molecular patterns (MAMPs). The host proteins that bind and recognize a PAMP/MAMP are called pattern recognition receptors (PRR). Binding of a PAMP to a PRR launches a basal immune response, which is classified as PAMP-triggered immunity (PTI). A well-studied example of PTI involves the binding of PAMP flg22, which is a component of bacterial flagellin, by *Arabidopsis thaliana* FLS2, an LRR-receptor kinase (14). The second level of defense is activated by the *R* gene-mediated recognition of pathogen effectors, and termed as effector-triggered immunity (ETI). Unlike PAMPs, which generally do not play a major role in pathogenicity, microbial effectors often serve as virulence factors and enter host cell to facilitate the establishment of disease. Effectors often work by suppressing PTI, as seen in the *Arabidopsis-Pseudomonas syringae* pathosystem. The *P. syringae* effector AvrPto inhibits the kinase activity of FLS2, which therefore inhibits PTI (15). However, pathogen effectors may be specifically recognized by particular *R* gene in a gene-for-gene type interaction and act negatively as avirulence factors. As a result, the suppression of PTI by AvrPto is in turn recognized by the host R protein Pto, leading to the activation of hypersensitive response and a strong resistance phenotype. The specificity of the receptor has implications for the spectra of resistance, as well as the effectiveness. Compared to PTI, ETI is a highly specialized interaction, involving mechanisms that vary greatly from pathosystem to pathosystem, and generally results in a stronger defense signal by the host.

2.1. Pattern-Recognition Receptors

The rice receptors classified as pattern-recognition receptors (PRRs) typically contain transmembrane and extracellular domains that involve in interacting with PAMPs or sense external stimuli. The most common transmembrane receptors are those of the receptor-like kinase (RLK) superfamily. RLKs are involved in the regulation of various developmental- and defense-related processes in plants. They consist of an extracellular perception domain, which binds and recognizes specific environmental stimuli, and an intracellular kinase domain, to transmit the signal downstream.

An example of a defense-related putative RLK in rice is OsWAK1, which belongs to the wall-associated protein kinase subfamily of RLKs. OsWAK1 transcripts are significantly induced by *M. oryzae* infection, and constitutive overexpression of OsWAK1 in rice results in higher levels of resistance to blast infection (16).

The largest subfamily of RLKs consists of the leucine-rich repeat RLKs (LRR-RLKs). (17). The defining characteristic of LRR-RLKs is the presence of an extracellular leucine-rich motif arranged in a specific number of tandem repeats and the level of specificity for any given elicitor is often determined by the LRR domain (18). LRR-RLKs are implicated in the perception of diverse stimuli, such as hormones, wounding, light, and steroids (19–22). There are also numerous examples of LRR-RLKs involvement in pathogen perception. Two somatic embryogenesis receptor kinase (SERK)-type LRR-RLKs have been implicated in perception and resistance to *M. oryzae*. *OsSERK1* was identified as a potential defense-related gene through a differential screening technique (23) and its constitutive overexpression in rice resulted in heightened levels of resistance (24). A second SERK-type LRR-RLK with 85% homology to *OsSERK1*, called *OsBISERK1* (*BTH-induced SERK1*) is also implicated in playing a role in resistance to *M. oryzae*. Like the name implies, *OsBISERK1* is upregulated strongly after treatment with the disease resistance-inducing compound benzothiadiazole (BTH), as well as after inoculation by an incompatible strain of *M. oryzae* (25). Another LRR-RLK implicated in disease resistance is OsBRR1, which was discovered through screening of an RNAi population with *M. oryzae*. Constitutive overexpression of *OsBRR1* resulted in heightened resistance against several compatible isolates of *M. oryzae* (26). Xa21, which confers resistance to *X. oryzae* pv. *oryzae*, contains an LRR domain and an intracellular kinase domain (27), *Xa21* was once classified as an *R* gene, however, it has been recently reclassified as a PRR after characterization of *X. oryzae* pv. *oryzae* AvrXa21, now renamed Ax21. Ax21 is highly conserved within *X. oryzae* pv. *oryzae* strains as well as other *Xanthomonas* species. Additionally, *Xylella fastidiosa* and the human pathogen *Stenotrophomonas maltophilia* contain putative Ax21 orthologs with 48% and 61% similarity, respectively (28). Transfer of *Xa21* into previously susceptible rice lines resulted in resistance to 29 out of 32 tested isolates of *X. oryzae* pv. *oryzae* (29), making *Xa21* a good candidate for broad-spectrum resistance to rice bacterial blight.

Chitin oligosaccharides from the cell wall of *M. oryzae* were found to bind with high affinity to a plasma membrane-bound protein, CEBiP (chitin oligosaccharide elicitor-binding protein) (30). Unlike proteins from the receptor-like kinase family, CEBiP is predicted to lack intracellular domains, and may require a chitin elicitor receptor kinase (OsCERK1) as a co-receptor for subsequent signal transduction (31). Chitin is conserved cell wall component of fungi, and acts as a general elicitor of basal plant defense. Kaku et al. (32) used RNA interference techniques to suppress *CEBiP* in rice cell cultures. Treatment of *CEBiP*-suppression cells with chitin oligosaccharides resulted in an attenuation of reactive oxygen species production as compared to nontransformed cells. In addition to this, microarray data showed differential regulation of defense-related genes. Upon elicitor treatment, 71% of upregulated genes and 80% of downregulated genes that were differentially expressed in the control rice cells remained at, or close to the basal level in CEBiP-suppressing cells. This suggests that CEBiP is necessary for perception of chitin and relay of the downstream defense signal.

2.2. Resistance Proteins

Of the five classes of resistance (R) proteins, the most common is the nucleotide-binding, leucine-rich repeat (NBS-LRR) family (18). Unlike the receptor-like kinases, NBS-LRR proteins are presumed to be localized to the cytoplasm. A meta-analysis shows that out of 73 rice blast resistance genes, 80% of them encode NBS-LRR proteins (33). To date, of the 13 cloned and characterized blast *R* genes, all but one (*Pi-d2*, a receptor kinase) encode NBS-LRR proteins (34). There have not been many studies done on the efficacy of transforming NBS-LRR-type *R* genes in rice (35). Qu et al. (10) transformed the NBS-LRR resistance gene *Pi9* into a blast-susceptible cultivar under the control of its native promoter. The transgenic *Pi9* lines retained the original levels of resistance to the same spectra of *M. oryzae* isolates as compared to the donor cultivar, showing that direct transfer of an NBS-LRR gene can be a useful tool in protecting against rice disease.

To date, only two genes (*Xa3/Xa26* and *Xa21*) encoding LRR-RLKs that are involved in race-specific resistance have been characterized (36). These genes were identified through their mediation of resistance to specific isolates of *X. oryzae* pv. *oryzae*, however the levels of resistance has been shown to fluctuate depending on the growth stage. Rice seedlings carrying *Xa21* are susceptible to *X. oryzae* pv. *oryzae*, however the level of resistance increases as the plant matures (37). This adult plant resistance pattern is also seen in rice plants carrying *Xa3/Xa26* (38). The level of resistance may also vary depending on the genetic background of the host. Japonica rice carrying transgenic *Xa3/Xa26* behind the native promoter display higher levels of resistance to *X. oryzae* pv. *oryzae* than in the donor indica cultivars (39).

2.3. Modification of Resistance Genes

Because of the efficacy of major resistance genes in preventing infection and subsequent loss of yield, it is desirable to utilize and/or modify these genes to avoid breakdown of resistance as well as to protect against a wider spectra of microbial pathogens. One method used in conventional and maker-based breeding approaches is gene pyramiding. Gene pyramiding refers to the combining of more than one gene of interest in a single cultivar to create an additive effect. When pyramiding *R* genes for broad-spectrum resistance, *R* genes recognizing different pathogen effectors are characterized, mapped, and then introgressed into an agronomically superior cultivar. There are a few examples of using a marker-based approach to pyramid *R* genes conferring resistance to *M. oryzae*, such as the combination of *Pi1*, *Piz-5*, and *Pi-ta* (40) and *Pish* and *Pib* (41) that have been successful, however, there is a lack of reports on using transformation-based methods to pyramid rice *R* genes. This is due to a lack of detailed knowledge on the mechanisms of successful *R* gene-mediated resistance. Theoretically, the best targets for transformation-mediated *R* gene pyramiding are those that not only recognize a corresponding *Avr* gene product but also transduce the defense signal without a need of specific partner proteins.

Other methods to improve the recognition spectrum of *R* genes are through modification of its various components. One example of this is through modification of *R* gene expression levels. A recent study (38) done with *Xa3/Xa26* suggest that the race-specific nature of this *R* gene is dependent on the expression level. Japonica cultivars carrying transgenic *Xa3/Xa26* driven by a strong constitutive ubiquitin promoter showed enhanced resistance to a wider spectrum of *X. oryzae* pv. *oryzae* isolates, along with rapid induction of defense-related genes *OsWRKY13* and *NH1*. Correspondingly, *Xa3/Xa26* driven by the weak pathogen-inducible *OsWRKY13* promoter showed weaker levels of resistance, and were resistant to fewer isolates of *X. oryzae* pv. *oryzae* as compared to transgenic plants using the native promoter. The authors suggest that the increased spectra of resistance could be explained by a differing affinity of the LRR domain for different effectors. An increased amount of *Xa3/Xa26* transcripts may increase the likelihood of a lower-affinity binding of Xa3 to otherwise virulent *X. oryzae* pv. *oryzae* effectors. A second explanation is that the enhanced resistance is due to the amplification of downstream defense signaling as indicated by the increased levels of defense-related *OsWRKY13* and *NH1*. This study suggests that race-specificity can be altered through adjustment or change of the promoter region and has implications for engineering broad-spectrum resistance.

Resistance genes may also be modified through alterations of the domains involved in effector recognition. To date, this approach has not yet been attempted in rice, however, it was shown that random mutagenesis of the LRR domain for Rx, which confers race-specific resistance against Potato virus X (PVX), resulted in recognition of additional races of PVX, as well as to a race of distantly related poplar mosaic virus PoMV (42). As numerous *R* genes encode proteins containing an LRR domain, this approach may be used to expand the recognition specificity of various rice *R* genes.

3. Enhancing Host Defense Signaling and Response

After detection of a potential pathogen by a host receptor or sensor, the defense signal is transduced downward, involving components such as reactive oxygen species, protein kinases and phosphatases, transcription factors, secondary messengers, hormones, and subsequent induction of pathogenesis-related (PR) genes. As a result, the outcome of the rice–pathogen interaction is greatly influenced by the timing and intensity of downstream defense signaling, as well as crosstalk between various signaling components and pathways.

3.1. Reactive Oxygen Species

One of the earliest physiological reactions of the host following pathogen recognition is the rapid, local production of reactive oxygen species (ROS) such as the superoxide anion (O_2^-), hydrogen peroxide (H_2O_2), and the hydroxyl ion (OH^-). This rapid production is known as the oxidative burst, and is a conserved immune response across kingdoms. O_2^- is produced primarily by NADPH oxidase (43) and is converted into the more stable H_2O_2 by superoxide dismutase (SOD). H_2O_2 catalyzes the crosslinking of cell wall components, which protects against penetration by a potential pathogen. Additionally, the oxidative burst has been shown to induce the expression of defense-related genes, and in high-enough concentrations, can trigger the hypersensitive response (HR) (44). For these reasons, the oxidative burst is a crucial step in the defense signaling.

In plants, members of the Rac/Rop small GTPase family act as switches for various signaling events. In rice, OsRac1 has been shown to interact with the N-terminal domain of NADPH oxidase (45). Upon activation, OsRac1 induces production of ROS and subsequent cell death in rice cell suspension cultures (46). Transgenic rice plants constitutively expressing active *OsRac1* acquire high levels of resistance to a virulent isolate of *M. oryzae*, and develop HR-like lesions reminiscent of an incompatible interaction along with biosynthesis of momilactone A, a common phytoalexin in rice (47). These lines of evidence suggest that enhancement of the oxidative burst through constitutive expression of *OsRac1* are an effective means of improving disease resistance in rice.

Germin-like proteins (GLPs), previously known as oxalate oxidase-like genes, are conserved in higher plants. GLPs are related to the barley HvGER4 subfamily, which has been shown to have superoxide dismutase activity (48). Superoxide dismutases catalyze the conversion of superoxide to hydrogen peroxide, which is essential for basal defense responses such as fortification of cell wall and production of papillae. GLPs are implicated in defense signaling responses, as many members are upregulated after exposure to pathogens, insect feeding, and exogenous application of defense-related compounds such as hydrogen peroxide, salicylic acid, and ethylene. In rice, a major QTL on chromosome 8 was found to contain a cluster of 12 GLPs. RNAi-mediated silencing of several of these GLPs resulted in increased susceptibility to both *M. oryzae* and *R. solani*, suggesting that these genes confer broad-spectrum resistance (49).

3.2. Mitogen-Activated Protein Kinases

Mitogen-activated protein kinase (MAPK) cascades are involved in many aspects of plant growth and development, along with responses to biotic and abiotic stresses (50). In rice, there are 17 MAPKs and a half of them are induced after infection by *M. oryzae* (51, 52). OsMPK12 (also known as BWMK1) has been shown to phosphorylate transcription factors that in turn induce expression of defense-related genes (53, 54). Transgenic studies have been conducted in a number of rice MPKs, including *OsMPK5* and *OsMPK6*. OsMPK5 has been recognized as a negative regulator of blast resistance (55) and suppression of *OsMPK5* through RNAi resulted in enhanced resistance to *M. oryzae*. On the opposite side, OsMPK5 has also been implicated as a positive regulator of resistance to *Cochliobolus miyabeanus*, which causes brown spot disease. *OsMPK5* suppression lines showed an increase in symptom development as compared to wild-type lines (56).

Like OsMPK5, the role of OsMPK6 in disease resistance also depends on the specific rice–pathogen interactions. It has been suggested that OsMPK6 plays a negative role in rice bacterial blight resistance, as RNAi-mediated suppression resulted in enhanced resistance to multiple races of *X. oryzae* pv. *oryzae*, which are associated with more rapid expression of *PR5, PR10/PBZ1,* phenylalanine ammonia lyase (*PAL*) and *NH1*, along with elevated levels of salicylic acid (57). However, OsMPK6 is also induced by fungal sphingolipid elicitors and has been shown to form a complex with OsRac1, RAR1, SGT1, Hsp90, and RACK1 (52). This protein complex was demonstrated to have an important role in basal immunity and transmission of the defense signal to downstream components. Additionally, OsMPK6 was found to be necessary for the production of diterpenoid phytoalexins (58), suggesting a positive role in resistance for OsMPK6.

3.3. Hormones

Pathogen infection often triggers the biosynthesis and signal transduction of several plant hormones which may serve as the local and/or systemic defense signals. The major defense-related hormones are salicylic acid, jasmonic acid, and ethylene. In addition, abscisic acid and auxin have also been shown to play roles in various plant–pathogen interactions. Depending on specific rice–pathogen interactions, these hormones may play a positive or negative role in disease resistance. In addition, hormone pathways may act either synergistically or antagonistically to mediate rice defense responses.

3.3.1. Salicylic Acid

Salicylic acid (SA) is a phenolic compound that has been well-documented to mediate disease resistance in many plant–pathogen interactions. However, rice has basal SA levels far exceeding (about ten times higher) the elevated SA levels in infected *Arabidopsis* or tobacco. Furthermore, there is little or no change in free SA levels in rice after pathogen infection (59). Using the salicylic acid-deficient *NahG* rice, Yang et al. (60) showed that SA does not appear to act as a defense signal, rather it plays an antioxidative role to protect rice plants from oxidative damage caused by pathogen infection.

Although SA itself may not serve as an effective defense signal, other similar phenolics may serve as signals that are mediated by NPR1 (*Non-expressor of PR1*), an integral component of SA signaling pathway. Studies done using *Arabidopsis* reveal that NPR1 exists as a cytosolic oligomer under normal conditions and is reduced to monomeric forms after accumulation of SA. NPR1 monomers enter and accumulate in the nucleus, where they activate the transcription of various *PR* genes (61). Overexpression of *Arabidopsis NPR1* or its rice homolog *OsNH1* in transgenic rice resulted in enhanced resistance to multiple isolates of *X. oryzae* pv. *oryzae* (62, 63), including the increased expression of defense genes such as *PR1b*, *PBZ1/PR10*, phenylalanine ammonia lyase (*PAL*), and *POX* (a peroxidase) (63). An in vitro assay showed that OsNH1 strongly interacted with the rice rTGA2.2, a bZIP transcription factor, suggesting that the downstream signaling mechanism of OsNH1 is similar to NPR1 in *Arabidopsis* (63).

Another method of improving disease resistance through alteration of SA signaling was revealed in studies involving a potential negative regulator. In *Arabidopsis*, *SSI2* encodes a fatty acid desaturase which acts as a suppressor of NPR1. Silencing of rice *SSI2* ortholog results in spontaneous lesion formation, an increase in free SA levels, elevated expression of SA-inducible genes, and enhanced resistance to both *M. oryzae* and *X. oryzae* pv. *oryzae*, (64).

Extensive studies done with dicotyledonous plants have shown that SA mediates the development of systemic acquired resistance (SAR). This phenomenon has not been well-studied in monocots, however, there have been a few studies showing a potential role for SAR in rice disease resistance. Probenazole (3-allyloxy-1,2-benzisothiazole-1,1-dioxide) is a chemical inducer of SAR and has been shown to increase SA levels and induce resistance in adult rice plants to *M. oryzae* infection (65). Umemura et al. (66) identified and characterized a probenazole-inducible *OsSGT1*, which encodes a UDP-glucose:SA glucosyltransferase that converts free salicylic acid into SA *O*-β-glucoside (SAG). SAG is a physiologically inactive form of free salicylic acid (67), which has been shown to increase in leaves during establishment of SAR, along with free SA (68). RNAi-mediated suppression of *OsSGT1* in rice resulted in an inhibition of probenazole-mediated resistance to rice blast, suggesting that its overexpression could potentially confer probenazole-independent resistance to rice disease (66).

3.3.2. Jasmonic Acid

A second stress and defense-related hormone that has been shown to play a role in numerous plant–pathogen interactions is jasmonic acid (JA). Jasmonates are lipid-derived molecules that in addition to mediating stress response, also play roles in plant growth and development, as well as senescence-related processes (69). In monocots, application of exogenous JA or methyl jasmonate (MeJA) results in the induction of several *PR* genes, such as *PR1a*, *PR1b*, *PR2*, *PR3*, *PR5*, and *PBZ1/PR10*, as well as the rice phytoalexins sakuranctin and momilactone A (70). Inhibition of JA biosynthesis by tetcyclacis results in the inhibition of *PR1* accumulation (71). These lines of evidence suggest that JA plays a role in rice defense response. Allene oxide synthase (AOS), a key enzyme in the JA biosynthesis pathway, is induced upon inoculation by *M. oryzae*. Overexpression of *OsAOS2* resulted in increased accumulation of JA, expression of *PR1a*, *PR1b*, *PR2*, *PR3*, *PR5*, and *PBZ1/PR10* and increased resistance to *M. oryzae* infection (72).

3.3.3. Ethylene

Ethylene (ET) is a gaseous hormone that plays diverse roles in plant growth and development as well as biotic and abiotic stress responses. The role of ethylene in disease resistance is considered to be mixed, depending on the types of plant–pathogen interactions. There are several lines of evidence suggesting that ET plays a role in rice resistance to *M. oryzae*. ET was first implicated as a positive regulator of resistance against blast disease through observational data involving the flooding of rice fields. Submergence causes a buildup of endogenous ET within the aerenchyma cells of rice plants, and this is correlated with a reduction in blast disease incidence. This effect was mimicked in non-flooded fields with application of ethephon, an ET-releasing chemical (73). Endogenous ET levels increase within 48 h post-inoculation with an avirulent isolate of *M. oryzae*, along with simultaneous induction of ET biosynthetic genes *OsACS1*, *OsACS2*, and *OsACO7* (74). As expected, knockout lines of both *OsACS2* and *OsACO7* showed increased susceptibility to *M. oryzae* (75). Transgenic studies involving the rice *EIN2* (*Ethylene insensitive 2*), a positive regulator of ET signaling have added to the hypothesis of ET as a positive regulator of resistance, as rice lines suppressing *OsEIN2* were more susceptible to *M. oryzae* (76). However, *OsEIN2*-supression lines showed increased resistance to *C. miyabeanus*, demonstrating a negative role of ET in resistance against brown spot disease (56).

3.3.4. Abscisic Acid

Abscisic acid (ABA) is a terpenoid phytohormone that plays a variety of roles throughout the life of a plant, such as in seed maturation and dormancy, inhibition of germination, photoregulation, senescence, flowering inhibition, and tolerance to abiotic stresses such as salinity, drought, and cold temperatures. ABA has also been shown to play a mixed role in response to biotic stress, including callose deposition, induction of stomatal closure, and suppression of ROS generation.

Most notably, ABA interacts in an antagonistic manner with SA and JA/ET signaling (77). In rice, the effects of ABA signaling on both SA and JA/ET-mediated resistance are seen in the rice-*M. oryzae* interaction. Application of exogenous ABA has been shown to result in increased blast disease severity (76, 78) and decreased expression of *WRKY45* and *OsNH1*, which are SA signaling components. This effect was negated in both *OsWRKY45*- and *OsNH1*-overexpressing lines (78). Along with an increase in blast symptoms, ABA application results in a decrease in endogenous ET levels. This decrease in ET was found to be mediated by OsMPK5. *OsMPK5* is induced by ABA, and confers tolerance to abiotic stresses drought, salinity, and cold temperatures, but negatively regulates ET levels and *PR* gene expression (55). On the other hand, ABA mediates increased resistance to *C. miyabeanus* through antagonism with ET signaling, as mentioned above. Other lines of evidence include the observation that ABA treatment results in enhanced resistance to *C. miyabeanus*, whereas *OsMPK5*-suppression lines show increased susceptibility (56).

3.3.5. Auxins

Auxin is a plant hormone that is implicated in many processes related to plant growth and development, such as formation of lateral roots, apical dominance, vascular differentiation, and flower development. Auxin-related physiological responses occur through polar transport from cell to cell, forming an auxin gradient. Indole acetic acid (IAA) is the most common form of auxin and is involved in several processes related to plant growth and development, but has been shown to have a negative role in disease resistance (79). This could be through antagonistic cross-talk with SA signaling, as auxin inhibits the expression of several *PR* genes (80) and SA-deficient (*NahG)* Arabidopsis showed increased levels of IAA (81).

Two members of the *GH3* gene family in rice, *OsGH3.1* and *OsGH3.8* are implicated as positive regulators of disease resistance. GH3 proteins catalyze the conjugation of IAA to amino acids, rendering it inactive. Overexpression of *OsGH3.1* resulted in a depletion of free IAA in cells, an increase in the expression of several *PR* genes, and a decrease in disease severity after inoculation by two different isolates of *M. oryzae* (82). Similarly, overexpression of *OsGH3.8* prevented the accumulation of free IAA and promoted enhanced resistance to *X. oryzae pv. oryzae* (83).

3.4. Transcription Factors

Transcription factors, also known as sequence-specific DNA binding proteins, bind to specific DNA sequences and activate transcription. This central role in gene regulation makes transcription factors an integral part in signaling processes. Transcription factors are classified into different families based on the presence of DNA-binding domains and conserved sequences. The WRKY family of transcription factors is a major group of plant-specific transcription factors involved in plant defense response and disease resistance. However, other transcription factors have also been shown to play a role in defense signaling and resistance response.

3.4.1. WRKY Transcription Factors

In rice, several transcription factors from the WRKY superfamily have been implicated in resistance against both *M. oryzae* and *X. oryzae* pv. *oryzae*. Overexpression studies have shown that three WRKY-type transcription factors, OsWRKY13, OsWRKY45, and OsWRKY71 are involved in resistance to *M. oryzae* and *X. oryzae* pv. *oryzae* (84, 85). *OsWRKY13* was identified as a potential defense-related gene through a candidate gene approach, which combines data from linkage map analyses, expression profiles, and functional complementation experiments (85). *OsWRKY13* was shown to be induced after inoculation by *M. oryzae* or *X. oryzae* pv. *oryzae* (86). In addition, mapping data showed that *OsWRKY13* localized to the same region in chromosome 1 as a resistance QTL (87, 88). Constitutive expression of *OsWRKY13* in a susceptible rice cultivar led to increased resistance to both *M. oryzae* and *X. oryzae* pv. *oryzae*, and increased salicylic acid-dependent signaling, but decreased jasmonic acid signaling (89). *OsWRKY45* was identified as a potential defense-related gene from a microarray screen of BTH-responsive genes (90). *OsWRKY45*-supression lines treated with BTH showed increased susceptibility to *M. oryzae* as compared to BTH-treated wild-type plants, suggesting that OsWRKY45 is a necessary component in BTH-mediated resistance. Constitutive overexpression of *OsWRKY45* in susceptible rice showed increased levels of resistance to *M. oryzae*, along with increased expression of *PR1b* and *PR2* (90). *OsWRKY71* is induced by MeJA, SA, wounding and pathogen infection (84). Overexpression of *OsWRKY71* resulted in increased expression of *OsNH1* and *PR1b*, along with increased resistance to *X. oryzae pv. oryzae* (84).

3.4.2. Other Classes of Transcription Factors

The rice transcription factor OsNAC6 has been implicated in mediating biotic and abiotic stress signaling in plants (91, 92). OsNAC6 is a member of the NAC transcription factor family, which binds specifically to the CATGTG motif (93). *OsNAC6* is induced after exposure to dehydration, high salinity, wounding, and treatment with ABA, MeJA, and *M. oryzae* (92). Constitutive overexpression of *OsNAC6* in rice resulted in higher levels of tolerance to drought and high salt conditions. The same *OsNAC6*-overexpression lines infected with *M. oryzae* showed lesions of a similar length as compared to wild-type plants, however, the lesion width was smaller, suggesting that the fungus was blocked from entering leaf veins (92). The marginal increase of resistance to *M. oryzae*, along with higher levels of tolerance to abiotic stresses such as drought and salinity suggest that overexpression of *OsNAC6* may be a useful tool for improvement of rice cultivars.

The alteration of a TGA-class transcription factor of the basic leucine zipper (bZIP) family has been shown to enhance resistance in rice. Rice transcription factor TGA2.1 is activated by interaction with OsNH1 (63). Activation of TGA transcription factors by NPR1 or its homologs results in activation of *PR* genes, which results in higher levels of resistance. Rice rTGA2.1 binds to the *PR1* gene promoter, along with the promoter of *RCH10*, a rice chitinase (63, 94). Additionally, two different rice mutants, the first carrying a suppression of *rTGA2.1* and the second carrying an *rTGA2.1* mutant that is unable to bind DNA, showed higher levels of resistance to *X. oryzae* pv. *oryzae*, manifested through smaller lesions. Additionally, the mutant lines showed higher expression of *PR* genes, suggesting that rTGA2.1 is a negative factor in resistance to *X. oryzae* pv. *oryzae* (94).

3.5. Pathogenesis-Related Genes

The basic definition of PR proteins is any pathogen-inducible, defense-related protein (95). The induction of *PR* genes is a common hallmark of resistance, having been described in many plant–pathogen interactions involving bacteria, fungi, oomycetes, and viruses. PR proteins are currently classified into 17 groups, and typically encode proteins with direct antimicrobial properties (95). For this reason, overexpression of *PR* genes can be considered a straightforward method in enhancing disease resistance in plants. For example, *PR3* encodes a chitinase which catalyzes the hydrolysis of β-1,4-linkage of the *N*-acetylglucosamine polymer of chitin, a cell wall component of fungi. Constitutive overexpression of a rice *PR3* resulted in enhanced resistance in multiple rice cultivars to *Rhizoctonia solani* (96). Another example involves *PR5*, which encodes a thaumatin-like protein. These belong to a large protein family that includes permatins, which create pores in fungal cell membranes (97). Thaumatin-like proteins are commonly induced in response to infection by *M. oryzae*, as well as by various stress and developmental signals. Overexpression of *PR5* in rice resulted in enhanced resistance to *R. solani*, manifested through a reduction in lesion size as compared to wild-type plants (98).

PR genes can also be transferred from one species to another. *PRms*, a fungal-inducible *PR* gene from maize (*Zea mays*) was transformed into rice, where it enhanced resistance to a number of rice diseases, including blast, bakanae disease (*Fusarium verticillioides*), foot rot (*Erwinia chrysanthemi*), and brown spot (99). In addition, the transgenic rice lines showed either constitutive or earlier induction of defense-related genes *OsPR1a*, *OsPR1b*, *PR5*, *PBZ1/PR10*, and *OsNH1*.

3.6. Phytoalexins

Phytoalexins are defined as small molecular weight, antimicrobial secondary metabolites that are induced upon pathogen infection. In rice, the majority of phytoalexins are diterpenoid, including momilactones. Shimura et al. (100) identified a cluster of five rice genes involved in momilactone biosynthesis, which include two diterpene cylcase genes, *OsKS4* and *OsCyc1*, along with two putative microsomal cytochrome P450 monooxygenases and one dehydrogenase. A follow-up study done by Okada et al. (101) showed that the expression of these five genes is controlled by a chitin oligosaccharide elicitor-inducible basic leucine zipper (bZIP) transcription factor OsTGAP1. Knockout *Ostgap1* mutants had negligible expression of all five genes listed, along with greatly decreased production of momilactones after treatment with chitin-derived elicitors. As expected, *OsTGAP1*-overexpression lines showed enhanced expression of all five genes, along with hyperaccumulation of momilactones. This result suggests that OsTGAP1 can potentially be exploited as a master switch for the increased production of antimicrobial compounds after pathogen detection, and thereby may result in enhanced resistance to fungal diseases.

4. Interference with Virulence Targets and Suppression of Susceptibility Genes

The third main technique in improving disease resistance in plants is to remove host components that promote susceptibility, known as susceptibility factors (102). Numerous recent studies have suggested that the main role of pathogen effectors is in the suppression of basal immunity, or PTI through interacting with specific host factors (13). Theoretically, knocking out, or modifying these factors would avoid the suppression of plant defense signaling, leading to resistance (103). Although some of these susceptibility factors have been identified and characterized (i.e., *Arabidopsis* RIN4), disease susceptibility factors in rice and their underlying mechanisms remain largely undiscovered to date.

Susceptibility genes, or *S* genes, are defined as dominant genes that when mutated, result in recessively inherited resistance (102, 103). In rice, there are three well-characterized examples of this, *Xa5*, *Xa13*, and *Pi21*. The recessive forms of *Xa5* and *Xa13* (denoted *xa5* and *xa13*) confer *R* gene-type resistance against specific races of *X. oryzae* pv. *oryzae* (104, 105), whereas *pi21* confers durable resistance against *M. oryzae* (9). *Xa5* encodes the small subunit of transcription factor IIA (TFIIAγ). Substitution of two nucleotides resulting in an amino acid change from valine to glutamic acid in position 39 resulted in resistance to *X. oryzae* pv. *oryzae* (104, 106). Recessive *R* gene *xa13* confers race-specific resistance to *X. oryzae* pv. *oryzae* race 6. *xa13* and its functional form *Xa13* code for identical proteins; the difference between the resistant and susceptible forms is within the promoter region. Chu et al. (105) characterized eleven *xa13*-carrying lines and noted that all lines had an alteration in the −69 to −86 region of the *Xa13* promoter, suggesting that *Xa13* is a susceptibility factor that is induced by *X. oryzae* pv. *oryzae*. In this same study, it was also noted that the dominant *Xa13* is not only required for colonization of *X. oryzae* pv. *oryzae* but also for development of functional pollen, as *xa13* lines are male sterile.

pi21 was mapped and identified as a recessive gene conferring durable, broad-spectrum resistance to *M. oryzae* (107). The resistance conferred by *pi21* does not result in HR, like many other blast resistance genes, but instead results in greatly reduced symptom development. However, transfer of *pi21* through marker-assisted breeding methods also resulted in poor eating quality. Further characterization of *pi21* by Fukuoka et al. (9) showed that *Pi21* encodes a proline-rich protein with a heavy metal-binding domain and protein–protein interaction motifs. The recessive *pi21* allele carries a deletion in the 18–48 bp region, within the consensus PxxPxxP motif, housed in a putative protein-protein interaction domain. In addition to this, transformation studies show that the poor eating quality is not a pleiotropic effect of *pi21*, but instead is conferred by a gene that is closely linked with the *Pi21* locus. RNAi-mediated suppression of *Pi21* resulted in resistance to *M. oryzae*, and preservation of good eating quality (9).

5. Modification of Transgenes

In any organism, genes and their products are subject to regulation at numerous levels in response to environmental cues. Along these lines, current technology allows for modification of transgenes both at the expression level through the use of different promoters and at the posttranscriptional level through alteration of protein stability.

5.1. Choosing Appropriate Promoters

Theoretically, the strength of the promoter linked to a resistance or defense gene is correlated to the magnitude of resistance. As mentioned in this chapter, a number of genes linked to constitutive promoters such as ubiquitin (*Ubi*) or *CaMV35* have shown increased resistance. For example, race-specific *R* gene *Xa3/Xa27* recognized a broader spectrum of *X. oryzae* pv. *oryzae* races when constitutively expressed. However, this is a very simplistic view, and in many cases, the use of a constitutive or any other strongly expressed promoter may have negative effects. A number of studies have shown that there is a trade-off between the utilization of defense resources in the plant, and its reproductive capability and overall health (reviewed in ref. 108). For example, constitutive overexpression of *Arabidopsis* jasmonic acid methyltransferase (*AtJMT*) in rice resulted in enhanced stress tolerance, however the transgenic lines exhibited an inhibition in spikelet development and subsequent low yield (109). To avoid this trade-off, it may be beneficial to use a specialized or inducible promoter. A number of specialized promoters have been isolated or synthesized, including promoters that are tissue-specific, stress-induced, or induced by application of exogenous compounds (reviewed in ref. 110). Pathogen-inducible promoters may be used when constitutive expression of a resistance-related gene exerts a negative effect on the overall fitness of the plant. In this case, the transgene is only expressed upon exposure to the pathogen. An example of utilization of a pathogen-inducible promoter is in the overexpression of allene oxide synthase (*OsAOS2*), as mentioned earlier in this chapter. The use of the stress-inducible *PBZ1/PR10* promoter help enhance resistance to *M. oryzae* without sacrificing plant vigor (72).

5.2. Alteration of Protein Structure

The alteration of protein structure and function is another potential method of improving defense signaling in plants. This could occur through many strategies, depending on the specific properties of the protein, and the mechanisms by which it is involved in the signaling process. A common method is to increase the stability of a defense-related protein. For example, the rate-limiting step in the ethylene biosynthetic pathway is catalyzed by the enzyme 1-aminocyclopropane-1-carboxylic acid synthase (ACS), which has a very short half-life. In *Arabidopsis*, two ACS isoenzymes have been found to be ubiquitinated and degraded through the 26S proteasome pathway. Phosphorylation of the C-terminal ends of these enzymes inhibits degradation, thereby resulting in an increase in ethylene production and enhanced response to stress (111, 112). Changing the serine residues to aspartate in each C-terminal phosphorylation site mimics the effect of phosphorylation, and these ACS^{DDD} phosphomimicking mutants show increased accumulation, and an increase in ethylene production (112). In rice, the MAP kinase OsMKK4 activates MAP kinases OsMPK3 and OsMPK6, which are necessary for production of diterpenoid phytoalexins. A study using rice cell cultures showed that a phosphomimicking mutant of OsMKK4 (OsMKK4DD) resulted in an increase of various defense responses, including cell death, production of lignin, and biosynthesis of phytoalexins through increased activation of OsMPK3 and OsMPK6 (58).

Protein structure can also be altered to change or inhibit signaling mechanisms, which may be beneficial in cases where a factor is known to be a negative regulator of resistance. An example of this was through the alteration of the DNA-binding site of transcription factor rTGA2.1. Fitzgerald et al. (62) substituted nucleotides in the DNA-binding domain of *rTGA2.1*, resulting in a change from two alanine to proline residues, rendering the protein unable to bind DNA. Constitutive overexpression of *rTGA2.1* in rice resulted in enhanced resistance to *X. oryzae* pv. *oryzae*, and an increased expression of *PR* genes.

6. Perspectives and Conclusions

Rice, like all higher plants, is a complex organism, consisting of many regulatory mechanisms controlling a balance between growth, development, reproduction, and response to environmental stimuli. As been shown in this chapter, there are many benefits to using transgenic approaches to improve disease resistance in rice. However, when deciding which genetic components to alter, there are a number of factors to be considered.

First of all, many resistance mechanisms involve a signaling complex, or at least partner proteins. The basic principle behind using transformation to improve cultivars is to transform a single gene conferring the desired trait. However, more than one gene or the collective action of a gene complex may be required to carry out a sufficient function. There are a few well-studied examples of "partner genes" in plant–pathogen interactions, such as tomato *Prf* (113) and *Arabidopsis NDR1* (114). A recent study revealed that a gene closely linked with the blast resistance gene *Pi-ta* was required for full *Pi-ta*-mediated resistance. This gene, *Ptr(t)* is located within 9 megabases of *Pi-ta* (115). Similarly, *Pi5*-mediated blast resistance was found to be conferred by two CC-NB-LRR genes. A map-based cloning approach found two CC-NB-LRR genes, *Pi5-1* and *Pi5-2*. However, transformation of each gene into a susceptible cultivar failed to result in an incompatible phenotype. Crossing the two transgenic lines resulted in resistance to *M. oryzae*, suggesting that both *Pi5-1* and *Pi5-2* are required for *Pi5*-mediated resistance. Interestingly, the two genes show different expression levels. *Pi5-1* is induced after pathogen challenge, whereas *Pi5-2* is expressed constitutively. These two examples suggest that resistance signaling is often more complex, involving more than one gene. If this is the case, overexpression of just one gene many not have a significant effect (28).

Secondly, expression or silencing of genes may have secondary effects on other important aspects of plant health, such as germination, flowering, and seed development. An obvious example is seen with the recessive resistance gene *xa13*. *Xa13* is a modulator of disease susceptibility, which encourages colonization of *X. oryzae* pv. *oryzae*. However, it is also required for pollen development. *xa13* lines show high levels of resistance to *X. oryzae* pv. *oryzae*, however, the lines also display male sterility (105). Another very important aspect in maintaining plant health is the ability to response to and protection against the effects of abiotic stresses, such as drought, hypoxia, cold temperatures, and saline soils. A few of the defense-related strategies mentioned in this review have been found to have negative effects on tolerance to abiotic stresses. As mentioned previously, OsMPK5 acts as a negative regulator of resistance to *M. oryzae*, but a positive regulator of tolerance to drought, salinity, and cold temperatures. Silencing of *OsMPK5* results in a reduction in blast disease severity, however it is at the cost of tolerance to the aforementioned abiotic stresses (55). Other examples with potential negative effects on abiotic stress signaling include OsWRKY13 and AtNPR1 (reviewed in ref. 35).

Lastly, enhancement of resistance against one specific pathogen may result in an increase in susceptibility to another pathogen. Plant defense pathways often have similar signaling components, however, simultaneous stress signaling pathways may have antagonizing effects on one another. Plant pathogens are known to have different lifestyles, ranging from obligate biotrophs such as rust fungi (survives solely on living tissue) to necrotrophs such as *Botrytis cinerea* (kills host tissue before obtaining nutrients). In accordance with this, plants have different pathways responding to pathogen challenge. For example, generally salicylic acid signaling is deployed against biotrophic pathogens, whereas JA and ET signaling is deployed against necrotrophic pathogens (reviewed in ref. 116). In rice, antagonism between different signaling pathways conferring resistance to different pathogens has not been well characterized, however two known examples include the MAP kinases OsMPK5 and OsMPK6. OsMPK5 has been implicated in acting as a negative regulator of resistance to *M. oryzae*, but has been found to play a positive role in defense against *C. miyabeanus* possibly through mediation of abscisic acid signaling (56). OsMPK6 has been shown to play both positive and negative roles in resistance, however the detailed mechanisms are still unknown. *OsMPK6* has been found to be induced by chitin and is essential for synthesis of specific phytoalexins (58). In contrast, suppression of *OsMPK6* results in increased resistance to *X. oryzae* pv. *oryzae* (57). For these reasons, it is important to consider the effect of specific gene transformation on resistance to different types of rice pathogens.

Research done in the past decade has resulted in the discovery of novel genes and mechanisms involved in disease resistance as well as the development of new technologies to incorporate factors that confer resistance into agronomically superior cultivars. These two lines of knowledge should lead to more efficient and effective strategies for genetic improvement of rice disease resistance in the near future.

Acknowledgment

This work was supported by the grants from USDA/NRI (2008-35301-19028) and NSF Plant Genome Research Program (DBI-0605017 and DBI-0922747).

References

1. Skamnioti P, Gurr SJ (2009) Against the grain: safeguarding rice from blast disease. Trends Biotechnol 27:141–150
2. Wilson RA, Talbot NJ (2009) Under pressure: investigating the biology of plant infection by Magnaporthe oryzae. Nat Rev 7:185–195
3. Banniza S, Holderness M (2001) Pathogen biology and diversity. In: Sreenivasaprasad S, Johnson R (eds) Major fungal diseases of rice: recent advances. Kluwer Academic Publishers, The Netherlands. pp 201–211
4. Ou SH (1972) Rice diseases. Commonwealth Mycological Institute, Kew, Surrey, England
5. Niño-Liu DO, Ronald PC, Bogdanove AJ (2006) *Xanthomonas oryzae* pathovars: model pathogen of a model crop. Mol Plant Pathol 7:303–324
6. Baulcombe D (2004) RNA silencing in plants. Nature 431:356–363
7. Gust AA, Brunner F, Nürnberger T (2010) Biotechnological concepts for improving plant innate immunity. Curr Opin Biotechnol 21:204–210
8. Kou Y, Wang S (2010) Broad-spectrum and durability: understanding of quantitative disease resistance. Curr Opin Plant Biol 13:181–185
9. Fukuoka S, Saka N, Koga H, Ono K, Shimizu T, Ebana K, Hayashi N, Takahashi A, Hirochika H, Okuno K, Yano M (2009) Loss of function of a proline-containing protein confers durable disease resistance in rice. Science 325:998–1001
10. Qu S, Liu G, Zhou B, Bellizzi M, Zeng L, Dai L, Han B, Wang G-L (2006) The broad-spectrum blast resistance gene *Pi9* encodes a nucleotide-binding site leucine-rich repeat protein and is a member of a multigene family in rice. Genetics 172:1901–1914
11. Niks RE, Marcel TC (2009) Nonhost and basal resistance: how to explain specificity? New Phytol 182:817–828
12. Poland JA, Balint-Kurti PJ, Wisser RJ, Pratt RC, Nelson RJ (2008) Shades of gray: the world of quantitative disease resistance. Trends Plant Sci 14:21–29
13. Jones JDG, Dangl JL (2006) The plant immune system. Nature 444:323–329
14. Gomez-Gomez L, Boller T (2006) FLS2: an LRR receptor like kinase involved in the perception of the bacterial elicitor flagellin in *Arabidopsis*. Mol Cell 5:1003–1011
15. Göhre V, Spallek T, Haweker H, Mersmann S, Mentzel T, Boller T, de Torres M, Mansfield JW, Robatzke S (2008) Plant pattern-recognition receptor FLS2 is directed for degradation by the bacterial ubiquitin ligase AvrPtoB. Curr Biol 18:1824–1832
16. Li H, Zhou SY, Zhao WS, Su SC, Peng YL (2009) A novel wall-associated receptor-like kinase gene, *OsWAK1*, plays important roles in rice blast disease resistance. Plant Mol Biol 69:337–346
17. Shiu S-H, Karlowski WM, Pan R, Tzeng Y-H, Mayer KFX, Li W-H (2004) Comparative analysis of the receptor-like kinase family in *Arabidopsis* and Rice. Plant Cell 16:1220–1234
18. Dangl JL, Jones JDG (2001) Plant pathogens and integrated defense responses to infection. Nature 411:826–833
19. Carole LB, Michael LN, Reuben AC, Mangalathu SR (2000) Alternative transcript initiation and novel post-transcriptional processing of a leucine-rich repeat receptor-like protein kinase gene that responds to short-day photoperiodic floral induction in morning glory (*Ipomoea nil*). Plant Mol Biol 43:43–58
20. Nishiguchi M, Yoshida K, Sumizono T, Tazaki K (2002) A receptor-like protein kinase with a lectin-like domain from lombardy poplar: gene expression in response to wounding and characterization of phosphorylation activity. Mol Genet Genomics 267:506–514
21. Hong SW, Jon JH, Kwak JM, Nam HG (1997) Identification of a receptor-like protein kinase gene rapidly induced by abscisic acid, dehydration, high salt, and cold treatments in *Arabidopsis thaliana*. Plant Physiol 113:1203–1212

22. Li J, Wen J, Lease KA, Doke JT, Tax FE, Walker JC (2002) BAK1, an Arabidopsis LRR receptor-like protein kinase, interacts with BRI1 and modulates brassinosteroid signaling. Cell 110:213–222
23. Xiong L, Lee MW, Qi M, Yang Y (2001) Identification of defense-related genes by suppression subtractive hybridization and differential screening. Mol Plant Microbe Interact 14:685–692
24. Hu H, Xiong L, Yang Y (2005) Rice SERK1 gene positively regulates somatic embryogenesis of cultured cell and host defense response against fungal infection. Planta 222:107–117
25. Song D, Li G, Song F, Zheng Z (2008) Molecular characterization and expression analysis of OsBISERK1, a gene encoding a leucine-rich repeat receptor-like kinase, during disease resistance responses in rice. Mol Biol Rep 35:275–283
26. Peng H, Zhang Z, Li Y, Lei C, Zhai Y, Sun X, Sun D, Sun Y, Lu T (2009) A putative leucine-rich repeat receptor kinase, OsBRR1, is involved in rice blast resistance. Planta 230:377–385
27. Song WY, Wang G-L, Chen LL, Kim HS, Pi LY, Holsten T, Wang B, Zhai WX, Zhu H, Fauquet C, Ronald PC (1995) A receptor kinase-like protein encoded by the rice disease resistance gene, *Xa21*. Science 270: 1804–1806
28. Lee SW, Han SW, Sririyanum M, Park CJ, Seo YS, Ronald PC (2009) A type-I secreted, sulfated peptide triggers XA21-mediated innate immunity. Science 326:850–853
29. Wang G-L, Song WY, Ruan DL, Sideris S, Ronald PC (1996) The cloned gene, Xa21, confers resistance to multiple *X. oryzae pv. oryzae* isolates in transgenic plants. *Mol.* Plant-Microbe Interct 9:850–855
30. Ito Y, Kaku H, Shibuya N (2002) Identification of a high affinity binding protein for N-acetylchitooligosaccharide elicitor in the plasma membrane of suspension-cultured rice cells by affinity labeling. Plant J 12:347–356
31. Shimizu T, Nakano T, Takamizawa D, Desaki Y, Ishii-Minami N, Nishizawa Y, Minami E, Okada K, Yamane H, Kaku H, Shibuya N (2010) Two LysM receptor molecules, CEBiP and OsCERK1, cooperatively regulate chitin elicitor signaling in rice. Plant J 64:204–214
32. Kaku H, Nishizawa Y, Ishii-Minami N, Akimoto-Tomiyama C, Dohmae N, Takio K, Minami E, Shibuya N (2006) Plant cells recognize chitin fragments for defense signaling through a plasma membrane receptor. Proc Natl Acad Sci USA 103:11086–11091
33. Ballini E, Morel J-B, Droc G, Price A, Courtois B, Notteghem J-L, Tharreau D (2008) A genome-wide meta-analysis of rice blast resistance genes and quantitative trait loci provides new insights into partial and complete resistance. Mol Plant Microbe Interact 21:859–868
34. Liu J, Wang X, Mitchell T, Hu Y, Liu X, Dai L, Wang G-L (2010) Recent progress and understanding of the molecular mechanisms of the rice-*Magnaporthe oryzae* interaction. Mol Plant Pathol 11:419–427
35. Delteil A, Zhang J, Lessard P, Morel JB (2010) Potential candidate genes for improving rice disease resistance. Rice 3:56–71
36. Zhao J, Fu J, Li X, Xu C, Wang S (2009) Dissection of the factors affecting development-controlled and race-specific disease resistance conferred by leucine-rich repeat receptor kinase-type R genes in rice. Theor Appl Genet 119:231–239
37. Century KS, Lagman RA, Adkisson M, Morlan J, Tobias R, Schwartz K, Smith A, Love J, Ronald PC, Whalen MC (1999) Developmental control of Xa21- mediated disease resistance in rice. Plant J 20:231–236
38. Cao Y, Ding X, Cai M, Zhao J, In Y, Li X, Xu C, Wang S (2007) Expression pattern of a rice disease resistance gene *Xa3/Xa26* is differentially regulated by the genetic backgrounds and developmental stages that influence its function. Genetics 177:523–533
39. Sun X, Cao Y, Yang Z, Xu C, Li X, Wang S, Zhang Q (2004) *Xa26*, a gene conferring resistance to *Xanthomonas oryzae* pv. *oryzae* in rice, encodes an LRR receptor kinase-like protein. Plant J 37:517–527
40. Hittalmani S, Parco A, Mew TV, Zeigler RS, Huang N (2000) Fine mapping and DNA marker-assisted pyramiding of the three major genes for blast resistance in rice. Theor Appl Genet 100:1121–1128
41. Koide Y, Kawasaki A, Telebanco-Yanoria MJ, Hairmansis A, Nguyen NTM, Bigirimana J, Fujita D, Kobayashi N, Fukuta Y (2010) Development of pyramided lines with two resistance genes, *Pish* and *Pib*, for blast disease (*Magnaporthe oryzae* B.Couch) in rice (*Oryza sativa* L.). Plant Breed 129:670–675
42. Farnham G, Baulcombe DC (2006) Artificial evolution extends the spectrum of viruses that are targeted by a disease-resistance gene from potato. Proc Natl Acad Sci USA 103:18828–18833
43. Foreman J, Demidchik V, Bothwell JHF, Mylona P, Miedema H, Torres MA, Linstead P, Costa S, Brownlee C, Jones JDG, Davies JM, Dolan L (2003) Reactive oxygen species

produced by NADPH oxidase regulate plant cell growth. Nature 422:422–426

44. Levine A, Tenhaken R, Dixon R, Lamb C (1994) H_2O_2 from the oxidative burst orchestrates the plant hypersensitive disease resistance response. Cell 79:583–593
45. Wong HL, Pinontoan R, Hayashi K, Tabata R, Yaeno T, Hasegawa K, Kojima C, Yoshioka H, Iba K, Kawasaki T, Shimamoto K (2007) Regulation of rice NADPH oxidase by binding of Rac GTPase to its N-terminal extension. Plant Cell 19:4022–4034
46. Kawasaki T, Henmi K, Ono E, Hatakeyama S, Iwamo M, Satoh H, Shimamoto K (1999) The small GTP-binding protein Rac is a regulator of cell death in plants. Plant Cell 14:763–776
47. Ono E, Wong HL, Kawasaki T, Hasegawa M, Kodama O, Shimamoto K (2001) Essential role of the small GTPase Rac in disease resistance of rice. Proc Natl Acad Sci USA 98:759–764
48. Godfrey D, Able A, Dry I (2007) Induction of a grapevine germin-like protein (*HvGLP3*) gene is closely linked to the site of Erysiphe necator infection: a possible role in defense? Mol Plant Microbe Interact 20:1112–1125
49. Manosalva PM, Davidson RM, Liu B, Zhu X, Hulbert SH, Leung H, Leach JE (2009) A germin-like protein gene family functions as a complex quantitative trait locus conferring broad-spectrum disease resistance in rice. Plant Physiol 149:286–296
50. Rohila JS, Yang Y (2007) Rice mitogen-activated protein kinase gene family and its role in biotic and abiotic stress response. J Integr Plant Biol 49:751–759
51. Reyna NS, Yang Y (2006) Molecular analysis of the rice MAP kinase gene family in relation to *Magnaporthe grisea* infection. Mol Plant Microbe Interact 19:530–540
52. Lieberherr D, Thao NP, Nakashima A, Umemura K, Kawasaki T, Shimamoto K (2005) A sphingolipid elicitor-inducible mitogen-activated protein kinase is regulated by the small GTPase OsRac1 and heterotrimeric G-protein in rice. Plant Physiol 138:1644–1652
53. Cheong YH, Moon BC, Kim JK, Kim CY, Kim MC, Park CY, Kim JC, Park BO, Koo SC, Yoon HW, Chung WS, Lim CO, Lee SY, Cho MJ (2003) BWMK1, a rice mitogen-activated protein kinase, locates in the nucleus and mediates pathogenesis-related gene expression by activation of a transcription factor. Plant Physiol 132:1961–1972
54. He C, Fong SH, Yang D, Wang GL (1999) BWMK1, a novel MAP kinase induced by fungal infection and mechanical wounding in rice. Mol Plant Microbe Interact 12:1064–1073
55. Xiong L, Yang Y (2003) Disease resistance and abiotic stress tolerance in rice are inversely modulated by an abscisic acid-inducible mitogen-activated protein kinase. Plant Cell 15:745–759
56. Vleesschauwer D, Yang Y, Vera Cruz C, Höfte M (2010) Abscisic acid-induced resistance against the brown spot pathogen *Cochliobolus miyabeanus* in rice involves MAP kinase-mediated repression of ethylene signaling. Plant Physiol 152:2036–2052
57. Yuan B, Shen X, Li X, Xu C, Wang S (2007) Mitogen-activated protein kinase OsMPK6 negatively regulates rice disease resistance to bacterial pathogens. Planta 226:953–960
58. Kishi-Kaboshi M, Okada K, Kurimoto L, Murakami S, Umezawa T, Shibuya N, Yamane H, Miyao A, Takatsuji H, Takahashi A, Hirochika H (2010) A rice fungal MAMP-responsive MAPK cascade regulates metabolic flow to antimicrobial metabolite synthesis. Plant J 63:599–612
59. Silverman P, Seskar M, Kanter D, Schweizer P, Métraux J-P, Raskin I (1995) Salicylic acid in rice: biosynthesis, conjugation, and possible role. Plant Physiol 108:633–639
60. Yang Y, Qi M, Mei C (2004) Endogenous salicylic acid protects rice plants from oxidative damage caused by aging as well as biotic and abiotic stress. Plant J 40:909–919
61. Mou Z, Fan W, Dong X (2003) Inducers of plant systemic acquired resistance regulate NPR1 function through redox changes. Cell 113:935–944
62. Chern MS, Fitzgerald HA, Yadav RC, Canlas PE, Dong X, Ronald PC (2001) Evidence for a disease-resistance pathway in rice similar to the NPR1-mediated signaling pathway in Arabidopsis. Plant J 27:101–113
63. Chern MS, Fitzgerald HA, Canlas PE, Navarre DA, Ronald PC (2005) Overexpression of a rice NPR1 homolog leads to constitutive activation of defense response and hypersensitivity to light. Mol Plant Microbe Interact 18:511–526
64. Jiang C-J, Shimono M, Maeda S, Inoue H, Mori M, Hasegawa M, Sugano S, Takasuji H (2009) Suppression of the rice fatty-acid desaturase gene *OsSSI2* enhances resistance to blast and leaf blight diseases in rice. Mol Plant Microbe Interact 22:820–829
65. Iwai T, Seo S, Mitsuhara I, Ohashi Y (2007) Probenazole-induced accumulation of salicylic acid confers resistance to *Magnaporthe grisea* in adult rice plants. Plant Cell Physiol 48:915–924

66. Umemura K, Satou J, Iwata M, Uozumi N, Koga J, Kawano T, Koshiba T, Anzai H, Mitomi M (2009) Contribution of salicylic acid glucosyltransferase, OsSGT1, to chemically induced disease resistance in rice plants. Plant J 57:463–472
67. Seo S, Ishizuka K, Ohashi Y (1995) Induction of salicylic acid β-glucosidase in tobacco leaves by exogenous salicylic acid. Plant Cell Physiol 36:447–453
68. Chen Z, Malamy J, Henning J, Conrath U, Sanchez-Casas P, Silva H, Ricigliano J, Klessig K (1995) Induction, modification, and transduction of the salicylic acid signal in plant defense responses. Proc Natl Acad Sci USA 92:4134–4137
69. He Y, Fukushige H, Hildebrand DF, Gan S (2002) Evidence supporting a role of jasmonic acid in Arabidopsis leaf senescence. Plant Physiol 128:876–884
70. Tamogami S, Rakwal R, Kodama O (1997) Phytoalexin production elicited by exogenously applied jasmonic acid in rice leaves (*Oryza sativa* L.) is under the control of cytokinins and ascorbic acid. FEBS Lett 412:61–64
71. Schweizer P, Buchala A, Métraux J-P (1997) Gene-expression patterns and levels of jasmonic acid in rice treated with the resistance inducer 2, 6-dichloroisonicotinic acid. Plant Physiol 115:61–70
72. Mei C, Min Q, Sheng G, Yang Y (2006) Inducible overexpression of a rice allele oxide synthase gene increases the endogenous jasmonic acid level, PR gene expression and host resistance to fungal infection. Mol Plant Microbe Interact 19:1127–1137
73. Singh MP, Lee FN, Counce PA, Gibbons JH (2004) Mediation of partial resistance to rice blast through anaerobic induction of ethylene. Phytopathology 94:819–825
74. Iwai T, Miyasaka A, Seo S, Ohashi Y (2006) Contribution of ethylene biosynthesis for resistance to blast fungus infection in young rice plants. Plant Physiol 142:1202–1215
75. Seo S, Mitsuhara I, Feng J, Iwai T, Hasegawa M, Ohashi Y (2010) Cyanide, a coproduct of plant hormone ethylene biosynthesis, contributes to the resistance of rice to blast fungus. Plant Physiol 155:502–514
76. Bailey TA, Zhou X, Chen J, Yang Y (2009) Role of ethylene, abscisic acid and MAP kinase pathways in rice blast resistance. In: Wang G-L, Valent B (eds) Advances in genetics, genomics and control of rice blast disease. Springer pp 185–190
77. Mauch-Mani B, Mauch F (2005) The role of abscisic acid in plant-pathogen interactions. Curr Opin Plant Biol 8:409–414
78. Jiang C-J, Shimono M, Sugano S, Kojima M, Azawa K, Yoshida R, Inoue H, Hayashi N, Sakakibara H, Takatsuji H (2010) Abscisic acid interacts antagonistically with salicylic acid signaling pathway in rice-Magnaporthe grisea interaction. Mol Plant Microbe Interact 23:791–798
79. Kazan K, Manners JM (2009) Linking development to defense: auxin in plant- pathogen interactions. Trends Plant Sci 14:373–382
80. Shinshu H, Mohnen D, Meins F (1987) Regulation of a plant pathogenesis-related enzyme: inhibition of chitinase and chitinase mRNA accumulation in cultured tobacco tissues by auxin and cytokinin. Proc Natl Acad Sci USA 84:89–93
81. Abreu ME, Munne-Bosch S (2009) Salicylic acid deficiency in *NahG* transgenic lines and *sid2* mutants increases seed yield in the annual plant *Arabidopsis thaliana*. J Exp Bot 60:1261–1271
82. Ding X, Cao Y, Huang L, Zhao J, Xu C, Li X, Wang S (2009) Activation of the indole-3-acetic acid-amido synthetase GH3. 8 suppresses expansin expression and promotes salicylate- and jasmonate-independent basal immunity in rice. Plant Cell 20:228–240
83. Domingo C, Andrés F, Tharreau D, Iglesias DJ, Talón M (2009) Constitutive expression of *OsGH3.1* reduces auxin content and enhances defense response and resistance to a fungal pathogen in rice. Mol Plant Microbe Interact 22:201–210
84. Liu X, Bai X, Wang X, Chu C (2007) *OsWRKY71*, a rice transcription factor, is involved in rice defense response. J Plant Physiol 164:969–979
85. Hu K, Qiu D, Shen X, Li X, Wang S (2008) Isolation and manipulation of quantitative trait loci for disease resistance in rice using a candidate gene approach. Mol Plant Pathol 1:786–793
86. Wen N, Chu Z, Wang S (2003) Three types of defense-responsive genes are involved in resistance to bacterial blight and fungal blast diseases in rice. Mol Genet Genomics 269:331–339
87. Wang GL, Mackill DJ, Bonman M, McCouch SR, Champoux MC, Nelson RJ (1994) RFLP mapping of genes conferring complete and partial resistance to blast in a durably resistant rice cultivar. Genetics 136:1421–1434
88. Chen H, Wang S, Xing Y, Xu C, Hayes P, Zhang Q (2003) Comparative analyses of genomic locations and race specificities of loci for quantitative resistance to *Pyricularia grisea* in rice and barley. Proc Natl Acad Sci USA 100:2544–2549
89. Qiu D, Xiao J, Ding X, Xiong M, Cai M, Cao Y, Li X, Xu C, Wang S (2007) OsWRKY13

mediates rice disease resistance by regulating defense related genes in salicylate and jasmonate-dependent signaling. Mol Plant Microbe Interact 20:492–499

90. Shimono M, Sugano S, Nakayama A, Jiang CJ, Ono K, Toki S et al (2007) Rice WRKY45 plays a crucial role in benzothiadiazole-inducible blast resistance. Plant Cell 19:2064–2076
91. Ohnishi T, Sugahara S, Yamada T, Kikuchi K, Yoshiba Y, Hirano HY, Tsutsumi N (2005) OsNAC6, a member of the NAC gene family, is induced by various stresses in rice. Genes Genet Syst 80:135–139
92. Nakashima K, Tran L-SP, Nguyen DV, Fujita M, Maruyama K, Todaka D, Ito Y, Hayashi N, Shinozaki K, Yamaguchi-Shinozaki K (2007) Functional analysis of a NAC-type transcription factor OsNAC6 involved in abiotic and biotic stress-responsive gene expression in rice. Plant J 51:617–630
93. Tran L-SP, Nakashima K, Sakuma Y, Simpson SD, Fujita Y, Maruyama K, Fujita M, Seki M, Shinozaki K, Yamaguchi-Shinozaki K (2004) Isolation and functional analysis of *Arabidopsis* stress-inducible NAC transcription factors that bind to a drought-responsive *cis*-element in the *early responsive to dehydration stress 1* promoter. Plant Cell 16:2481–2498
94. Fitzgerald HA, Canlas PE, Chern MS, Ronald PC (2005) Alteration of TGA factor activity in rice results in enhanced tolerance to *Xanthomonas oryzae* pv. *oryzae*. Plant J 43:335–347
95. van Loon LC, Rep M, Pieterse CMJ (2006) Significance of inducible defense-related proteins in infected plants. Annu Rev Phytopathol 44:1–28
96. Datta K, Baisakh N, Maung T, Thet KM, Tu J, Datta S (2002) Pyramiding transgenes for multiple resistance in rice against bacterial blight, yellow stem borer and sheath blight. Theor Appl Genet 106:1–8
97. Anžlovar S, Dermastia M (2003) The comparative analysis of osmotins and osmotin-like PR-5 proteins. Plant Biol 5:116–124
98. Datta K, Velazhahan R, Oliva N, Ona I, Mew T, Khush GS, Muthukrishnan S, Datta SK (1999) Over-expression of the cloned rice thaumatin-like protein (PR-5) gene in transgenic rice plants enhances environmental friendly resistance to *Rhizoctonia solani* causing sheath blight disease. Theor Appl Genet 98:1138–1145
99. Gomez-Ariza J, Campo S, Rufat M, Estopa M, Messeguer J, Segundo BS, Coca M (2007) Sucrose-mediated priming of plant defense responses and broad-spectrum disease resistance of the maize pathogenesis-related PRms protein in rice plants. Mol Plant Microbe Interact 20:832–842
100. Shimura K, Okada A, Okada K, Jikumaru Y, Ko K-W, Toyomasu T, Sassa T, Hasegawa M, Kodama O, Shibuya N, Koga J, Nojiri H, Yamane H (2007) Identification of a biosynthetic gene cluster in rice for momilactones. J Biol Chem 282:34013–34018
101. Okada A, Okada K, Miyamoto K, Koga J, Shibuya N, Nojiri H, Yamane H (2009) OsTGAP1, a bZIP transcription factor, coordinately regulates the inductive production of diterpenoid phytoalexins in rice. J Biol Chem 284:26510–26518
102. Eckhardt NA (2002) Plant disease susceptibility genes? Plant Cell 14:1983–1986
103. Pavan S, Jacobsen E, Visser RGF, Bai Y (2010) Loss of susceptibility as a novel breeding strategy for durable and broad-spectrum resistance. Mol Breed 25:1–12
104. Iyer AS, McCouch SR (2004) The rice bacterial blight resistance gene *xa5* encodes a novel form of disease resistance. Mol Plant Microbe Interact 17:1348–1354
105. Chu Z, Yuan M, Yao J, Ge X, Yuan B, Xu C, Li X, Fu B, Li Z, Bennetzen JL, Zhang Q, Wang S (2006) Promoter mutations of an essential gene for pollen development result in disease resistance in rice. Genes Dev 20:1250–1255
106. Iyer-Pascuzzi AS, McCouch SR (2007) Recessive resistance genes and the *Oryza sativa-Xanthomonas oryzae* pv *oryzae* pathosystem. Mol Plant Microbe Interact 20:731–739
107. Fukuoka S, Okuno K (2001) QTL analysis and mapping of *pi21*, a recessive gene for field resistance to rice blast in Japanese upland rice. Theor Appl Genet 103:185–190
108. Brown JKM (2002) Yield penalties of disease resistance in crops. Curr Opin Plant Biol 5:1–6
109. Kim EH, Kim YS, Park S-H, Koo YJ, Choi YD, Chung Y-Y, Lee I-J, Kim J-K (2009) Methyl jasmonate reduces grain yield by mediating signals to alter spikelet development in rice. Plant Physiol 149:1751–1760
110. Potenza C, Aleman L, Sengupta-Gopalan C (2004) Targeting transgene expression in research, agricultural, and environmental applications: promoters used in plant transformation. In Vitro Cell Dev Biol 40:1–22
111. Liu Y, Zhang S (2004) Phosphorylation of 1-aminocyclopropane-1-carboxylic acid synthase by MPK6, a stress-responsive mitogen-activated protein kinase, induces

ethylene biosynthesis in *Arabidopsis*. Plant Cell 16:3386–3399

112. Joo S, Liu Y, Lueth A, Zhang S (2008) MAPK phosphorylation-induced stabilization of ACS6 protein is mediated by the non-catalytic C-terminal domain, which also contains the *cis*-determinant for rapid degradation by the 26S proteasome pathway. Plant J 54:129–140
113. Salmeron JM, Oldroyd GED, Rommens CMT, Scofield SR, Kim H-S, Lavelle DT, Dahlbeck D, Staskawicz BJ (1996) Tomato *Prf* is a member of the leucine- rich repeat class of plant disease resistance genes and lies embedded within the *Pto* kinase gene cluster. Cell 86:123–133
114. Century KS, Shapiro AD, Repetti PP, Dahlbeck D, Holub E, Staskawicz BJ (1997) NDR1, a pathogen-induced component required for *Arabidopsis* disease resistance. Science 278:1963
115. Jia Y, Martin R (2008) Identification of a new locus, *Ptr(t)*, required for rice blast resistance gene *Pi-ta*-mediated resistance. Mol Plant Microbe Interact 21:396–403
116. Lee S-K, Song M-Y, Seo Y-S, Kim H-K, Ko S, Cao P-J, Suh J-P, Yi G, Roh J-H, Lee S, An G, Hahn T-R, Wang G-L, Ronald P, Jeon J-S (2009) Rice *Pi5*-mediated resistance to *Magnaporthe oryzae* requires the presence of two coiled-coil-nucleotide-binding-leucine-rich-repeat genes. Genetics 181:1627–1638
117. Spoel SH, Johnson JS, Dong X (2007) Regulation of tradeoffs between plant defenses against pathogens with different lifestyles. Proc Natl Acad Sci USA 104:18842–18847

Chapter 22

Molecular Strategies to Engineer Transgenic Rice Seed Compartments for Large-Scale Production of Plant-Made Pharmaceuticals

Trevor Greenham and Illimar Altosaar

Abstract

The use of plants as bioreactors for the large-scale production of recombinant proteins has emerged as an exciting area of research. The current shortages in protein therapeutics due to the capacity and economic bottlenecks faced with modern protein production platforms (microbial, yeast, mammalian) has driven considerable attention towards molecular pharming. Utilizing plants for the large-scale production of recombinant proteins is estimated to be 2–10% the cost of microbial platforms, and up to 1,000-fold more cost effective than mammalian platforms (Twyman et al. Trends Biotechnol 21:570–578, 2003; Sharma and Sharma, Biotechnol Adv 27:811–832, 2009). In order to achieve an economically feasible plant production host, protein expression and accumulation must be optimized. The seed, and more specifically the rice seed has emerged as an ideal candidate in molecular pharming due to its low protease activity, low water content, stable protein storage environment, relatively high biomass, and the molecular tools available for manipulation (Lau and Sun, Biotechnol Adv 27:1015–1022, 2009).

Key words: Bioreactor, Recombinant protein, Pharmaceutical, Biologics, Transgenic rice, Molecular pharming

1. Introduction

Biopharmaceuticals are the fastest growing sector within the pharmaceutical industry, with a market value of $120 billion (US dollars) in 2009 (1–5). These biologics are mainly produced using recombinant technology and established production platforms such as microbial, yeast, or mammalian cell cultures (6). The effectiveness of different platforms is judged primarily on protein yield, posttranslational modifications, ease of downstream purification and the capital requirements needed for commercialization. *E. coli* was the first large-scale protein production host and has several advantages such as cheap fermentation runs,

Yinong Yang (ed.), *Rice Protocols*, Methods in Molecular Biology, vol. 956,
DOI 10.1007/978-1-62703-194-3_22, © Springer Science+Business Media, LLC 2013

short generation times, and high titers of recombinant protein. Mammalian cultures (CHO cells predominantly) were introduced to overcome some of the shortfalls of the microbial expression platforms such as the formation of inclusion bodies upon high titers, difficulty in purification due to endogenous contaminants, and most importantly their lack of eukaryotic posttranslational modifications (glycosylation, acylation, disulphide bridge formation, etc.) which are often required for protein folding and function (7). CHO cells can produce recombinant proteins with similar glycoprofiles as native proteins and innovations in target-gene insertion, culture media manipulation and apoptosis inhibition have improved titers to over 5 g/L (8). Currently, CHO cells are the most utilized production platform despite their high infrastructure and process costs (9).

The biopharmaceutical industry has seen considerable growth over the past 5 years and has a forecasted growth between 7 and 15% during the next decade (10). This is attributed to the increasing population of elders concomitant with innovations in drug discovery. The growth is apparent when comparing past and present market demands for biopharmaceuticals. Insulin for example had a global demand of 5,000–6,000 kg/year in 2009 and is expected to reach 16,000 kg/year by 2012 (2). The growth of this industry has caused extreme shortages in manufacturing capacity. The high capital requirements related with the aforementioned platforms have put an inhibitory effect on supplying adequate amounts of biopharmaceuticals, prompting other production strategies to be investigated.

With the advent of plant transformation technology (11–13), plants have proven to be feasible bioreactors for the large-scale production of recombinant proteins (14). Due to their lengthy life cycles (3–6 months for rice), the true economic advantages of plants are obtained when dealing with large-scale applications. These advantages are in terms of production costs, scalability, product safety, ease of storage and distribution, none of which can be matched by any current bacterial or mammalian production platform (15–17). It is estimated that using achieved yields, enough hepatitis B-antigen to vaccinate all the infants worldwide could be produced on approximately 200 acres of land (18).

2. Rice Seed as an Expression Host

The seed has emerged as one of the most prominent plant organs for recombinant protein production. For plants, the seed evolved to serve as a storage organ for the long-term maintenance of nutritional components required for the germination of the new sporophytic generation (19). Rice seeds are composed of 7–8%

protein and 92–93% starch. It has been shown that throughout the dormancy period of the rice seed, its storage proteins remain intact and functional (20), meaning it should be a suitable area for the stable deposition of recombinant proteins. This is a considerable advantage over plant platforms that accumulate protein in leafy tissue (e.g., tobacco), as they are more prone to hydrolytic activity and have drastically shorter shelf life. Rice crops are self-pollinating, which ensures that no genetic material is gained or lost, and the gene coding for the protein of interest remains present in each new generation. This provides an advantage from a regulatory standpoint as it reduces the chances of environmental contamination. The seed is also being investigated as an edible vaccine (21–26). Edible vaccines show economic promise because it eliminates the need for downstream purification, reducing capital requirements needed for commercialization by over 60%.

Rice is the staple food worldwide, cultivated in over 100 countries on more than 150 million hectares of land (19). The familiarity with the agronomy and nutritional values of rice, along with GRAS (generally recognized as safe) designation by the Food and Drug Administration make it a strong candidate for the large-scale production of biopharmaceuticals.

3. Strategies to Boost Recombinant Protein Expression in Rice Seeds

The endosperm is the main storage compartment for rice and accounts for over 80% of the total seed weight, thus it is the most enticing site for protein accumulation. The protein composition of the endosperm is composed of 60–70% glutelins, 25–30% prolamins, 5–10% globulins, and 0–5% others. The highest yields seen in rice seeds is a result of utilizing the parameters from natively high expressing seed storage proteins. Protein yields have gotten considerable attention as they play a key role in the capital performance of the production platform. The most common strategies to achieve higher yields of recombinant protein in rice seeds are by optimizing the promoter region, the untranslated regions, translation efficiency and subcellular localization/targeting of the target protein (Table 1).

3.1. Promoters

An early process in recombinant protein generation, transcription is one of the most important areas in determining protein accumulation. In industry, the DNA constructs used for transformation are synthesized de novo from oligonucleotides, providing complete control over its sequence.

Promoters are stretches of DNA found upstream of a gene's coding sequence that interacts with transcription factors and allows RNA polymerase II to bind. In plants, at least seven different

Table 1
Strategies used for expression of biomedical recombinant proteins in rice seeds

Rice expression strategy	Target protein	Accumulation level	References
Glutelin B-1 promoter Glutelin B-1 signal peptide ER-retention signal KDEL Glutelin B-1 terminator	Allergenic T-cell epitope peptides	60 µg/grain	Takagi et al. (37)
Globulin promoter NosTer	Glucagon-like peptide-1 (GLP-1)	50 µg/20 mg seed	Sugita et al. (83)
Glutelin B-1 promoter Glutelin A-2 5′UTR N-terminus GluA-2 acidic subunit fusion C-terminus GluA-2 basic subunit fusion Glutelin B-1 terminator	Cedar Pollen Allergen Cry j1	15% total soluble protein	Yang et al. (91)
Glutelin-1 promoter Glutelin-1 signal peptide NosTer	Human GM-CSF	1.3% total soluble protein	Sardana et al. (34)
AGPase large subunit promoter Glutelin B-1 signal peptide ER-retention signal KDEL NosTer	7 Crp peptide	10 µg/grain	Takaiwa et al. (92)
Glutelin-13A promoter Glutelin-13A signal peptide NosTer	Human IGF-1	6.8 ± 0.5% total soluble protein	Xie et al. (93)
Bx17 HMW glutenin endosperm-specific promoter First intron of rice actin1 gene Heat-labile enterotoxin B subunit bacterial signal peptide ER-retention signal (SEKDEL) NosTer	Cholera Toxin B subunit	2.1% total soluble protein	Oszvald et al. (94)

Glutelin A-2 promoter N-terminus GluA-2 fusion NosTer	Type II-collagen tolerogenic peptide	1 μg/seed	Hashizume et al. (82)
Glutelin B-1 promoter Glutelin B-1 signal peptide NosTer	Luminal binding protein	15-fold increase relative to endogenous levels	Yasuda et al. (95)
Glutelin B-1 promoter Glutelin B-1 signal peptide ER-retention signal (KDEL) Glutelin B-1 3′ UTR	*Ascaris suum* As14 fused to Cholera Toxin B subunit	1.5 μg/grain	Matsumoto et al. (96)
Glutelin-1 promoter Glutelin-1 signal peptide NosTer	Human Lactoferrin	1% seed dry weight	Zhang et al. (38)

transcription factors are required to transcribe a single gene and most of them are dependant on the nucleotide sequence of the promoter. The use of strong constitutive promoters such as the plant cauliflower mosaic virus 35S promoter (27), the rice ubiquitin (28), and actin promoters (29) have been used to drive expression in rice seeds. Although these promoters are known to be highly active in plants, they showed low expression in monocot seeds (under 5% total seed protein) (30, 31). Constitutive promoters also do not allow for much control over the deposition of the recombinant protein, which can negatively effect plant growth and development. Constitutive expression also reduces the opportunity to develop a more cost effective purification strategy that does not rely on prior art.

Seed-specific promoters, such as the ones driving the expression of the major storage proteins (glutelin, prolamin, globulin), have shown to provide higher seed expression levels than when using a strong constitutive promoter (32–34). Using high expressing seed-specific promoters is the simplest way to increase transgene expression. Replacing constitutive promoters with a seed-specific promoter can account for up to a 10-fold increase in seed expression (33). Dozens of rice seed-specific promoters have been characterized for their activity and among them the glutelin B1 (GluB1), glutelin B4 (GluB4), globulin-1 (Glb-1), and 10 and 16 kDa prolamin promoters have shown to give the highest seed expression levels ranging from 6 to 15% of total seed protein (1–2% total seed weight) (33, 35–38).

Engineering promoters, or adding synthetic components to promoters can increase seed-specific expression of the target gene. Conserved plant promoter elements are important regulators of transcription and should be considered when designing synthetic promoters. The most studied elements are the TATA concensus sequence, the transcription initiation site (TS), the 5′ untranslated region (5′UTR), and the context of the translational start codon (39). The design of synthetic promoters relies heavily on additions and/or modification to the already high expressing seed promoters.

The most common sequences added to promoters are transcriptional enhancer domains (cis elements). Cis elements are what determine, either independently or synergistically, the spatial and temporal expression of the promoter. The strength of these motifs is dependant on their copy number and location (40). The conserved motifs found in the 5′ flanking region of the 27 kDa gamma-zein genes have shown to be critical for endosperm-specific expression (41). These motifs include a 13 bp AACA motif, a 7 bp GCN4 motif, a 96 bp GCAA motif, and a prolamin box (TG(T/A/C)AAA(G/T)) which are all found within 245 bp upstream of the TS site in glutelin genes and act in concert to direct the expression of their gene(s) (42). The rice 13 kDa prolamin gene has similar upstream combinations of ACGT, GCN4, and AACA motifs that direct its endosperm specific expression. These proximal elements when fused to a truncated CaMV35S promoter showed endosperm expression levels comparable to a seed-specific promoter, proving their efficacy to direct endosperm expression. Glutelin genes also contain distal motifs found from −245 bp to −1,302 that account for approximately a sixfold enhancement of their promoter activity (42). The GluA-2 promoter in rice for example has one distal and six proximal cis elements that are responsible for its high expression activity (42). Loss and gain of function experiments have shown that at least three of these cis elements are involved in the endosperm expression of the glutelin genes (42). Cis elements found in other monocot species (e.g., coconut, oats) can be used to direct transgene expression to the rice endosperm, though they have not proven to be more effective than the native seed storage promoter cis acting elements (43).

Inserting global regulatory sequences flanking the transgene can increase expression levels by several fold (2, 44). Some commonly used sequences are those required for stress response or matrix attachment. These sequences place the transgene in a surrounding locus that is suitable for the recruitment of transcription factors. Matrix attachment regions, for example, are AT-rich regions that reduce positional effects by forming chromatin loops, keeping the DNA proximal to the transgene loosely packed and available for transcription (45).

Transacting factors can be utilized to increase transgene expression by either directly interacting with cis elements within the promoter or interacting with other transcription factors, recruiting them to the promoter. Most transacting factors identified in rice are involved in stress response. Some of these include the iron deficiency regulator OSBZ8 which interacts with an abscisic acid response element, the dehydration response factors (DREBs) that interact with DRE elements (40), the iron deficiency IDEF1 which binds the cis element IDE1(43), and infection response factor rice endosperm bZIP (REB). REB is a transcription factor that binds to promoters containing a REB-responsive element. Incorporation of a REB-responsive element (GCCACGT(A/C)AG) in promoter regions showed increased seed expression by two- to fourfold (46). The alpha-globulin gene promoter contains three of these repeats, each of which can bind REB (47). REB(s) can increase transcription synergistically as homodimers or heterodimers depending on binding site proximity and flanking motif configuration (39). In order for expression to be elevated by transgenic overexpression, the transacting factor(s) must be limiting natively in the plant.

The use of hybrid promoters or a combination of transcriptional elements from more than one promoter has shown to increase seed expression compared to when used independently (48). Combining elements from the wheat puroindoline-b gene promoter and elements from the rice glutelin-1 promoter showed double the expression of a reporter gene in rice seed (49). The addition of repeated promoter elements, if spaced correctly, can increase seed expression levels while concomitantly reducing the possibility of recombination and positional/silencing effects (50).

3.2. Effects of Untranslated Regions

3.2.1. 5′ Untranslated Region

The 5′UTR is located just upstream of the translation initiation start site and plays an important role in translation (51, 52). Modifying the 5′UTR of a transgene can increase expression levels, though it is difficult to precisely evaluate their efficiency since they work in tandem with many other factors (53). A conserved 5′UTR from the soybean lipoxygenase genes, when fused upstream of a beta-glucuronidase start codon showed an 11-fold increase in expression in rice (54). The 5′UTR region from a tobacco alcohol dehydrogenase gene (NtADH) can act as a translational enhancer for rice transgenes (55). When used upstream of a reporter gene, the three regions of the 5′UTR (F, NF, 221) of NtADH when added to a CaMV35S promoter increased expression levels by 5-, 1.5-, and 2-fold in rice relative to the CaMV35S promoter alone (55). A 5′UTR from *Vigna radiata* aminocyclopropane-1-carboxylate synthase (VR-ACS1) gene has been used as a translational enhancer in plants. This 5′UTR region increased GUS expression by two- to fivefold relative to the control (56). A 66 bp chlorophyll a/b binding gene 5′UTR region has been reported to enhance translation up to eightfold (57). The use of a 5′UTR region from a rice polyubiquitin gene (*rubi3*) has shown its efficacy in enhancing monocot gene expression (58). Rubi3 is an abundant protein involved in protein degradation and cell cycle control. The 1,140 bp 5′UTR of this gene enhances its promoter activity by 20-fold in transient expression assays (59, 60). 5′UTR regions can sometimes be made more active by the use of introns. Intron mediated enhancement (IME) was first defined in 1990 when investigating two maize introns (61). The mechanisms of IME activity are poorly understood, but it has proven to be a useful tool to increase translation of rice transgenes. The highly expressed rice alpha-tubulin genes (*Ostua*1, *Ostua*2 and *Ostua*3) have been isolated for expression analyses. Characterization of their upstream 5′UTR intron demonstrated that it is highly responsible for their expression (62). The 5′UTR intron in rice beta-tubulin gene *Ostub*16 has also been used to enhance expression levels in rice (63). The rubi3 5′UTR contains an intron that increases its translation when present (58). Removing the intron from the rubi3 5′UTR showed that it accounted for a twofold enhancement in transcription activity and a 20-fold enhancement in transcript stability, implying their important role of 5′UTRs in mRNA processing and export (64).

3.2.2. 3′ Untranslated Region

The 3′ untranslated region (3′UTR), located just downstream of the transcription stop codon, is responsible for pre-mRNA 3′end formation (cleavage and addition of the poly(A) tract) and helps stabilize the transcript. The poly(A) tail in particular plays an important role in determining transcript stability and function (65), and a poor 3′UTR can greatly reduce transcript stability (66). There are three primary elements that dictate the efficiency of poly(A) signals in plant 3′UTRs: the far upstream elements (FUE), the near upstream elements (NUE), and the cleavage site (67). Incorporating 3′UTRs harboring these elements can increase gene expression (68). Several studies have shown the efficacy of 3′UTRs for increasing expression levels in plants (69, 70). One study using rice as a host demonstrated that the rice glutelin, GluB-1 3′UTR, when used with a ubiquitin constitutive promoter to drive reporter gene expression, caused an increase in recombinant seed protein levels by 1.8- and 4-fold higher when compared to using a nopaline synthase terminator (19, 71). Very few plant 3′UTRs have been investigated for their impact on rice transgene expression, but it is apparent that the choice of a 3′UTR can significantly contribute to high mRNA stability and accumulation levels of a recombinant protein (72, 73).

3.3. Protein Accumulation and Storage

Once a messenger RNA transcript for a recombinant protein is translated, focus shifts towards its stable accumulation. The two main strategies for the stable accumulation of a recombinant protein are targeting it to a subcellular compartment or using a fusion partner (3). Targeting recombinant proteins to subcellular compartments in plants can increase accumulation levels by several fold (2, 16, 19, 74, 75). The deposition of a recombinant protein impacts not only its accumulation levels but also its assembly and posttranslational modifications. The rationale behind fusion proteins is if a high-expressing plant protein can be stably accumulated, then fusing it with a recombinant protein should also result in the same high and stable accumulation (3). If the natively fused protein component resides in a certain compartment, then it can be assumed that is where the fusion partner will reside, although this is not always the case. In plant cells, an expressed protein without a signal sequence or fusion partner will accumulate in the cytoplasm (76). Otherwise, the protein can be targeted for chloroplast, mitochondrial, or secretory pathway (ER, protein bodies, protein storage vacuoles, apoplasts) deposition. For seed-specific localization, targeting a protein towards the secretory pathway via the ER has proven to be the preferred method. The ER has numerous chaperone proteins (BiP, PDI, calnexin, caltreticulin) and an oxidizing environment that is suitable for most proteins (3, 76). Targeting a recombinant protein to the secretory pathway often results in over a tenfold increase in recombinant seed protein levels relative to cytoplasmic expression (3, 77–79).

The highest seed accumulation levels have been achieved by targeting recombinant proteins to the starchy endosperm. The endosperm is composed primarily of starch and storage proteins found in the form of protein bodies (PB) (80). There are two classes of PBs in rice, PBI and PBII. PBII mainly store glutelins and globulins that get sequestered via the Golgi network or vesicular pathway, and PBI store prolamins via ER-derived deposition (19). Recombinant proteins can be tailored with N or C terminal propeptides, internal sequence motifs, signal peptides, or fusions partners to direct targeting to a specific subcellular domain (49). The most utilized targeting sequences are those found within the major seed storage proteins.

Several strategies have been used for the localization of a recombinant protein towards the PBIIs. The N-terminus portion of the 26 kDa rice globulin protein fused to a green fluorescent protein, under the control of a seed-specific promoter resulted in PBII localization (81). The rice glutelin A (GluA) protein has been used as a fusion partner for PBII localization of a type II-collagen tolerogenic peptide. This peptide, when fused to the C-terminus of the GluA-2 deposited mainly in PBIIs with trace amounts found in the ER, and had a seed expression level of 1 μg/seed (82). It is also a common approach to direct recombinant proteins to PBIIs by inserting them within the hyper variable region of the large acidic subunit of rice glutelin (37, 83, 84). The latter has shown accumulation levels of 20–50 μg per seed (83). The N-terminus from glycinin, a soybean seed storage protein can also direct proteins to PBII in rice seeds, showing the conservation in protein sorting mechanisms between plant species (85).

Targeting recombinant proteins to PBIs has proven to be a valid strategy for their high accumulation. PBIs are viewed as one of the most prominent areas for the deposition of a protein tailored for oral delivery. It has been determined both in vitro and in vivo that prolamins, the main component of PBIs, are highly resistant to gastrointestinal digestion due in part to their hydrophobic properties (43). Directing recombinant proteins towards PBIs usually involves a fusion partner with prolamin or a C-terminus ER-retention signal (86). A green fluorescent protein when fused to the C-terminus of rice prolamin localized within the endosperm's PBIs (87). PBI localization was also observed when a hybrid of seven human T-cell epitopes (7Crp) was fused with GluA-2. The localization was unexpected, as the GluA-2 natively resides in PBIIs, however, the PBI localization was attributed to a free cysteine residue found within the 7Crp that caused it to interact with prolamins via disulfide bridges (86). ER-retention signals (H/KDEL) are involved in the retrieval between the Golgi and ER and localize proteins to PBIs. The ER-retention signal showed an increase in accumulation levels by up to fourfold when compared to the similar construct lacking the C-terminal signal (37, 88). The expression of a cholera toxin B subunit containing the KDEL signal accumulated to level of 30 μg/seed in PBIs. A mature dust mite allergen (Der p 1) was expressed in rice under the control of the GluB-1 promoter, GluB-1 signal sequence, and a KDEL retention signal. This resulted in PBI accumulation levels of 60 μg/seed, whereas a yield of 9.5 μg/seed was obtained when the signal sequence was used independently (86). Combining N-terminal signal peptides with C-terminal ER-retention signals will result in more efficient deposition of recombinant proteins into PBs (37). Targeting recombinant proteins to endosperm-specific organelles via the ER by signal sequence(s), retention signal(s), or fusion partner(s) increases seed accumulation levels by several fold.

There have been a few observations of targeting a recombinant protein within or to the surface of starch granules. Starch granules are indeed an attractive rice seed organelle for the deposition of recombinant proteins because they occupy over 90% of the seed endosperm and are a valuable commodity in several different industries. The promoter and signal sequence from puroindoline-b, a protein natively found on the surface of starch granules in wheat, were found to direct the expression and localization of human lysozyme in rice seed (49). The localization of the recombinant protein was expected to be to the starch granules, however, this was not the case as analyses showed only PBI and PBII deposition. These results are estimated to be attributed to possible cryptic sequences within the human lysozyme that triggers it to undergo deposition via a rice default pathway, or perhaps lysozyme contains internal PB localization sequences (49). These localization results are similar to those found in maize when the promoter and signal peptide of the maize storage protein (zein) was fused to the bacterial protein LT-B (41). The LT-B protein was deposited solely within the maize starch granules, suggesting that it must have a native internal starch targeting signal, as a C-terminal PB signal peptide was used (41). Although starch granules are the major occupants in rice endosperm, the highest recombinant protein yields recorded remain to be achieved through such PB deposition strategies (Table 1).

3.4. Multigene Engineering

Multigene engineering can be used for the production of multimeric proteins, manipulation of metabolic pathways, or introduction of foreign metabolic pathways. The most prominent example of multigene engineering in rice is "Golden rice," which involved the insertion of three genes (phytoene synthase, phytoene desaturase, and lycopene β-cyclase) that resulted in the production of β-carotene (43). The traditional methods used for multigene insertions are crossing two independent transgenic lines, or by performing sequential transformations (transforming a transgenic plant with another transgene). These processes, however, are time and labor intensive and are susceptible to segregation events in later generations due to the result of the different transgenes residing on different loci (86). To insert multiple genes in one generation and on the same locus, co-transformation methods have been developed (84, 87, 88). The most common co-transformation method for rice is using *Agrobacterium*-mediated transformations with the transgenes on a single T-DNA or inserted using multiple T-DNA fragments (86). Particle bombardment has also shown efficacy in multigene engineering, as it has been reported that 11 different plasmids, each containing a different marker and transgene were found on the same locus and expressed in the subsequent generation (89). Co-transforming plants with five to ten transgenes is now possible with the innovations in T-DNA-based vectors and artificial chromosomes but there are still issues involved with transgene silencing and unpredictable interactions with the host proteome.

4. Conclusion

Plant molecular pharming has undergone considerable advancements in recombinant protein production and is recognized as a promising opportunity to meet the future demands for biopharmaceuticals. The drastic shortage of protein-based therapeutics has pushed collaborations between industry giants and molecular pharming start-ups as proof-of-concept trials have proven the large-scale efficacy of plants as production hosts (18). Molecular pharming is still in its infancy as there has yet to be a plant-made pharmaceutical approved for commercialization. There are 21 plant-derived pharmaceuticals currently in phases I and II of clinical trials and eleven plant-derived pharmaceuticals that are either in phases III and IV, or are currently being marketed as a medical device or fine chemical (3, 90). The seed, and more specifically the rice seed has emerged as an ideal candidate for the large-scale production of protein therapeutics due to its high protein content, low protease activity, low water content, stable protein storage environment, relatively high biomass, and molecular tools available for manipulation (3). Due to the preservation capabilities of seeds, recombinant seed banks could be generated, providing an economically feasible, timely scale-up of therapeutics in response to changing market demands. With the increasing demand for protein therapeutics and the era of follow-on biologics upon us, molecular pharming as an industry will see considerable growth and attention over the next decade. To fully utilize the advantages of seeds, advancements in protein expression and downstream purification will get the most consideration as they are the key factors that determine the efficacy of a platform's economic performance (50).

References

1. Twyman RM et al (2003) Molecular farming in plants: host systems and expression technology. Trends Biotechnol 21:570–578
2. Sharma AK, Sharma MK (2009) Plants as bioreactors: recent developments and emerging opportunities. Biotechnol Adv 27: 811–832
3. Lau OS, Sun S (2009) Plant seeds as bioreactors for recombinant protein production. Biotechnol Adv 27:1015–1022

4. IMS Health (2010) Accurate global assessment of a dynamic and complex niche market. IMS global biologics perspectives. IMS Health Incorporated
5. Walsh G (2010) Biopharmaceutical benchmarks 2010. Nat Biotechnol 28:917–924
6. Boehm R (2007) Bioproduction of therapeutic proteins in the 21st century and the role of plants and plant cells as production platforms. Ann NY Acad Sci 1102:121–134
7. Karg S, Kallio P (2009) The production of biopharmaceuticals in plant systems. Biotechnol Adv 27:879–894
8. Parl J et al (2010) Cellular engineering for the high-level production of recombinant proteins in mammalian cell systems. Kor J Chem Eng 27:1042–1048
9. Wurm FM (2004) Production of recombinant protein therapeutics in cultivated mammalian cells. Nat Biotechnol 11:1393–1398
10. Hiller A (2009) Fast growth foreseen for protein therapeutics. GEN 29:153–155
11. Cheng XY et al (1998) *Agrobacterium*-transformed rice plants expressing synthetic *cry1A(b)* and *cry1A(c)* genes are highly toxic to striped stem borer and yellow stem borer. Proc Natl Acad Sci USA 6:2767–2772
12. Ma JK et al (2003) The production of recombinant pharmaceuticals in plants. Nat Rev Genet 4:794–805
13. De Zoeten G et al (1989) The expression, localization, and effect of a human interferon in plants. Virology 172:213–222
14. Hood EE et al (1997) Commercial production of avidin from transgenic maize: characterization of transformant, production, processing, extraction and purification. Mol Breed 3:291–306
15. Twyman RM et al (2005) Transgenic plants in the biopharmaceutical market. Expert Opin Emerg Drugs 10:185–218
16. Boothe J et al (2010) Seed-based expression systems for plant molecular farming. Plant Biotechnol J 8:588–606
17. Houdebine LM (2009) Production of pharmaceutical proteins by transgenic animals. Comp Immunol Microbiol Infect Dis 32:107–121
18. Basaran P, Rodriguez-Cerezo E (2008) Plant molecular farming: opportunities and challenges. Crit Rev Biotechnol 28:153–172
19. Yang L et al (2008) Biopharming to increase bioactive peptides in rice seed. J AOAC Int 91:957–963
20. Golovina EA et al (2008) Sustained expression of human cytomegalovirus glycoprotein B (UL55) in the seeds of homozygous rice plants. Mol Biotechnol 40:1–12
21. Tackaberry ES et al (2008) Sustained expression of human cytomegalovirus glycoprotein B (UL55) in the seeds of homozygous rice plants. Mol Biotechnol 40:1–12
22. Alli Z et al (2002) Pharming vaccines for hepatitis and cytomegalovirus: towards the development of multivalent and subunit vaccines for oral delivery of antigens. Phytochem Rev 1:55–66
23. Ganz PR et al (1996) Expression of human blood proteins in transgenic plants: the cytokine GM-CSF as a model protein. In: Transgenic plants: a production system for industrial and pharmaceutical proteins, ch 7–4. pp 281–297
24. Ganz PR et al (1995) Plants as factories for producing human blood proteins. Can Soc Transfus Med Bull 7:112–121
25. Blais DR, Altosaar I (2006) Human CD14 expressed in seeds of transgenic tobacco displays similar proteolytic resistance and bioactivity with its mammalian-produced counterpart. Transgenic Res 15:151–164
26. Blais DR, Altosaar I (2007) Humanizing infant milk formula to decrease postnatal HIV transmission. Trends Biotechnol 25:376–384
27. Stoger E et al (2000) Cereal crops as viable production and storage systems for pharmaceutical scFv antibodies. Plant Mol Biol 4: 583–590
28. Wang J, Oard H (2003) Rice ubiquitin promoters: deletion analysis and potential usefulness in plant transformation systems. Plant Cell Rep 22:129–134
29. Huang Z et al (2006) Rapid, high-level production of hepatitis B core antigen in plant leaf and its immunogenicity in mice. Vaccine 24:2506–2513
30. Stoger E et al (2002) Practical considerations for pharmaceutical antibody production in different crop systems. Mol Breed 9:149–158
31. Torres E et al (1999) Rice cell culture as an alternative production system for functional diagnostic and therapeutic antibodies. Transgenic Res 8:441–449
32. Hood E et al (2003) Criteria for high-level expression of a fungal laccase gene in transgenic maize. Plant Biotechnol J 1:129–140
33. Qu L, Takaiwa F (2004) Evaluation of tissue specificity and expression strength of rice seed component gene promoters in transgenic rice. Plant Biotechnol J 2:113–125
34. Sardana R et al (2007) Biologically active human GM-CSF produced in the seeds of transgenic rice plants. Transgenic Research 16:713–721
35. Wu CY et al (1998) Promoters of rice seed storage protein genes direct endosperm-

specific gene expression in transgenic rice. Plant Cell Physiol 39:885–889
36. Wakasa Y et al (2006) High accumulation of bioactive peptide in transgenic rice seeds by expression of introduced multiple genes. Plant Biotechnol J 4:499–510
37. Takagi H et al (2005) Oral immunotherapy against pollen allergy using a seed-based peptide vaccine. Plant Biotechnol J 3:521–533
38. Zhang D et al (2010) Expression, purification, and characterization of recombinant human transferring from rice (*Oryza sativa* L.). Protein Expr Purif 74:69–79
39. Sawant S et al (2001) Designing of an artificial expression cassette for the high-level expression of transgenes in plants. Theor Appl Genet 102:635–644
40. Rushton PJ et al (2002) Synthetic plant promoters containing defined regulatory elements provide novel insights into pathogen and wound-induced signaling. Plant Cell 14:749–762
41. Chikwamba RK et al (2003) Localization of a bacterial protein in starch granules of transgenic maize kernels. Proc Natl Acad Sci USA 100:11127–11132
42. Wu CY et al (1998) The GCN4 motif in a rice glutelin gene is essential for endosperm-specific gene expression and is activated by Opaque-2 in transgenic rice plants. Plant J 14:673–683
43. Ye X et al (2000) Engineering the provitamin A (beta-carotene) biosynthetic pathway into (carotenoid-free) rice endosperm. Science 287:303–305
44. Li X et al (2002) Influence of matrix attachment regions from maize on transgene expression level in tobacco. Acta Botanica Sinica 44:804–808
45. Tetko I et al (2006) Spatiotemporal expression control correlates with intragenic matrix attachment regions (S/MARs) in *Arabidopsis thaliana*. PLoS Comput Biol 2:136–145
46. Yang D et al (2001) Expression of the REB transcriptional activator in rice grains improves the yield of recombinant proteins whose genes are controlled by a Reb-responsive promoter. Proc Natl Acad Sci USA 98:11438–11443
47. Nakase M et al (1997) Characterization of a novel rice bZIP protein which binds to the α-globulin promoter. Plant Mol Biol 33:513–522
48. Comai L et al (1990) Novel and useful properties of a chimeric plant promoter combining CaMV 35S and MAS elements. Plant Mol Biol 15:373–381
49. Hennegan K et al (2005) Improvement of human lysozyme expression in transgenic rice grain by combining wheat *Triticum aestivum* puroindoline b and rice *Oryza sativa* Gt1 promoters and signal peptides. Transgenic Res 14:583–592
50. Streatfield S (2007) Approaches to achieve high-level heterologous protein production in plants. Plant Biotechnol J 5:2–15
51. Futterer J, Hohn T (1996) Translation in plants—rules and exceptions. Plant Mol Biol 32:159–189
52. Kawaguchi R, Bailey-Serres J (2002) Regulation of translation initiation in plants. Curr Opin Plant Biol 5:460–465
53. Mauro VP et al (2002) The ribosome filter hypothesis. Proc Natl Acad Sci USA 99: 12031–12036
54. Kato T et al (1991) A modified B-glucuronidase gene: sensitive detection of plant promoter activities in suspension cultured cells of tobacco and rice. Plant Mol Biol Rep 9:333–339
55. Satoh J et al (2004) The 5′-untranslated region of the tobacco alcohol dehydrogenase gene functions as an effective translational enhancer in plant. J Biosci Bioeng 98:1–8
56. Wever W et al (2010) The 5′ untranslated region of the VR-ACS1 mRNA acts as a strong translational enhancer in plants. Transgenic Res 19:667–674
57. Danthinne X, Emmelo J (1990) Studies on the translational properties of STNV RNA non-coding regions. In: 42nd international symposium on crop protection. Gent, Belgium
58. Sivamani E et al (2009) Sequence analysis of rice rubi3 promoter gene expression cassette for improved transgene expression. Plant Sci 177:549–556
59. Sivamani E, Qu R (2006) Expression enhancement of a rice polyubiquitin gene promoter. Plant Mol Biol 60:225–239
60. Lu J et al (2008) Activity of the 5′ regulatory regions of the rice polyubiquitin rubi3 gene in transgenic rice plants as analyzed by both GUS and GFP reporter genes. Plant Cell Rep 27:1587–1600
61. Mascarenhas D et al (1990) Intron-mediated enhancement of heterologous gene expression in maize. Plant Mol Biol 15:913–920
62. Fiume E et al (2004) Introns are key regulatory elements of rice tubulin expression. Planta 218:693–703
63. Giani S et al (2009) In transgenic rice, a- and b-tubulin regulatory sequences control GUS amount and distribution through intron mediated enhancement and intron dependant spatial expression. Transgenic Res 18:151–162
64. Samadder P et al (2008) Transcriptional and post-transcriptional enhancement of gene

expression by the 5'UTR intron of rice rubi3 gene in transgenic rice cells. Mol Genet Genomics 279:429–439

65. Hunt AG (2008) Messenger RNA 3' end formation in plants. Nuclear pre-mRNA processing in plants, Curr Top Microb Immunol 326, 151–177
66. Green PJ (1993) Control of mRNA stability in higher plants. Plant Physiol 102:1065–1070
67. Shen Y et al (2008) Genome level analysis of rice mRNA 3'-end processing signals and alternative polyadenylation. Nucleic Acids Res 36:3150–3161
68. Dong H et al (2006) An exploration of 3'-end processing signals and their tissue distribution in *Oryza sativa*. Gene 389:107–113
69. Ali S, Taylor W (2001) The 3' non-coding region of a C4 photosynthesis gene increases transgene expression when combined with heterologous promoters. Plant Mol Biol 46:325–333
70. Knirsch L, Clerch LB (2000) A region in the 3′ UTR of MnSOD RNA enhances translation of a heterologous RNA. Biochem Biophy Res Commun 272:164–168
71. Yang L et al (2009) The 3′-untranslated region of rice glutelin GluB-1 affects accumulation of heterologous protein in transgenic rice. Biotechnol Lett 31:1625–1631
72. Mishra S et al (2006) Ubiquitin fusion enhances cholera toxin B subunit expression in transgenic plants and the plant-expressed protein binds GM1 receptor more efficiently. J Biotechnol 127:95–108
73. Kang TJ et al (2004) Modification of the cholera toxin B subunit coding sequence to enhance expression in plants. Mol Breed 13: 143–153
74. Ma JK et al (2005) Plant-derived pharmaceuticals—the road forward. Trends Plant Sci 10:580–585
75. Stoger E et al (2005) Sowing the seeds of success: pharmaceutical proteins from plants. Curr Opin Biotechnol 16:167–173
76. Vitale A, Pedrazzini E (2005) Recombinant pharmaceuticals from plants: the plant endomembrane system as bioreactor. Mol Interv 5:216–225
77. Avesani L et al (2003) Improved in planta expression of the human islet autoantigen glutamic acid decarboxylase (GAD65). Transgenic Res 12:203–212
78. Garbarino JE et al (1995) Isolation of a polyubiquitin promoter and its expression in transgenic potato plants. Plant Physiol 109:1371–1378
79. Benchabane M et al (2008) Preventing unintended proteolysis in plant protein biofactories. Plant Biotechnol J 6:633–648
80. Jiang L, Sun SS (2002) Membrane anchors for vacuolar targeting: applications in plant bioreactors. Trends Biotechnol 20:99–102
81. Kawagoe Y et al (2005) The critical role of disulfide bond formation in protein sorting in the endosperm of rice. Plant Cell 17:1141–1153
82. Hashizume F et al (2008) Development and evaluation of transgenic rice seeds accumulating a type II-collagen tolerogenic peptide. Transgenic Res 17:1117–1129
83. Sugita K et al (2005) Genetically modified rice seeds accumulating GLP-1 analogue stimulate insulin secretion from a mouse pancreatic beta-cell line. FEBS Lett 579:1085–1088
84. Wu G et al (2005) Stepwise engineering to produce high yields of very long-chain polyunsaturated fatty acids in plants. Nat Biotechnol 23:1013–1017
85. Takaiwa F et al (2008) Co-expression of soybean glycinins A1aB1b and A3B4 enhances their accumulation levels in transgenic rice seed. Plant Cell Physiol 49:1589–1599
86. Naqvi S et al (2009) When more is better: multigene engineering in plants. Trends Plant Sci 15:48–56
87. Abbadi A et al (2004) Biosynthesis of very-long-chain polyunsaturated fatty acids in transgenic oilseeds: constraints on their accumulation. Plant Cell 16:2734–2748
88. Wada N et al (2009) Bioactive beads-mediated transformation of rice with large DNA fragments containing *Aegilops tauschii* genes. Plant Cell Rep 28:759–768
89. Chen L et al (1998) Expression and inheritance of multiple transgenes in rice plants. Nat Biotechnol 16:1059–1064
90. Obembe O et al (2011) Advances in plant molecular farming. Biotechnol Adv 29: 210–222
91. Yang L et al (2007) Development of transgenic rice seed accumulating a major Japanese cedar pollen allergen (Cry j 1) structurally disrupted for oral immunotherapy. Plant Biotechnol J 5:815–826
92. Takaiwa F et al (2007) Endosperm tissue is good production platform for artificial recombinant proteins in transgenic rice. Plant Biotechnol J 5:84–92
93. Xie T et al (2008) A biologically active rhIGF-1 fusion accumulated in transgenic rice seeds can reduce blood glucose in diabetic mice via oral delivery. Peptides 29:1862–1870

94. Oszvald M et al (2008) Expression of cholera toxin B subunit in transgenic rice endosperm. Mol Biotechnol 40:261–268
95. Yasuda H et al (2009) Overexpression of BiP has inhibitory effects on the accumulation of seed storage proteins in endosperm cells of rice. Plant Cell Physiol 50:1532–1543
96. Matsumoto Y et al (2009) Oral immunogenicity and protective efficacy in mice of transgenic rice plants producing a vaccine candidate antigen (As16) of *Ascaris suum* fused with cholera toxin B subunit. Transgenic Res 18:185–192

Appendix

Genomic and Bioinformatic Resources for Rice Research

Wenhua Liu, Kabin Xie, and Yinong Yang

Abstract

During the past decade, there has been an explosive accumulation of "omics" data due to the development of advanced and efficient genome technology. Both japonica and indica rice genomes have been completely sequenced, annotated, and made publicly available. Transcriptome profiling has been conducted on a large scale by microarray and deep-sequencing to chart the atlas of rice gene expression under normal or stressed conditions. In addition, yeast two-hybrid, tandem affinity purification, and other proteomic techniques have accelerated the identification of rice proteins and the elucidation of complex networks of protein–protein interactions. In combination with extensive resources of natural rice variants, deletion and insertional mutants, these genomic, transcriptomic, and proteomic information will greatly facilitate functional characterization of rice genes and networks in a genetic, biological, and evolutionary context. With exquisite bioinformatic tools, many user-friendly databases such as the ones described below have been built to integrate large data sets and allow us to more effectively explore the treasure troves of rice genomic and genetic resources.

Key words: Bioinformatics; Genome annotation; Gene expression; Transcriptomics; Proteomics; Protein–protein interaction; Rice mutants

Yinong Yang (ed.), *Rice Protocols*, Methods in Molecular Biology, vol. 956,
DOI 10.1007/978-1-62703-194-3, © Springer Science+Business Media, LLC 2013

Table A.1
Useful Web sites for rice research

Database	Web address	Description
Integrated rice genome annotation		
RAP-DB (Rice Annotation Project Database)	http://rgp.dna.affrc.go.jp/E/IRGSP/	Genome annotation
RGAP (Rice Genome Annotation Project)	http://rice.plantbiology.msu.edu/	Genome annotation (1)
RiceGE: Rice Functional Genomic Express Database	http://signal.salk.edu/cgi-bin/RiceGE	Genome annotation, expression data and mutant resources
GRAMENE	http://www.gramene.org/	Comparative grass genomics
BGI-RIS	http://rice.genomics.org.cn/rice/	Beijing Genomics Institute (BGI) rice information system (2)
KOME (Knowledge-based Oryza Molecular Biological Encyclopedia)	http://cdna01.dna.affrc.go.jp/cDNA/	Rice full-length cDNA collection, gene ontology and expression database (3)
RICD (Rice Indica cDNA Database)	http://www.ncgr.ac.cn/ricd/	Rice (indica) full-length cDNA collection (4)
Rice HaplotypeMap Project	http://www.ncgr.ac.cn/RiceHapMap/	Genetic map-based SNPs (5)
RiceNet	http://www.functionalnet.org/ricenet	Functional association of genes in rice genome (6)
PlaNet	http://aranet.mpimp-golm.mpg.de	Whole-genome coexpression networks for *Arabidopsis* and six important plant and crop species (7)
DRTF (Database of Rice Transcription Factors)	http://drtf.cbi.pku.edu.cn/	Transcription factor database (8)
PlnTFDB	http://plntfdb.bio.uni-potsdam.de/v3.0/	Transcription factor database (9)
Genetic resources		
Rice *Tos17* Insertion Mutant Database	http://tos.nias.affrc.go.jp/~miyao/pub/tos17/	Rice mutants induced by rice retrotransposon *Tos17*
Rice Transposon Flanking Sequence Tag Database	http://sundarlab.ucdavis.edu/rice/blast/blast.html	Insertion mutants of the maize *En/Spm* or *Ac-Ds* transposons in *japonica* rice cv. Nipponbare
Oryza Tag Line	http://urgi.versailles.inra.fr/OryzaTagLine/	T-DNA insertion mutants
OryGenesDB	http://orygenesdb.cirad.fr/	T-DNA and Ds flanking sequence tags derived from the genomics initiative Genoplante (Oryza Tag Line) and the European consortium Cereal Gene Tags (10)

TRIM (Taiwan Rice Insertional Mutants Database)	http://trim.sinica.edu.tw/home	T-DNA insertion mutants
RISD (Rice T-DNA Insertion Sequence Database)	http://www.postech.ac.kr/life/pfg/risd/	T-DNA insertion and activation tagging lines in *japonica* rice cv. Dongjin or Hwayoung.
RMD (Rice Mutant Database)	http://rmd.ncpgr.cn/	T-DNA insertion mutants generated by an enhancer trap system (11)
National Plant Germplasm System	http://www.ars-grin.gov/npgs/	Crop germplasm collection
IRRI Genetic Resource Center	http://beta.irri.org/seeds/	Rice knowledge and collection of natural variants and chemical mutants
International Rice Information System	http://rice.generationcp.org/germplasm/	A rice portal for global information on genetic resource and rice cultivars
Gene expression		
Genevestigator	https://www.genevestigator.com/gv/	Collection of public microarray data and analysis of rice gene expression (12, 13)
CREP (Collection of Rice Gene Expression Profiles)	http://crep.ncpgr.cn/	A database of the rice transcriptome project
Yale Virtual Center for Cellular Expression Profiling of Rice	http://bioinformatics.med.yale.edu/riceatlas/	Whole-genome transcriptional profiles of rice cell types
Rice Next-Gen Sequence Database	http://mpss.udel.edu/rice/	Rice mRNA and small RNA database generated by next-generation sequencing (14)
ROAD (Rice Oligonucleotide Array Database)	http://www.ricearray.org	A public resource for rice gene expression and coexpression analysis
DIURNAL	http://diurnal.cgrb.oregonstate.edu/	Circadian/diurnal gene expression data for Arabidopsis, rice, and poplar genes (15)
RED (Rice Expression Database)	http://cdna02.dna.affrc.go.jp/RED/	Rice gene expression database
OryzaExpress (Gene Expression Database for Rice)	http://riceball.lab.nig.ac.jp/oryzaexpress/	Rice gene expression database (16)

(continued)

Table A.1 (continued)

Database	Web address	Description
CSRDB: Cereal Small RNAs Database	http://sundarlab.ucdavis.edu/smrnas	Small RNAs database for cereal crops (17)
RiceXPro	http://ricexpro.dna.affrc.go.jp/	Collection of microarray data during the entire life cycle of rice plants
RiceFox	http://ricefox.psc.riken.jp/	Overexpression of rice full-length cDNAs in *Arabidopsis*
Proteomic database		
RPD (Rice Proteome Database)	http://gene64.dna.affrc.go.jp/RPD	2D-PAGE and mass-spectrometer analysis of rice proteome (18)
OryzaPG-BD	http://oryzapg.iab.keio.ac.jp/	Rice proteome database based on shotgun proteogenomics (19)
IntAct	http://www.ebi.ac.uk/intact/	General database for protein interactome (20)
RKD (Rice Kinase Database)	http://phylomics.ucdavis.edu/kinase/	Rice kinase database (21)
Bioinformatic tools		
PLACE (Plant Cis-acting Regulatory DNA Elements)	http://www.dna.affrc.go.jp/PLACE/	Analysis of plant promoters and *cis*-acting DNA elements (22)
WMD3 (Web MicroRNA Designer)	http://wmd3.weigelworld.org/	Artificial miRNA design, small RNA targets prediction (23)
Osiris	http://www.bioinformatics2.wsu.edu/Osiris	Rice promoter analysis (24)
PhosPhAt	http://phosphat.mpimp-golm.mpg.de/	Phosphorylation site prediction (25)

References

1. Ouyang S, Zhu W, Hamilton J, Lin H, Campbell M, Childs K, Thibaud-Nissen F, Malek RL, Lee Y, Zheng L et al (2007) The TIGR rice genome annotation resource: improvements and new features. Nucleic Acids Res 35:D883–D887
2. Zhao W, Wang J, He X, Huang X, Jiao Y, Dai M, Wei S, Fu J, Chen Y, Ren X et al (2004) BGI-RIS: an integrated information resource and comparative analysis workbench for rice genomics. Nucleic Acids Res 32:D377–D382
3. Kikuchi S, Satoh K, Nagata T, Kawagashira N, Doi K, Kishimoto N, Yazaki J, Ishikawa M, Yamada H, Ooka H et al (2003) Collection, mapping, and annotation of over 28,000 cDNA clones from japonica rice. Science 301:376–379
4. Lu T, Huang X, Zhu C, Huang T, Zhao Q, Xie K, Xiong L, Zhang Q, Han B (2008) RICD: a rice indica cDNA database resource for rice functional genomics. BMC Plant Biol 8:118
5. Huang X, Wei X, Sang T, Zhao Q, Feng Q, Zhao Y, Li C, Zhu C, Lu T, Zhang Z et al (2010) Genome-wide association studies of 14 agronomic traits in rice landraces. Nat Genet 42:961–967
6. Lee I, Seo Y, Coltrane D, Hwang S, Oh T, Marcotte E, Ronald P (2011) Genetic dissection of the biotic stress response using a genome-scale gene network for rice. Proc Natl Acad Sci U S A 108:18548–18553
7. Mutwil M, Klie S, Tohge T, Giorgi F, Wilkins O, Campbell M, Fernie A, Usadel B, Nikoloski Z, Persson S (2011) PlaNet: combined sequence and expression comparisons across plant networks derived from seven species. Plant Cell 23:895–910
8. Gao G, Zhong Y, Guo A, Zhu Q, Tang W, Zheng W, Gu X, Wei L, Luo J (2006) DRTF: a database of rice transcription factors. Bioinformatics 22:1286–1287
9. Riano-Pachon DM, Ruzicic S, Dreyer I, Mueller-Roeber B (2007) PlnTFDB: an integrative plant transcription factor database. BMC Bioinformatics 8:42
10. Droc G, Ruiz M, Larmande P, Pereira A, Piffanelli P, Morel JB, Dievart A, Courtois B, Guiderdoni E, Perin C (2006) OryGenesDB: a database for rice reverse genetics. Nucleic Acids Res 34:D736–D740
11. Zhang J, Li C, Wu C, Xiong L, Chen G, Zhang Q, Wang S (2006) RMD: a rice mutant database for functional analysis of the rice genome. Nucleic Acids Res 34:D745–D748
12. Zimmermann P, Laule O, Schmitz J, Hruz T, Bleuler S, Gruissem W (2008) Genevestigator transcriptome meta-analysis and biomarker search using rice and barley gene expression databases. Mol Plant 1:851–857
13. Hruz T, Laule O, Szabo G, Wessendorp F, Bleuler S, Oertle L, Widmayer P, Gruissem W, Zimmermann P (2008) Genevestigator v3: a reference expression database for the meta-analysis of transcriptomes. Adv Bioinformatics 2008:420747
14. Nobuta K, Venu RC, Lu C, Belo A, Vemaraju K, Kulkarni K, Wang W, Pillay M, Green PJ, Wang GL et al (2007) An expression atlas of rice mRNAs and small RNAs. Nat Biotechnol 25:473–477
15. Mockler TC, Michael TP, Priest HD, Shen R, Sullivan CM, Givan SA, McEntee C, Kay SA, Chory J (2007) The DIURNAL project: DIURNAL and circadian expression profiling, model-based pattern matching, and promoter analysis. Cold Spring Harb Symp Quant Biol 72:353–363
16. Hamada K, Hongo K, Suwabe K, Shimizu A, Nagayama T, Abe R, Kikuchi S, Yamamoto N, Fujii T, Yokoyama K et al (2011) OryzaExpress: an integrated database of gene expression networks and omics annotations in rice. Plant Cell Physiol 52:220–229
17. Johnson C, Bowman L, Adai AT, Vance V, Sundaresan V (2007) CSRDB: a small RNA integrated database and browser resource for cereals. Nucleic Acids Res 35:D829–D833
18. Komatsu S, Kojima K, Suzuki K, Ozaki K, Higo K (2004) Rice proteome database based on two-dimensional polyacrylamide gel electrophoresis: its status in 2003. Nucleic Acids Res 32:D388–D392
19. Helmy M, Tomita M, Ishihama Y (2011) OryzaPG-DB: rice proteome database based on shotgun proteogenomics. BMC Plant Biol 11:63
20. Aranda B, Achuthan P, Alam-Faruque Y, Armean I, Bridge A, Derow C, Feuermann M, Ghanbarian AT, Kerrien S, Khadake J et al (2010) The IntAct molecular interaction database in 2010. Nucleic Acids Res 38: D525–D531
21. Dardick C, Chen J, Richter T, Ouyang S, Ronald P (2007) The rice kinase database. A phylogenomic database for the rice kinome. Plant Physiol 143:579–586

22. Higo K, Ugawa Y, Iwamoto M, Korenaga T (1999) Plant cis-acting regulatory DNA elements (PLACE) database: 1999. Nucleic Acids Res 27:297–300
23. Schwab R, Ossowski S, Riester M, Warthmann N, Weigel D (2006) Highly specific gene silencing by artificial microRNAs in *Arabidopsis*. Plant Cell 18:1121–1133
24. Morris RT, O'Connor TR, Wyrick JJ (2008) Osiris: an integrated promoter database for *Oryza sativa* L. Bioinformatics 24:2915–2917
25. Durek P, Schmidt R, Heazlewood JL, Jones A, MacLean D, Nagel A, Kersten B, Schulze WX (2009) PhosPhAt: the *Arabidopsis thaliana* phosphorylation site database. An update. Nucleic Acids Res 38:D828–D834

Index

Yinong Yang (ed.), *Rice Protocols*, Methods in Molecular Biology, vol. 956,
DOI 10.1007/978-1-62703-194-3, © Springer Science+Business Media, LLC 2013

MIX
Papier aus verantwortungsvollen Quellen
Paper from responsible sources
FSC® C105338

If you have any concerns about our products,
you can contact us on
ProductSafety@springernature.com

In case Publisher is established outside the EU,
the EU authorized representative is:
Springer Nature Customer Service Center GmbH
Europaplatz 3, 69115 Heidelberg, Germany

Printed by Libri Plureos GmbH
in Hamburg, Germany